Lectures on Morse Homology

Kluwer Texts in the Mathematical Sciences

VOLUME 29

A Graduate-Level Book Series

The titles published in this series are listed at the end of this volume.

SPRINGER SCIENCE+BUSINESS MEDIA, B.V.

Lectures on Morse Homology

by

Augustin Banyaga

The Pennsylvania State University,
University Park, PA, U.S.A.

and

David Hurtubise

The Pennsylvania State University,
Altoona, P.A., US.A.

 SPRINGER SCIENCE+BUSINESS MEDIA, B.V.

A C.I.P. Catalogue record for this book is available from the Library of Congress.

ISBN 978-90-481-6705-0   ISBN 978-1-4020-2696-6 (eBook)
DOI 10.1007/978-1-4020-2696-6

Printed on acid-free paper

Contents

Preface   ix

1. Introduction   1
    1.1   Overview   1
    1.2   Algebraic topology   3
    1.3   Basic Morse theory   4
    1.4   Stable and unstable manifolds   5
    1.5   Basic differential topology   6
    1.6   Morse-Smale functions   7
    1.7   The Morse Homology Theorem   9
    1.8   Morse theory on Grassmann manifolds   10
    1.9   Floer homology theories   11
    1.10 Guide to the book   11

2. The CW-Homology Theorem   15
    2.1   Singular homology   15
    2.2   Singular cohomology   20
    2.3   CW-complexes   21
    2.4   CW-homology   23
    2.5   Some homotopy theory   31

3. Basic Morse Theory   45
    3.1   Morse functions   45
    3.2   The gradient flow of a Morse function   58
    3.3   The CW-complex associated to a Morse function   63
    3.4   The Morse Inequalities   73

	3.5	Morse-Bott functions	80
4.	The Stable/Unstable Manifold Theorem		93
	4.1	The Stable/Unstable Manifold Theorem for a Morse function	93
	4.2	The Local Stable Manifold Theorem	98
	4.3	The Global Stable/Unstable Manifold Theorem	111
	4.4	Examples of stable/unstable manifolds	116
5.	Basic Differential Topology		127
	5.1	Immersions and submersions	127
	5.2	Transversality	131
	5.3	Stability	132
	5.4	General position	134
	5.5	Stability and density for Morse functions	137
	5.6	Orientations and intersection numbers	143
	5.7	The Lefschetz Fixed Point Theorem	148
6.	Morse-Smale Functions		157
	6.1	The Morse-Smale transversality condition	157
	6.2	The λ-Lemma	165
	6.3	Consequences of the λ-Lemma	171
	6.4	The CW-complex associated to a Morse-Smale function	175
7.	The Morse Homology Theorem		195
	7.1	The Morse-Smale-Witten boundary operator	196
	7.2	Examples using the Morse Homology Theorem	201
	7.3	The Conley index	207
	7.4	Proof of the Morse Homology Theorem	211
	7.5	Independence of the choice of the index pairs	219
8.	Morse Theory On Grassmann Manifolds		227
	8.1	Morse theory on the adjoint orbit of a Lie group	228
	8.2	A Morse function on an adjoint orbit of the unitary group	235
	8.3	An almost complex structure on the adjoint orbit	243
	8.4	The critical points and indices of $f_A : U(n+k) \cdot x_0 \to \mathbb{R}$	246
	8.5	A Morse function on the complex Grassmann manifold	249
	8.6	The gradient flow lines of $f_A : G_{n,n+k}(\mathbb{C}) \to \mathbb{R}$	252
	8.7	The homology of $G_{n,n+k}(\mathbb{C})$	257

   8.8   Further generalizations and applications   260

9. An Overview of Floer Homology Theories   269
   9.1   Introduction to Floer homology theories   269
   9.2   Symplectic Floer homology   272
   9.3   Floer homology for Lagrangian intersections   280
   9.4   Instanton Floer homology   281
   9.5   A symplectic flavor of the instanton homology   284

Hints and References for Selected Problems   287

Bibliography   309

Symbol Index   317

Index   321

Preface

This book is based on the lecture notes from a course we taught at Penn State University during the fall of 2002. The main goal of the course was to give a complete and detailed proof of the Morse Homology Theorem (Theorem 7.4) at a level appropriate for second year graduate students. The course was designed for students who had a basic understanding of singular homology, CW-complexes, applications of the existence and uniqueness theorem for O.D.E.s to vector fields on smooth Riemannian manifolds, and Sard's Theorem.

We would like to thank the following students for their participation in the course and their help proofreading early versions of this manuscript: James Barton, Shantanu Dave, Svetlana Krat, Viet-Trung Luu, and Chris Saunders. We would especially like to thank Chris Saunders for his dedication and enthusiasm concerning this project and the many helpful suggestions he made throughout the development of this text.

We would also like to thank Bob Wells for sharing with us his extensive knowledge of CW-complexes, Morse theory, and singular homology. Chapters 3 and 6, in particular, benefited significantly from the many insightful conversations we had with Bob Wells concerning a Morse function and its associated CW-complex.

Augustin Banyaga and David Hurtubise
The Pennsylvania State University, 2004

Chapter 1

Introduction

1.1 Overview

A Morse-Smale function $f : M \to \mathbb{R}$ on a finite dimensional compact smooth Riemannian manifold (M, g) gives rise to a chain complex $(C_*(f), \partial_*)$, called the Morse-Smale-Witten chain complex (Definition 7.2), whose chains are generated by the critical points of f and whose boundary operator is defined by counting gradient flow lines (with sign).

The Morse Homology Theorem (Theorem 7.4) says that the homology of the Morse-Smale-Witten chain complex $(C_*(f), \partial_*)$ is isomorphic to the singular homology of M with integer coefficients.

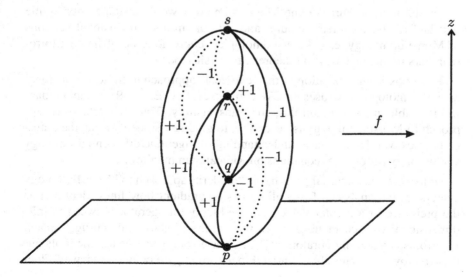

As an illustration of the Morse Homology Theorem, consider the preceeding picture of a torus that has been tilted slightly from its vertical position. (The reason the torus needs to be tilted slightly will be explained in Chapter 6 when we discuss the Morse-Smale transversality condition.) The height function in this picture has 4 critical points p, q, r, and s, and each of these critical points has an associated **index**, namely 0, 1, 1, and 2 (see Definition 3.1). Using the orientation conventions discussed in Chapter 7 we assign $+1$ or -1 to each of the gradient flow lines connecting critical points of relative index one.

The Morse-Smale-Witten chain complex $(C_*(f), \partial_*)$ is defined to be

$$
\begin{array}{ccccccc}
C_2(f) & \xrightarrow{\partial_2} & C_1(f) & \xrightarrow{\partial_1} & C_0(f) & \longrightarrow & 0 \\
\updownarrow{\scriptstyle\approx} & & \updownarrow{\scriptstyle\approx} & & \updownarrow{\scriptstyle\approx} & & \\
<s> & \xrightarrow{\partial_2} & <q,r> & \xrightarrow{\partial_1} & <p> & \longrightarrow & 0
\end{array}
$$

where the boundary operator is determined by counting (with sign) the number of gradient flow lines between critical points of relative index one. Since these gradient flow lines come in pairs with opposite signs we see that $\partial_* \equiv 0$, and thus,

$$
H_k((C_*(f), \partial_*)) = \begin{cases} \mathbb{Z} & \text{if } k = 0, 2 \\ \mathbb{Z} \oplus \mathbb{Z} & \text{if } k = 1 \\ 0 & \text{otherwise.} \end{cases}
$$

Hence, as guaranteed by the Morse Homology Theorem, the homology of the Morse-Smale-Witten chain complex $(C_*(f), \partial_*)$ agrees with the singular homology of T^2.

The basic ideas surrounding Morse homology were developed during the first half of the twentieth century, and various infinite dimensional versions of Morse homology, e.g. Floer homologies, continue to be of interest to researchers in mathematics and theoretical physics today.

This book is an exposition of the "classical" approach to finite dimensional Morse homology. It focuses on CW-complexes, O.D.E.s on Riemannian manifolds, stable/unstable manifolds, and transversality. Thus, the "classical" approach to Morse homology is accessible to graduate students in mathematics or physics who have a basic understanding of algebraic/differential topology and ordinary differential equations on Riemannian manifolds.

Unlike the "classical" approach, the "modern" approach to Morse homology centers on the analysis of moduli spaces of gradient flow lines identified as the preimage of a regular value of some Fredholm operator between infinite dimensional Banach manifolds. In the infinite dimensional setting, Banach manifolds, Fredholm operators, P.D.E.s, and Sobolev spaces are the tools one uses to prove theorems about Morse/Floer homology. For an overview of Floer homology theories we refer the reader to Chapter 9, and for an exposition of the "modern" approach to Morse homology applied to finite dimensional manifolds we refer the reader to the excellent book by M. Schwarz [131].

This book presents in great detail all the results one needs to prove the Morse Homology Theorem (Theorem 7.4) using the "classical" approach. Most of these results can be found scattered throughout the literature dating from the mid to late 1900's in some form or other, but often the results are proved in different contexts with a multitude of different notations and different goals. In this book we put all these results together into a single reference with complete and detailed proofs.

We have chosen to present what we consider to be the most accessible versions of these results rather than the most general. References to the literature are provided throughout the book to both the original proofs and the more general versions.

We assume that the reader is familiar with singular homology, CW-complexes, applications of the existence and uniqueness theorem for O.D.E.s to vector fields on smooth Riemannian manifolds, and Sard's Theorem. Excellent references for this material include [30], [77], [91], and [138], but there are of course many other references available.

1.2   Algebraic topology

In Chapter 2 we begin by reviewing the definitions and fundamental properties of singular homology and cohomology. We do not derive any of the fundamental properties of singular homology. For the proofs of the fundamental properties the reader can refer to any number of books on algebraic topology (see for instance [30], [39], [66], [109], or [138]). Next we define a CW-complex and prove the CW-Homology Theorem (Theorem 2.15). The CW-Homology Theorem says that if X is a CW-complex and $X^{(n)}$ is the n-skeleton of X, then the singular homology of X with coefficients in a commutative ring with unit Λ can be computed from the CW-chain complex $(\underline{C}_*(X; \Lambda), \underline{\partial}_*)$ whose n^{th} chain group $\underline{C}_n(X; \Lambda)$ is the singular homology group $H_n(X^{(n)}, X^{(n-1)}; \Lambda)$. The boundary operator $\underline{\partial}_n$ in the CW-chain complex can be defined either in terms of the connecting homomorphism in the homology exact sequence of the triple $(X^{(n)}, X^{(n-1)}, X^{(n-2)})$ or in terms of the degrees of the relative attaching maps. We end Chapter 2 by proving some basic theorems from homotopy theory. In particular, we prove that if $A \subseteq X$ is closed and the inclusion $A \hookrightarrow X$ is a cofibration, then $H_k(X, A; \Lambda) \approx H_k(X/A, *; \Lambda)$ for all k (Corollary 2.31).

1.3   Basic Morse theory

We begin Chapter 3 by recalling that a **Morse function** $f : M \to \mathbb{R}$ on a smooth manifold M is a smooth function whose critical points are all **non-degenerate**, i.e. at every critical point the Hessian is non-degenerate. The **index** λ_p of a critical point $p \in M$ is the number of negative eigenvalues of the Hessian matrix $M_p(f)$. We give several examples of Morse functions, including Bott's perfect Morse functions on the complex projective spaces (Example 3.7), and we prove that on any finite dimensional smooth manifold Morse functions always exist. Next we prove the Morse Lemma (Lemma 3.11) which says that at every non-degenerate critical point there exists a coordinate chart such that locally the Morse function f is a quadratic form. Our proof of the Morse Lemma follows Palais [111] and is based on the Moser "path method" [107].

After proving the Morse Lemma we choose a Riemannian metric g and study the gradient flow lines of a Morse function $f : M \to \mathbb{R}$. The gradient flow lines are the flow lines generated by minus the gradient vector field $-\nabla f$, where ∇f is defined by the condition $g(\nabla f, V) = df(V)$ for all smooth vector fields V on M. We prove that every smooth function decreases along its gradient flow lines (Proposition 3.18), and we prove that every gradient flow line on a compact smooth Riemannian manifold M begins and ends at a critical point (Proposition 3.19).

These simple results lead to the first significant result in Morse theory (Theorem 3.20) which says that the homotopy type of the space

$$M^t = f^{-1}((-\infty, t]) = \{x \in M \mid f(x) \leq t\}$$

only changes when t passes through a **critical value**, i.e. a number $c \in \mathbb{R}$ such that $f^{-1}(c)$ contains a critical point. When t passes through a critical value c the homotopy type of M^t changes by the attachment of one k-cell for every critical point in $f^{-1}(c)$ of index k (Theorem 3.25 and Remark 3.26). This leads to the most important result in Chapter 3 (Theorem 3.28) which says that if $f : M \to \mathbb{R}$ is a Morse function on a finite dimensional smooth Riemannian manifold M such that M^t is compact for all $t \in \mathbb{R}$, then M has the homotopy type of a CW-complex with one cell of dimension k for each critical point of f of index k. We prove Theorem 3.28 in complete detail, including the proofs of some lemmas regarding the homotopy type of a space with a k-cell attached (Lemma 3.29 and Lemma 3.30).

Combined with the CW-Homology Theorem (Theorem 2.15), this shows that the total number of critical points of a Morse function on a compact manifold is bounded below by the sum of the Betti numbers; it is this observation that motivated the Arnold Conjecture (Conjecture 9.1). More generally, in Section 3.4 we prove the Morse Inequalities (Theorem 3.33). The results in Section 3.4 are not needed in the rest of the book, but they are of historical significance and are included for completeness. As an elegant application of the Morse Inequalities we prove that the Euler characteristic $\mathcal{X}(M)$ of a compact manifold M of odd dimension is zero (Theorem 3.42). The last section of Chapter 3 is an introduction to Morse-Bott functions, including the Morse-Bott Lemma (Lemma 3.51) and the Morse-Bott Inequalities (Theorem 3.53).

1.4 Stable and unstable manifolds

If $f : M \to \mathbb{R}$ is a smooth function on a finite dimensional smooth Riemannian manifold (M, g) and $p \in M$ is a non-degenerate critical point of f, then the **stable manifold** of p is defined to be

$$W^s(p) = \{x \in M | \lim_{t \to \infty} \varphi_t(x) = p\}$$

where φ_t is the 1-parameter group of diffeomorphisms generated by minus the gradient vector field, i.e. $-\nabla f$. Similarly, the **unstable manifold** of p is defined to be

$$W^u(p) = \{x \in M | \lim_{t \to -\infty} \varphi_t(x) = p\}.$$

The Stable/Unstable Manifold Theorem for a Morse Function (Theorem 4.2) says that if $f : M \to \mathbb{R}$ is a Morse function and M is compact, then the tangent space at a critical point p splits as

$$T_p M = T_p^s M \oplus T_p^u M$$

where the Hessian is positive definite on $T_p^s M$ and negative definite on $T_p^u M$. Moreover, the stable and unstable manifolds of p are the surjective images of smooth embeddings

$$E^s : T_p^s M \to W^s(p) \subseteq M$$
$$E^u : T_p^u M \to W^u(p) \subseteq M.$$

Hence, $W^s(p)$ is a smoothly embedded open disk of dimension $m - \lambda_p$, and $W^u(p)$ is a smoothly embedded open disk of dimension λ_p where m is the dimension of M and λ_p is the index of the critical point p.

The proof of Theorem 4.2 takes up most of Chapter 4. In the first section we derive some local formulas for ∇f, $d\nabla f|_p$, and $d\varphi_t|_p$ (Lemmas 4.3, 4.4, and 4.5), and we use these formulas to show that a non-degenerate critical point of a smooth function $f : M \to \mathbb{R}$ is a **hyperbolic** fixed point of $\varphi_t : M \to M$ for any $t \neq 0$, i.e. $d\varphi_t|_p : T_p M \to T_p M$ has no eigenvalues of length 1. In the second section we prove the Local Stable Manifold Theorem (Theorem 4.11) following the proof of M.C. Irwin [82] [133]. Irwin's proof is based on the Lipschitz Inverse Function Theorem (Theorem 4.10), and it describes the local stable manifold of a hyperbolic fixed point as the graph of a differentiable map between Banach spaces of Cauchy sequences. Section 4.2 contains some fairly difficult real analysis. It is included for completeness, but the chapter is written so that readers can skip Section 4.2 without hindering their understanding of the rest of the book.

In the third section we use the Local Stable Manifold Theorem (Theorem 4.11) to prove the Global Stable and Unstable Manifold Theorems for a hyperbolic fixed point of a smooth diffeomorphism $\varphi : M \to M$ (Theorem 4.15 and Theorem 4.17). The Global Stable and Unstable Manifold Theorems show that the stable and unstable manifolds are the images of certain smooth injections whose construction is due to Smale [136]. Next, we consider the case of a diffeomorphism generated by a vector field on M. We show that if M is compact and the diffeomorphism is generated by the gradient vector field of a Morse function, then the stable and unstable manifolds are smoothly embedded open disks (Lemma 4.20).

In the last section of Chapter 4 we give some examples of stable and unstable manifolds. Our examples include stable and unstable manifolds of Morse functions, gradient vector fields, vector fields, and diffeomorphisms.

1.5 Basic differential topology

In Chapter 5 we prove some results on transversality, general position, and orientation that are used in subsequent chapters. We show that every immersion is locally equivalent to an inclusion (Theorem 5.2), and every submersion is locally equivalent to a projection (Theorem 5.5). As a corollary to Theorem 5.5 we obtain the Preimage Theorem (Corollary 5.9) which says that the preimage of a regular value is a submanifold.

If $f : M \to N$ and $g : Z \to N$ are smooth maps between smooth manifolds M, N, and Z, then f is **transverse** to g, $f \pitchfork g$, if and only if whenever $f(x) = g(z) = y$ we have

$$df_x(T_xM) + dg_z(T_zZ) = T_yN.$$

As a generalization of Corollary 5.9 we prove the Inverse Image Theorem (Theorem 5.11) which says that if $i : Z \hookrightarrow N$ is an immersed submanifold and $f \pitchfork i$, then $f^{-1}(Z)$ is a submanifold of M whose codimension in M is the same as the codimension of Z in N, i.e.

$$\dim M - \dim f^{-1}(Z) = \dim N - \dim Z.$$

As a corollary we note that if M and Z are immersed submanifolds of N that intersect transversally, then $M \cap Z$ is an immersed submanifold of N of dimension $\dim M + \dim Z - \dim N$ (Corollary 5.12).

Theorem 5.16 shows that transversality is a **stable** property, i.e. transverse maps remain transverse under small perturbations, and Theorems 5.17 and 5.19

show that smooth maps and embeddings can always be put in **general position**. More specifically, Theorems 5.17 and 5.19 show that if $f : M \to N$ and $g : Z \to N$ are smooth maps between smooth manifolds M, N, and Z, then we can perturb g by an arbitrarily small smooth homotopy $g_t : Z \to N$ such that the perturbed map g_1 is transverse to f. If $g : Z \to N$ is an embedding, then the smooth homotopy g_t can be chosen so that $g_t : Z \to N$ remains an embedding for all t.

Next we show that the class of Morse functions on a finite dimensional smooth manifold M is locally stable (Corollary 5.24) and dense as a subspace of the space of all smooth functions on M with the uniform topology (Theorem 5.27). We also show that the set of Morse functions on M is an open and dense subspace of the space of all smooth functions on M with the smooth topology (Theorem 5.31).

In the last two sections of Chapter 5 we define orientations and intersection numbers, and we prove the Lefschetz Fixed Point Theorem (Theorem 5.50). We end the chapter by showing that if $f : M \to M$ is a smooth map whose fixed points are non-degenerate and f is homotopic to the identity, then the Euler characteristic $\mathcal{X}(M)$ of M is a lower bound for the number of fixed points of f (Theorem 5.56), i.e.

$$\#\mathrm{fix}(f) \geq \sum_{j=0}^{m}(-1)^j \dim H_j(M;\mathbb{Q})$$

where m is the dimension of M.

1.6   Morse-Smale functions

A Morse function $f : M \to \mathbb{R}$ on a finite dimensional smooth Riemannian manifold (M, g) is said to satisfy the **Morse-Smale transversality condition** if and only if the stable and unstable manifolds of f intersect transversally, i.e.

$$W^u(q) \pitchfork W^s(p)$$

for all $p, q \in \mathrm{Cr}(f)$ (the set of all critical points of f). A Morse function that satisfies the Morse-Smale transversality condition is called **Morse-Smale**. A result due to Kupka [92] and Smale [135] implies that the set of Morse-Smale functions is a dense subspace of the space of all smooth functions on a finite dimensional compact smooth Riemannian manifold (M, g) (Theorem 6.6).

If $f : M \to \mathbb{R}$ is a Morse-Smale function, then for any two critical points p and q of f such that $W^u(q) \cap W^s(p) \neq \emptyset$

$$W(q,p) \stackrel{\mathrm{def}}{=} W^u(q) \cap W^s(p)$$

is an embedded submanifold of M of dimension $\lambda_q - \lambda_p$ (Proposition 6.2). Thus for a Morse-Smale function, the index of the critical points is strictly decreasing along the function's gradient flow lines (Corollary 6.3). In particular, this means that if $\lambda_q \leq \lambda_p$, then $W(q, p) = \emptyset$. Additional properties of the space $W(q, p)$ follow as consequences of Palis' λ-Lemma.

The λ-Lemma (Theorem 6.17) is one of the crucial theorems in the theory of smooth dynamical systems. A version of this theorem, as well as several important corollaries, were announced in a paper by Smale in 1960 [136]. Smale referenced the proof of the theorem to a preprint "On structural stability" which never appeared in print. The proof of the λ-Lemma and the proofs of the corollaries from [136] did not appear in print until 1969 when Palis, one of Smale's students, published his doctoral thesis [112]. In Chapter 6 we include a complete and detailed proof of the λ-Lemma following [114] and [143].

Following the proof of the λ-Lemma we prove several of Smale's corollaries (Corollaries 6.19, 6.20, 6.23, and 6.26). In particular, we show if $f : M \to \mathbb{R}$ is a Morse-Smale function on a finite dimensional compact smooth Riemannian manifold (M, g), then there is a partial ordering on the set of critical points $\mathrm{Cr}(f)$ (Definition 6.22). If $p, q \in \mathrm{Cr}(f)$, then we say that q is **succeeded** by p, $q \succeq p$, if and only if $W(q, p) = W^u(q) \cap W^s(p) \neq \emptyset$, i.e. there exists a gradient flow line from q to p. Using this notation we can state the most important result of Chapter 6 (Corollary 6.28) as follows: the closure of $W(q, p) \subseteq M$ consists of the points in M which lie on piecewise gradient flow lines from q to p. That is, if p and q are critical points of $f : M \to \mathbb{R}$ such that $q \succeq p$, then

$$\overline{W(q, p)} = \bigcup_{q \succeq \tilde{q} \succeq \tilde{p} \succeq p} W(\tilde{q}, \tilde{p}).$$

In particular, Corollary 6.29 says that if q and p are critical points of relative index one, i.e. if $\lambda_q - \lambda_p = 1$, then

$$\overline{W(q, p)} = W(q, p) \cup \{p, q\}$$

and $W(q, p)$ has finitely many components. That is, the number of gradient flow lines from q to p is finite when $\lambda_q - \lambda_p = 1$.

In the last section of Chapter 6 we present a couple of results due to Franks [59] that relate the stable and unstable manifolds of a Morse-Smale function $f : M \to \mathbb{R}$ to the cells and attaching maps in the CW-complex X determined by f (see Theorem 3.28). Franks' results are not needed for the proof of the Morse Homology Theorem (Theorem 7.4), but they are of interest because they show how the Morse-Smale transversality condition (and an additional assumption on the Riemannian metric) can be used to strengthen the results in Chapter 3 concerning the CW-complex associated to a Morse function.

The first of Franks' results (Theorem 6.34) says that if $f : M \to \mathbb{R}$ is a Morse-Smale function on a finite dimensional compact smooth Riemannian manifold (M, g) that satisfies a certain generic condition with respect to the metric g near its critical points, then there is a homotopy equivalence $h : M \to X$ that sends a k-dimensional unstable manifold of f into the base of a unique k-cell in the CW-complex X determined by f. The second result (Theorem 6.40) says that if p and q are critical points of f such that p is an immediate successor of q, then the homotopy class of the relative attaching map of the cell associated to q to the cell associated to p corresponds to the framed submanifold $W(q,p)/\mathbb{R}$ under the Thom-Pontryagin construction.

1.7 The Morse Homology Theorem

In Chapter 7 we construct the Morse-Smale-Witten chain complex of a Morse-Smale function $f : M \to \mathbb{R}$ on a compact smooth Riemannian manifold (M, g), and we and prove that the homology of the Morse-Smale-Witten chain complex coincides with the singular homology of M (Theorem 7.4: Morse Homology Theorem).

The kth chain group in the Morse-Smale-Witten chain complex $(C_*(f), \partial_*)$ is defined to be the free abelian group generated by $\mathrm{Cr}_k(f)$, the critical points of index k. The Morse-Smale-Witten boundary operator ∂_* is defined as

$$\partial_{k+1}(q) = \sum_{p \in \mathrm{Cr}_k(f)} n(q,p)p$$

where $q \in \mathrm{Cr}_{k+1}(f)$ is a critical point of index $k+1$ and the sum is taken over all critical points p of index k. The number $n(q,p) \in \mathbb{Z}$ is the algebraic sum of signed gradient flow lines from q to p. It is also the intersection number of the "unstable sphere" of q and the "stable sphere" of p at a non-critical level between p and q.

Since the Morse-Smale-Witten boundary operator is defined by counting gradient flow lines (with sign), it has a geometric flavor and is easy to visualize. In Section 7.2 we compute the homology of some manifolds using the Morse-Smale-Witten chain complex. In each example we draw the gradient flow lines and show the orientations used to determine the coefficients in the Morse-Smale-Witten boundary operator, i.e. $n(q,p) \in \mathbb{Z}$.

In Sections 7.3, 7.4, and 7.5 we prove the Morse Homology Theorem (Theorem 7.4) using the Conley index and Conley's connection matrix. The Conley index is a pointed homotopy class associated to an isolated compact invariant set of a dynamical system that remains the same under small perturbations of the system. In Section 7.3 we review the basic facts of the Conley index theory and describe examples of index pairs which are pertinent to the proof of the main theorem. In particular, we observe that the Conley index of the critical set $\mathrm{Cr}_k(f)$ agrees with the homotopy class of $X^{(k)}/X^{(k-1)}$, where $X^{(k)}$ denotes the k-skeleton of the CW-complex X associated to the Morse function.

The proof we give in Section 7.4 of the Morse-Homology Theorem is due to Salamon [128], and it describes the Morse-Smale-Witten boundary operator as a special case of Conley's connection matrix [35] [60] [61]. The main step in the proof is Lemma 7.21 which relates the Morse-Smale-Witten boundary operator to a connecting homomorphism in the homology exact sequence of a triple. Lemma 7.21 is proved using the fact that we can choose "nice" index pairs for the critical points and the Conley index is independent of the choice of the index pairs representing it.

Once we prove Lemma 7.21 the proof of the Morse Homology Theorem follows easily from arguments similar to those used to prove the CW-homology Theorem (Theorem 2.15). Section 7.5 is devoted to proving that the Conley index is independent of the choice of index pairs representing it [35] [127].

1.8   Morse theory on Grassmann manifolds

In Chapter 8 we show how Bott's perfect Morse functions (discussed in Example 3.7) are examples of a more general class of Morse-Smale functions defined on the complex Grassmann manifolds.

The Morse-Smale functions $f_A : G_{n,n+k}(\mathbb{C}) \to \mathbb{R}$ are defined by first fixing an embedding of the complex Grassmann manifold $G_{n,n+k}(\mathbb{C})$ into the Lie algebra of skew-Hermitian matrices $\psi : G_{n,n+k}(\mathbb{C}) \hookrightarrow \mathfrak{u}(n + k)$ and then taking the inner product of $x \in G_{n,n+k}(\mathbb{C})$ with a matrix $A \in \mathfrak{u}(n + k)$ using the negative of the trace form, i.e.

$$f_A(x) \stackrel{\text{def}}{=} < \psi(x), A > = -\text{trace}(\psi(x)A)$$

where $\psi(x)A$ denotes matrix multiplication. We prove in Theorem 8.12 that whenever A has distinct eigenvalues f_A is a Morse function.

In Section 8.3 we describe an almost complex structure on $G_{n,n+k}(\mathbb{C})$ in terms of the Lie bracket on $\mathfrak{u}(n + k)$, and we compute the gradient vector field of the function f_A in terms of the almost complex structure (Theorem 8.28). In Section 8.4 we specialize to a specific matrix $A \in \mathfrak{u}(n + k)$, and we analyze the critical points and gradient flow lines of f_A. We give a simple formula for computing the index of the critical points of f_A (Theorem 8.32) which shows that the critical points of f_A are all of even index, and we prove that there is a bijection between the critical points of f_A of index λ and the Schubert cells of $G_{n,n+k}(\mathbb{C})$ of dimension λ for all $\lambda \in \mathbb{Z}_+$ (Corollary 8.33). In Section 8.6 we show that the unstable manifolds of f_A are the Schubert cells of $G_{n,n+k}(\mathbb{C})$ (Theorem 8.40), and we use this fact to show that the function f_A satisfies the Morse-Smale transversality condition (Theorem 8.45).

Theorem 8.32 shows that the boundary operators in the Morse-Smale-Witten chain complex of f_A are all trivial. Thus, the homology of $G_{n,n+k}(\mathbb{C})$ can be computed using either the Morse Homology Theorem (Theorem 7.4) or from the CW-Homology Theorem (Theorem 2.15) and Theorem 3.28. This is discussed in Section 8.7.

In the last section of Chapter 8 we briefly outline further generalizations and applications to the theory of Lie groups and symplectic geometry.

1.9   Floer homology theories

In Chapter 9 we give a brief introduction to Floer homology theories: these are attempts to build Morse-Smale-Witten chain complexes $(C_*(\mathcal{F}), \partial_*)$, where the chains $C_*(\mathcal{F})$ are free abelian groups generated by the critical points of some function $\mathcal{F}$ defined on some infinite dimensional manifold $\mathcal{M}$, and to identify the resulting Floer homology.

In the first theory $\mathcal{M}$ is the space of smooth contractible free loops on a symplectic manifold (M, ω), and $\mathcal{F}$ is the action functional. Floer invented this theory as a tool to solve the Arnold conjecture, which asserts that if all the fixed points of a Hamiltonian diffeomorphism of (M, ω) are non-degenerate, then their number is bounded from below by the sum of the Betti numbers of M.

In the second theory the space $\mathcal{M}$ is the space of smooth paths in (M, ω) starting at a Lagrangian submanifold L and ending at another Lagrangian submanifold L'. There is a function on $\mathcal{M}$ whose critical points are the intersection points of L and L'. This theory is called the Floer homology for Lagrangian intersections, and it was invented to solve the problem of finding a lower bound on the number of intersection points of L and L'. It is a generalization of the first theory since fixed points of a symplectic diffeomorphism (M, ω) coincide with intersection of the graph of the symplectic diffeomorphism and the diagonal in $M \times M$, which are both Lagrangian submanifolds of $(M \times M, \omega \ominus \omega)$.

In the third theory, $\mathcal{M}$ is the space of all $SO(3)$ (or $SU(2)$) connections on a principal $SO(3)$ (or $SU(2)$) bundle over a compact oriented 3-dimensional manifold, and the function $\mathcal{F}$ is the Chern-Simons functional, whose critical points coincide with flat connections. This homology is called the instanton homology.

In the case where the 3-manifold M is a homology sphere and the bundle is the trivial bundle $M \times SU(2)$, the "Euler characteristic" of the instanton homology is an invariant of the 3-manifold M equal to twice the Casson invariant of M.

The last section of Chapter 9 shows that the instanton homology, like the first two homologies, also has a symplectic flavor.

1.10   Guide to the book

This book has several uses, both as a textbook and as a reference.

One semester course

A one semester course on Morse homology for students who have already completed a course on algebraic topology might look something like this:

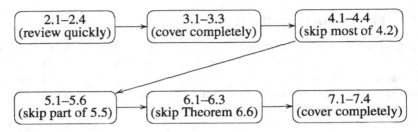

Most students should also find the results in Section 6.4 interesting, but those results rely on an additional assumption on the Riemannian metric that is not needed for the proof of the Morse Homology Theorem given in Chapter 7.

Two semester course

A two semester course on Morse homology might look something like this:

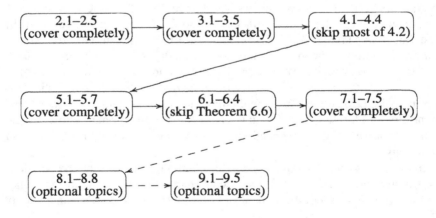

A two semester course could cover Chapters 2 and 3 completely at a leisurely pace. Chapter 4 could then be covered fairly quickly as long as Section 4.2 is skipped. Chapters 5, 6, and 7 could be covered completely, with the exception of the proof of Theorem 6.6: Kupka-Smale Theorem (which most students should skip). The topics in Chapters 8 and 9 could also be covered as optional topics.

For the expert

For the expert in algebraic and differential topology who wants to get to the proof of the Morse Homology Theorem as quickly as possible we recommend the following:

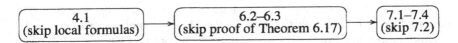

Note that Smale's corollaries to the λ-Lemma (Theorem 6.17) found in Section 6.3 are **essential** for the proof of the Morse Homology Theorem given in Chapter 7. However, the results in Section 6.4 are **not needed** to prove the Morse Homology Theorem. There is an additional assumption on the Riemannian metric used to prove the results in Section 6.4 that is **not necessary** for the proof of the Morse Homology Theorem given in Chapter 7.

Notes on the proofs of the Morse Homology Theorem

There are at least four different ways to prove the Morse Homology Theorem:

1. Using the infinite dimensional techniques of Floer homology.

2. Using techniques from cobordism theory, i.e. handlebody decompositions.

3. Using Franks' Connecting Manifold Theorem (Theorem 6.40).

4. Using the Conley index and Conley's connection matrix (Chapter 7).

The first approach is the "modern" one found in the book **Morse Homology** by M. Schwarz [131]. In that book Schwarz constructs the Morse-Smale-Witten chain complex and shows that the resulting Morse homology theory satisfies the Eilenberg-Steenrod axioms. This proves that there exists (a non-explicit) isomorphism between Morse homology and singular homology.

The second approach is implicit in Section 6 of the book **Lectures on the h-cobordism Theorem** by J. Milnor [102]. However, the results in Milnor's book, as written, apply only to "self-indexing" Morse functions. The "self-indexing" assumption is not necessary for the Morse Homology Theorem.

The third approach uses the results of Franks [59], discussed in Section 6.4 of this book. This approach first appeared explicitly in [34]. Franks' results apply only in cases where the Riemannian metric is "compatible with the Morse charts" for the function (see Definition 6.30). In the literature, this additional assumption is sometimes expressed by saying that the Riemannian metric is "nice", the vector field is in "standard form" [59], or the gradient vector field is "Special Morse" [20]. Another similar assumption is that the function has "split Morse singularities" [19]. Although these sorts of extra assumptions greatly simplify some of the proofs in the theory, they are not needed for the Morse Homology Theorem.

The fourth approach, which we give in Chapter 7, is due to Salamon [128]. There are several advantages to this approach. First of all, it does not require any superfluous assumptions on either the Morse function or the Riemannian metric, i.e. it works for any Morse-Smale function. Secondly, the proof produces an explicit isomorphism between the Morse homology groups and the singular homology groups (see the last line of the proof of Theorem 7.4). And finally, the proof uses only elementary "classical" techniques from algebraic topology and homotopy theory.

Chapter 2

The CW-Homology Theorem

In this chapter we introduce singular homology, and we prove the CW-Homology Theorem. The CW-Homology Theorem (Theorem 2.15) states that the singular homology $H_*(X, A; \Lambda)$ is isomorphic to the homology of the CW-chain complex $(\underline{C}_*(X, A; \Lambda), \underline{\partial}_*)$, and it gives a formula for computing the boundary operator $\underline{\partial}_*$ in the CW-chain complex in terms of the degrees of the attaching maps. We also prove some basic theorems from homotopy theory. In particular, we prove that if $A \subseteq X$ is closed and the inclusion $A \hookrightarrow X$ is a cofibration, then $H_k(X, A; \Lambda) \approx H_k(X/A, *; \Lambda)$ for all k (Corollary 2.31).

2.1 Singular homology

For any $k \in \mathbb{Z}_+$, the **standard k-simplex** Δ^k is the subspace of $\mathbb{R}^{k+1}$ consisting of $(k+1)$-tuples $(t_0, t_1, \ldots, t_k)$ with $t_i \geq 0$ and $t_0 + t_1 + \cdots + t_k = 1$. There are **face maps** $F_i^k : \Delta^{k-1} \to \Delta^k$ defined by

$$F_i^k(t_0, \ldots, t_{k-1}) = (t_0, \ldots, t_{i-1}, 0, t_i, \ldots, t_{k-1}) \subset \Delta^k$$

for $0 \leq i \leq k$. A **singular k-simplex** of X is a continuous map $\sigma : \Delta^k \to X$. For $k \geq 0$ and Λ a commutative ring with unit, we denote by $C_k(X; \Lambda)$ the free Λ-module with generators the singular k-simplices, i.e. an element of $C_k(X; \Lambda)$ is a formal sum $\sum_{i \in I} \lambda_i \sigma_i$ where $\lambda_i \in \Lambda$, the σ_i are k-simplices, and λ_i is non-zero for only a finite number of $i \in I$. For $k < 0$ we define $C_k(X; \Lambda)$ to be the trivial module.

There is a boundary operator

$$\partial_k : C_k(X; \Lambda) \to C_{k-1}(X; \Lambda)$$

defined on a generator $\sigma \in C_k(X; \Lambda)$ by

$$\partial_k(\sigma) = \sigma \circ F_0^k - \sigma \circ F_1^k + \cdots + (-1)^k \sigma \circ F_k^k.$$

∂_k extends to a Λ-module homomorphism that satisfies

$$\partial_k \circ \partial_{k+1} = 0,$$

and therefore the kernel of $\partial_k : C_k(X; \Lambda) \to C_{k-1}(X; \Lambda)$, which is denoted $Z_k(X; \Lambda)$, contains the image of $\partial_{k+1} : C_{k+1}(X; \Lambda) \to C_k(X; \Lambda)$, which is denoted $B_k(X; \Lambda)$.

Definition 2.1 *The quotient Λ-module*

$$H_k(X; \Lambda) = Z_k(X; \Lambda)/B_k(X; \Lambda)$$

*is called the kth **singular homology group** of X with coefficients in Λ.*

We will denote the graded Λ-module

$$H_*(X; \Lambda) = \bigoplus_{k \geq 0} H_k(X; \Lambda)$$

and write $H_*(X)$ when $\Lambda = \mathbb{Z}$ or when Λ is understood from the context.

The relative groups

If A is a subspace of the topological space X (the subspace A may be empty), then we call (X, A) a **topological pair**. The inclusion $i : A \to X$ induces a Λ-module homomorphism $i_* : C_k(A; \Lambda) \to C_k(X; \Lambda)$ by $i_*(\sigma) = i \circ \sigma$. We denote by

$$C_k(X, A; \Lambda) = C_k(X; \Lambda)/C_k(A; \Lambda)$$

the quotient module of $C_k(X; \Lambda)$ by the sub Λ-module $C_k(A; \Lambda)$. Clearly $\partial_k(C_k(A; \Lambda)) \subseteq C_{k-1}(A; \Lambda)$, and hence $\partial_k : C_k(X; \Lambda) \to C_{k-1}(X; \Lambda)$ induces a map

$$\overline{\partial}_k : C_k(X, A; \Lambda) \to C_{k-1}(X, A; \Lambda)$$

that satisfies $\overline{\partial}^2 = 0$. Hence, we can define the relative homology group

$$H_k(X, A; \Lambda) = \frac{\ker \overline{\partial}_k : C_k(X, A; \Lambda) \to C_{k-1}(X, A; \Lambda)}{\operatorname{im} \overline{\partial}_{k+1} : C_{k+1}(X, A; \Lambda) \to C_k(X, A; \Lambda)}.$$

Observe that $C_k(X; \Lambda) = C_k(X, \emptyset; \Lambda)$. As above, we will write $H_k(X, A)$ when $\Lambda = \mathbb{Z}$ or when Λ is understood from the context.

Fundamental properties

A map $f : (X, A) \to (Y, B)$ between topological pairs is a continuous function $f : X \to Y$ such that $f(A) \subseteq B$. Such an f induces a Λ-module homomorphism

$$f_* : H_*(X, A) \to H_*(Y, B)$$

that preserves the grading. If $g : (Y, B) \to (Z, C)$ is also map of pairs, then the induced maps satisfy $(f \circ g)_* = f_* \circ g_* : H_*(A, X) \to H_*(Z, C)$. For instance, we have a homomorphism $i_* : H_*(A) \to H_*(X)$ induced by the inclusion $i : A \to X$ and a homomorphism $j_* : H_*(X) \to H_*(X, A)$ induced by the inclusion $j : (X, \emptyset) \to (X, A)$.

Suppose now that we are given a triple (X, B, A) where $A \subseteq B \subseteq X$. The inclusions $(B, A) \subseteq (X, A)$ and $(X, A) \subseteq (X, B)$ induce homomorphisms which we will also denote by i_* and j_*

$$i_* : H_*(B, A) \ \to \ H_*(X, A)$$
$$j_* : H_*(X, A) \ \to \ H_*(X, B),$$

and these maps satisfy $(j \circ i)_* = j_* \circ i_*$.

The singular homology groups defined above have the following fundamental properties.

(P1) (**Homotopy property.**) If $f, g : (X, A) \to (Y, B)$ are homotopic as maps of pairs, then

$$f_* = g_* : H_*(X, A) \to H_*(Y, B).$$

It follows that if $(X, A) \simeq (Y, B)$, then $H_*(X, A) \approx H_*(Y, B)$.

(P2) (**Exactness property.**) For any $A \subseteq X$ there is a **connecting homomorphism**

$$\delta_k : H_k(X, A) \to H_{k-1}(A)$$

for all k which fits into the following exact sequence.

$$\cdots \longrightarrow H_k(A) \xrightarrow{i_*} H_k(X) \xrightarrow{j_*} H_k(X, A) \xrightarrow{\delta_k} H_{k-1}(A) \longrightarrow \cdots$$

Moreover, δ_k is a **natural transformation**, i.e. if $f : (X, A) \to (Y, B)$ is a map of pairs, then the following diagram commutes for all k.

$$
\begin{array}{ccc}
H_k(X, A) & \xrightarrow{\delta_k} & H_{k-1}(A) \\
\downarrow{\scriptstyle f_*} & & \downarrow{\scriptstyle f_*} \\
H_k(Y, B) & \xrightarrow{\delta_k} & H_{k-1}(B)
\end{array}
$$

Note: A sequence of abelian groups and homomorphisms $G_1 \xrightarrow{\alpha} G_2 \xrightarrow{\beta} G_3$ is said to be **exact** if and only if im $\alpha = \ker \beta$.

As a consequence of the exactness property one can show that for a triple $A \subseteq B \subseteq X$ the connecting homomorphism induces a homomorphism

$$\delta_* = j_* \circ \delta_k : H_k(X, B) \xrightarrow{\delta_k} H_{k-1}(B) \xrightarrow{j_*} H_{k-1}(B, A)$$

where the second map is induced from the inclusion $j : (B, \emptyset) \to (B, A)$. This homomorphism fits into the following exact sequence.

$$\cdots \longrightarrow H_k(B, A) \xrightarrow{i_*} H_k(X, A) \xrightarrow{j_*} H_k(X, B) \xrightarrow{\delta_*} H_{k-1}(B, A) \longrightarrow \cdots$$

(P3) (**Excision property.**) If $U \subseteq A \subseteq X$ and $\overline{U} \subseteq \text{int}\,(A)$, then there is an isomorphism

$$H_*(X - U, A - U) \approx H_*(X, A)$$

induced by the excision map $e : (X - U, A - U) \to (X, A)$.

(P4) (**Dimension property.**) If P is a one point space, then

$$H_k(P) = \begin{cases} \Lambda & \text{if } k = 0 \\ 0 & \text{otherwise.} \end{cases}$$

The properties (P1) through (P4) completely "characterize" singular homology for connected topological spaces and the disjoint union of finitely many connected topological spaces. For spaces that are an infinite disjoint union of connected topological spaces we need to add the following property.

(P5) (**Additivity property.**) For the disjoint union $X = \coprod_\alpha X_\alpha$ the inclusions $i_\alpha : X_\alpha \hookrightarrow X$ induce for all k an isomorphism

$$\oplus (i_\alpha)_* : \bigoplus_\alpha H_k(X_\alpha) \xrightarrow{\approx} H_k(X).$$

For more details and proofs of the fundamental properties see [30], [66], [109], or [138].

Although the singular homology groups are defined for any topological space, they are difficult to compute directly from the definition. Properties (P1) through (P5) can often be used to compute the homology of a topological space, but computations involving properties (P1) through (P5) can sometimes be non-trivial. As an example, we will now compute the homology of the n-sphere S^n using only the definition of singular homology and properties (P1) through (P4). In Section 2.4 we will see that the homology of S^n is trivial to compute using the CW-chain complex.

Example 2.2 If $D^n = \{x \in \mathbb{R}^n | \, |x| \leq 1\}$ is the n-disk and ∂D^n is its boundary $S^{n-1} = \{x \in \mathbb{R}^n | \, |x| = 1\}$, then we have the following for $n > 0$.

$$H_k(S^n) = \begin{cases} \Lambda & \text{if } k = n \text{ or } k = 0 \\ 0 & \text{otherwise} \end{cases}$$

$$H_k(D^n, \partial D^n) = \begin{cases} \Lambda & \text{if } k = n \\ 0 & \text{otherwise} \end{cases}$$

When $k = 0$ this follows easily from the definition of singular homology. For $k > 0$ consider the long exact sequence of the pair (D^n, S^{n-1}).

$$\cdots \longrightarrow H_k(D^n) \xrightarrow{j_*} H_k(D^n, S^{n-1}) \xrightarrow{\delta_k} H_{k-1}(S^{n-1}) \xrightarrow{i_*} H_{k-1}(D^n) \longrightarrow$$

Since D^n is homotopic to a point, $H_k(D^n) = 0$ if $k > 0$, and hence we have $H_k(D^n, S^{n-1}) \approx H_{k-1}(S^{n-1})$ if $k > 1$. When $k = 1$ we have

$$0 \longrightarrow H_1(D^n, S^{n-1}) \xrightarrow{\delta_1} H_0(S^{n-1}) \xrightarrow{i_*} H_0(D^n) \longrightarrow 0,$$

and hence $H_1(D^n, S^{n-1}) \approx \ker(H_0(S^{n-1}) \xrightarrow{i_*} \Lambda)$ where i_* is surjective. Since $H_0(S^{n-1}) \approx \Lambda$ if $n > 1$ and $H_0(S^{n-1}) \approx \Lambda \oplus \Lambda$ if $n = 1$, we see that $H_1(D^n, S^{n-1}) = 0$ if $n > 1$ and $H_1(D^1, S^0) \approx \Lambda$. More explicitly, when $n = 1$ we have $H_0(S^0) = C_0(-1; \Lambda) \oplus C_0(1; \Lambda) \approx \Lambda \oplus \Lambda$ and $i_*([\sigma_{-1}], [\sigma_1]) = [\sigma_{-1}] + [\sigma_1] \in H_0(D^1)$. So, $\delta_1(H_0(D^1, S^0)) = \ker(H_0(S^0) \xrightarrow{i_*} \Lambda)$ is generated by $(-[\sigma_{-1}], [\sigma_1])$ where $\sigma_{-1} : \Delta^0 \to -1 \in D^1$ and $\sigma_1 : \Delta^0 \to 1 \in D^1$ are unique maps to points.

Now consider the long exact sequence of the pair (S^n, D^n_+) where D^n_+ is the upper hemisphere.

$$\cdots \longrightarrow H_k(D^n_+) \xrightarrow{i_*} H_k(S^n) \xrightarrow{j_*} H_k(S^n, D^n_+) \xrightarrow{\delta_k} H_{k-1}(D^n_+) \longrightarrow \cdots$$

Using the same reasoning as before we see that $H_k(S^n) \approx H_k(S^n, D^n_+)$ for $k > 1$. When $k = 1$ and $n > 0$ we have

$$\cdots \longrightarrow 0 \longrightarrow H_1(S^n) \xrightarrow{j_*} H_1(S^n, D^n_+) \xrightarrow{\delta_1} H_0(D^n_+) \xrightarrow{i_*} H_0(S^n) \longrightarrow 0$$

since $H_0(S^n, D^n_+) = 0$ (as seen from the definition of relative singular homology). Since $i_* : \Lambda \to \Lambda$ is an isomorphism, $\delta_1 = 0$, and hence j_* is also an isomorphism. Therefore, $H_k(S^n) \approx H_k(S^n, D^n_+)$ for all $k > 0$ if $n > 0$.

Finally, if U is a small open neighborhood of the north pole in S^n we can apply excision and the homotopy invariance of homology to conclude that

$$H_k(S^n, D^n_+) \approx H_k(S^n - U, D^n_+ - U) \approx H_k(D^n, S^{n-1})$$

for all k. Putting this all together we conclude that for $n > 0$ we have

$$H_k(S^n) \overset{k \geq 0}{\approx} H_k(S^n, D^n_+) \approx H_k(D^n, S^{n-1}) \overset{k \geq 1}{\approx} H_{k-1}(S^{n-1}).$$

Starting with

$$H_1(S^n) \approx H_1(D^n, S^{n-1}) = \begin{cases} \Lambda & n = 1 \\ 0 & n > 1 \end{cases}$$

the result now follows by induction.

2.2   Singular cohomology

With the same notation as above consider

$$C^k(X; \Lambda) \stackrel{\text{def}}{=} \text{Hom}_\Lambda(C_k(X; \Lambda), \Lambda),$$

the group of homomorphisms over Λ of the Λ-module $C_k(X; \Lambda)$ into Λ. Let $< c, a > \in \Lambda$ denote the value of $c \in C^k(X; \Lambda)$ on $a \in C_k(X; \Lambda)$, and define the coboundary operator

$$d_k : C^k(X; \Lambda) \to C^{k+1}(X; \Lambda)$$

as follows. If $c \in C^k(X; \Lambda)$, then $d_k c \in C^{k+1}(X; \Lambda)$ is the cochain whose value over a chain $a \in C_{k+1}(X; \Lambda)$ is

$$< d_k c, a > = < c, \partial_{k+1} a > .$$

Clearly $d_k \circ d_{k-1} = 0$.

Definition 2.3 *The kth **singular cohomology group** of X with coefficients in the commutative ring Λ is defined to be*

$$H^k(X; \Lambda) = Z^k(X; \Lambda)/B^k(X; \Lambda)$$

where

$$Z^k(X; \Lambda) = \ker d_k : C^k(X; \Lambda) \to C^{k+1}(X; \Lambda)$$

and

$$B^k(X; \Lambda) = \text{im } d_{k-1} : C^{k-1}(X; \Lambda) \to C^k(X; \Lambda).$$

As with the homology groups, we will write $H^k(X)$ when $\Lambda = \mathbb{Z}$ or when Λ is understood from the context.

In the same fashion we can define the relative cohomology groups, denoted by $H^k(X, A; \Lambda)$, starting with the relative cochain group

$$C^k(X, A; \Lambda) = \text{Hom}_\Lambda(C_k(X, A), \Lambda).$$

The following theorem gives a relation between homology and cohomology. (See for instance Theorem A.1 of [104].)

Theorem 2.4 *Suppose that $H_{k-1}(X)$ is zero or is a free Λ-module. Then $H^k(X)$ is canonically isomorphic to the Λ-module $\text{Hom}_\Lambda(H_k(X), \Lambda)$.*

2.3   CW-complexes

The CW-complexes are a nice category of topological spaces introduced by J.C.H. Whitehead [151]. The "C" in CW stands for "closure finite" and the "W" stand for "weak topology". A CW-complex is built step by step by successive operations called **attaching cells**. Let D^n be the unit n-disk in $\mathbb{R}^n$, i.e. $D^n = \{x \in \mathbb{R}^n \mid |x| \leq 1\}$, and S^{n-1} the unit $(n-1)$-sphere $S^{n-1} = \partial D^n = \{x \in \mathbb{R}^n \mid |x| = 1\}$. If $f_\partial : S^{n-1} \to X$ is a continuous map into a topological space X, we denote by

$$X \cup_{f_\partial} D^n$$

the quotient space of the disjoint union $X \amalg D^n$ where $x \in \partial D^n = S^{n-1}$ is identified with $f_\partial(x) \in X$. We say that $X \cup_{f_\partial} D^n$ is obtained from X by **attaching an n-cell** and f_∂ is called the **attaching map**.

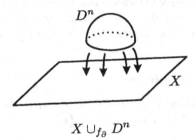

$$X \cup_{f_\partial} D^n$$

Definition 2.5 *A topological space X has a **CW-structure** if there are subspaces $X^{(n)}$ with*

$$X^{(0)} \subseteq X^{(1)} \subseteq \cdots \subseteq X = \bigcup_{n \in \mathbb{Z}_+} X^{(n)}$$

such that

1. $X^{(0)}$ *is a discrete set of points.*

2. $X^{(n+1)}$ *is obtained from $X^{(n)}$ by attaching $(n+1)$-cells for all $n \geq 0$.*

3. X *has the **weak topology**. This means that a subspace of X is open if and only if its intersection with $X^{(n)}$ is open in $X^{(n)}$ for all $n \in \mathbb{Z}_+$.*

A space X with a specified CW-structure is called a **CW-complex**, and the subspace $X^{(n)}$ is called the n-**skeleton** of the CW-complex X. An attaching map $f_\partial : S^{n-1} \to X^{(n-1)}$ extends to a map $f : D^n \to X^{(n)}$ called the **characteristic map**. We will call the image of D^n under f a **closed cell** in X and the image of $D^n - \partial D^n$ an **open cell** in X. Note that an open cell is open as a subspace of $X^{(n)}$, but it is usually not open as a subspace of X. By a

subcomplex of a CW-complex X we mean a union of closed cells in X which is itself a CW-complex with the same attaching maps. The following theorem gives the fundamental properties of CW-complexes (see Proposition IV.8.1 and Theorem IV.8.2 of [30]).

Theorem 2.6 *For any CW-complex X we have the following.*

1. *If $A \subseteq X$ has no two points in the same open cell, then A is closed and discrete.*

2. *If $K \subseteq X$ is compact, then K is contained in a finite union of open cells.*

3. *Every closed cell of X is contained in a finite subcomplex of X.*

4. *Any compact subset of X is contained in a finite subcomplex.*

As a consequence of property (3) we see that the image of an attaching map $f_\partial : S^{n-1} \to X^{(n-1)}$ lies in a finite subcomplex, and hence the closure of every open cell lies in a finite subcomplex. This is the origin of the term "closure finite".

Examples of CW-complexes

Example 2.7 (The n-sphere) The n-sphere S^n is a CW-complex with $X^{(0)} = \{*\}$, $X^{(n)} = \{*\} \cup_{f_\partial} D^n$, and $f_\partial : \partial D^n \to \{*\}$.

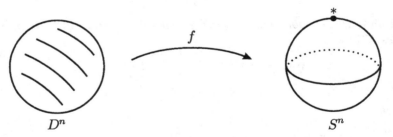

$$D^n \qquad\qquad\qquad\qquad\qquad\qquad S^n$$

Example 2.8 (Projective spaces) For any $n \in \mathbb{Z}_+$, the real projective space $\mathbb{R}P^n$ is the space of all lines in $\mathbb{R}^{n+1}$. The space $\mathbb{R}P^n$ is topologized by noting that there is a free $\mathbb{R}^*$-action

$$\begin{aligned}
\mathbb{R}^* \times \mathbb{R}^{n+1} - \{0\} &\to \mathbb{R}^{n+1} - \{0\} \\
(\lambda, (x_0, \ldots, x_n)) &\mapsto (\lambda x_0, \ldots, \lambda x_n),
\end{aligned}$$

where $\mathbb{R}^* = \mathbb{R} - \{0\}$, and identifying $\mathbb{R}P^n \approx (\mathbb{R}^{n+1} - \{0\})/\mathbb{R}^*$, where $(\mathbb{R}^{n+1} - \{0\})/\mathbb{R}^*$ is given the quotient topology. One can show that there is homeomorphism

$$\mathbb{R}P^n \approx \mathbb{R}P^{n-1} \cup_{f_\partial} D^n$$

where $f_\partial : S^{n-1} \to \mathbb{R}P^{n-1}$ is the map assigning to each point $p \in S^{n-1}$ the line through p and the origin. By induction this determines a CW-complex structure on $\mathbb{R}P^n$ for all $n \in \mathbb{Z}_+$.

Analogously, for the complex projective space $\mathbb{C}P^n \approx (\mathbb{C}^{n+1} - \{0\})/\mathbb{C}^*$ there is a homeomorphism

$$\mathbb{C}P^n \approx \mathbb{C}P^{n-1} \cup_{f_\partial} D^{2n}$$

where $f_\partial : S^{2n-1} \to \mathbb{C}P^{n-1}$ is the map assigning to each point $p \in S^{2n-1} \subseteq \mathbb{C}^n$ the complex line through p and the origin. By induction this determines a CW-complex structure on $\mathbb{C}P^n$ for all $n \in \mathbb{Z}_+$.

Both $\mathbb{R}P^n$ and $\mathbb{C}P^n$ are examples of a more general class of spaces known as Grassmann manifolds which are discussed in more detail in Chapter 8.

Example 2.9 (Simplicial complexes) Any simplicial complex X is a CW-complex. The n-cells of the CW-complex are just the n-simplices.

Example 2.10 (Compact manifolds) One of the basic theorems from Morse theory (Theorem 3.28) says that any Morse function (see Definition 3.1) on a compact smooth Riemannian manifold M determines a CW-complex which is homotopic to M.

2.4 CW-homology

A topological pair (X, A) where X is a CW-complex and A is a subcomplex of X is called a **CW-pair**. Throughout this section we fix a CW-pair (X, A) and let $X^{(n)}$ denote the n-skeleton of X union the entire subcomplex A. For every σ let D^n_σ be a copy of D^n and let S^{n-1}_σ a copy of S^{n-1} where σ ranges over the n-cells of X not in A. Let $\{f_{\partial\sigma}\}$ be the attaching maps $f_{\partial\sigma} : S^{n-1}_\sigma \to X^{(n-1)}$ and $\{f_\sigma\}$ the characteristic maps $f_\sigma : D^n_\sigma \to X^{(n)}$.

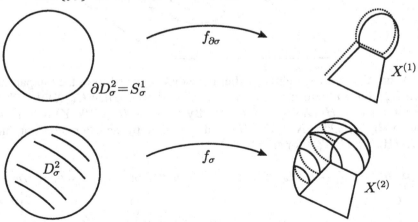

Attaching a 2-cell (Figure adapted from [30].)

Lemma 2.11 *For any commutative ring with unit Λ,*

$$H_k(X^{(n)}, X^{(n-1)}; \Lambda) \approx \begin{cases} \underline{C}_n(X, A; \Lambda) & \text{for } k = n \\ 0 & \text{otherwise.} \end{cases}$$

where

$$\underline{C}_n(X, A; \Lambda) \approx \bigoplus_\sigma H_n(D_\sigma^n, \partial D_\sigma^n; \Lambda) \approx \bigoplus_\sigma \Lambda$$

is the free Λ-module generated by the n-cells of X not in A. Moreover, the map

$$\bigoplus_\sigma f_{\sigma*} : \bigoplus_\sigma H_n(D_\sigma^n, \partial D_\sigma^n; \Lambda) \to H_n(X^{(n)}, X^{(n-1)}; \Lambda)$$

is an isomorphism.

Proof:

Each closed cell D_σ^n is a subset of $X^{(n)}$ with identifications on its boundary given by the attaching map $f_\sigma : D_\sigma^n \to X^{(n)}$.

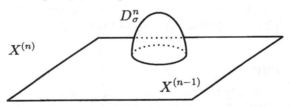

Let U be $X^{(n)}$ with a closed disk C_σ removed from the top part of each $f_\sigma(D_\sigma^n)$, and let $N_\sigma = f_\sigma(D_\sigma^n) - C_\sigma$.

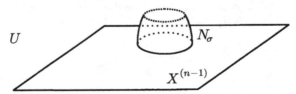

Let Y be $X^{(n)}$ with a slightly smaller open disk removed from the top part of each D_σ^n. Since Y deformation retracts to $X^{(n-1)}$, we see that $H_k(X^{(n)}, X^{(n-1)})$ is isomorphic to $H_k(X^{(n)}, Y)$ for all k. By excision, $H_k(X^{(n)}, Y)$ is isomorphic to $H_k(\amalg_\sigma(D_\sigma^n - N_\sigma), Y - U)$, and by homotopy the latter is isomorphic to $H_k(\amalg_\sigma(D_\sigma^n, \partial D_\sigma^n))$ for all k.

$$\begin{array}{ccccccc}
H_k(\amalg_\sigma(D_\sigma^n, \partial D_\sigma^n)) & \xrightarrow[(\text{id})_*]{=} & H_k(\amalg_\sigma(D_\sigma^n, \partial D_\sigma^n)) & \xrightarrow[\text{excision}]{\approx} & H_k(\amalg_\sigma(D_\sigma^n, \partial D_\sigma^n)) & \xrightarrow[(\text{id})_*]{=} & H_k(\amalg_\sigma(D_\sigma^n, \partial D_\sigma^n)) \\
\downarrow{(\text{id})_*} & & \downarrow & & \downarrow{\amalg_\sigma f_\sigma} & & \downarrow{\amalg_\sigma f_\sigma} \\
H_k(\amalg_\sigma(D_\sigma^n, \partial D_\sigma^n)) & \xrightarrow{\approx} & H_k(\amalg_\sigma(D_\sigma^n - N_\sigma), Y - U) & \xrightarrow[\text{excision}]{\approx} & H_k(X^{(n)}, Y) & \xrightarrow{\approx} & H_k(X^{(n)}, X^{(n-1)})
\end{array}$$

The result now follows from the additivity property (P5) and Example 2.2.

$\square$

Note: We will denote $\underline{C}_n(X, A; \Lambda)$ by $\underline{C}_n(X, A)$ when $\Lambda = \mathbb{Z}$ or when Λ is understood from the context.

Lemma 2.12

i) $H_k(X^{(n)}, A) \xrightarrow{i_*} H_k(X, A)$ *is an isomorphism for all $k < n$.*

ii) $H_k(X^{(n)}, A) = 0$ *for all $k > n$.*

iii) $H_n(X^{(n)}, X^{(n-2)}) \approx H_n(X^{(n)}, A)$.

Proof:

Proof of i) Let $k < n$ and consider the exact sequence of the triple $A \subseteq X^{(n)} \subseteq X^{(n+1)}$.

$$H_{k+1}(X^{(n+1)}, X^{(n)}) \xrightarrow{\delta_*} H_k(X^{(n)}, A) \xrightarrow{i_*} H_k(X^{(n+1)}, A) \xrightarrow{j_*} H_k(X^{(n+1)}, X^{(n)})$$

The end terms are zero by Lemma 2.11 (since $k + 1 < n + 1$ and $k < n + 1$). Hence, $H_k(X^{(n)}, A) \approx H_k(X^{(n+1)}, A)$, and by induction, $H_k(X^{(n)}, A) \approx H_k(X^{(N)}, A)$ for all $N > n$. By Theorem 2.6, the image of every singular k-simplex $\Delta^k \to X$ must lie in a finite subcomplex. Hence, every singular chain in $H_k(X, A)$ is a singular chain in $H_k(X^{(N)}, A)$ for some $N \in \mathbb{Z}_+$, and $H_k(X^{(n)}, A) \xrightarrow{i_*} H_k(X, A)$ is an isomorphism for all $k < n$.

Proof of ii) We will use induction on n. When $n = 0$ it is clear that $H_k(X^{(0)}, A)$ is zero for all $k > 0$ since $X^{(0)}$ is a disjoint union of points union A. Now assume that $H_k(X^{(n-1)}, A) = 0$ for all $k > n - 1$ and consider the following exact sequence coming from the triple $A \subseteq X^{(n-1)} \subseteq X^{(n)}$.

$$H_k(X^{(n-1)}, A) \to H_k(X^{(n)}, A) \to H_k(X^{(n)}, X^{(n-1)})$$

For all $k > n$, $H_k(X^{(n-1)}, A) = 0$ by the induction hypothesis and the fact that $H_k(X^{(n)}, X^{(n-1)}) = 0$ by Lemma 2.11. Hence, $H_k(X^{(n)}, A) = 0$.

Proof of iii) Consider the exact sequence of the triple $A \subseteq X^{(n-2)} \subseteq X^{(n)}$.

$$H_n(X^{(n-2)}, A) \to H_n(X^{(n)}, A) \to H_n(X^{(n)}, X^{(n-2)}) \to H_{n-1}(X^{(n-2)}, A)$$

The end groups are zero by part ii). Hence,

$$H_n(X^{(n)}, A) \approx H_n(X^{(n)}, X^{(n-2)}).$$

$\square$

Remark 2.13 In the proof of the Morse Homology Theorem (Theorem 7.4) we will apply similar arguments to a filtration:

$$\emptyset = N_{-1} \subseteq N_0 \subseteq N_1 \subseteq \cdots \subseteq N_m = M$$

that satisfies $H_j(N_k, N_{k-1}) = 0$ for all $j \neq k$.

Let $X^{(n)}/X^{(n-1)}$ be the space obtained by identifying $X^{(n-1)}$ to a single point when $n > 0$, and let $X^{(n)}/X^{(n-1)}$ be $X^{(0)}$ union a disjoint basepoint when $n = 0$, i.e. $X^{(n)}/X^{(n-1)} = X^{(0)} \amalg \{*\}$ when $n = 0$. Then $X^{(n)}/X^{(n-1)} \approx \bigvee_\sigma S_\sigma^n$ is a bouquet of n-spheres unioned at the basepoint $*$. For each σ let $p_\sigma : X^{(n)} \to S_\sigma^n$ be the composition $X^{(n)} \to X^{(n)}/X^{(n-1)} \to S_\sigma^n$ where in the second map we collapse S_τ^n to the basepoint if $\tau \neq \sigma$. $p_\sigma : (X^{(n)}, X^{(n-1)}) \to (S_\sigma^n, *)$ is a map of pairs, and clearly $p_\tau \circ f_\sigma : D^n \to S_\sigma^n$ is a constant map to the base point if $\tau \neq \sigma$.

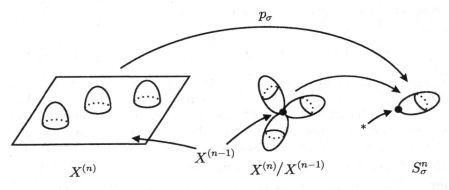

Denote by $\Psi_n : \underline{C}_n(X, A; \Lambda) \to H_n(X^{(n)}, X^{(n-1)}; \Lambda)$ the isomorphism given by Lemma 2.11. That is,

$$\Psi_n\left(\sum_\sigma n_\sigma \sigma\right) = \sum_\sigma n_\sigma f_{\sigma*}[D^n]$$

where $[D^n]$ is the generator of $H_n(D^n, \partial D^n)$.

Proposition 2.14 *The inverse of Ψ_n is the map*

$$\Phi_n : H_n(X^{(n)}, X^{(n-1)}; \Lambda) \to \underline{C}_n(X, A; \Lambda)$$

given by

$$\Phi_n(\alpha) = \sum_\sigma \phi_n(p_{\sigma*}\alpha)\sigma$$

where the map $\phi_n : H_n(S^n, *; \Lambda) \to \Lambda$ is the unique homomorphism such that $\phi_n[S^n] = 1$ and $[S^n]$ is the fundamental class of $(S^n, *)$.

Note: When $n = 0$ we will take as a fundamental homology class $[S^0]$ of $(S^0, *)$ the homology class generated by $\{1\}$, and we take $\{-1\}$ as the basepoint $*$. For more details on orientation conventions see Section IV.9 of [30] and Sections II.2 and II.6 of [150].

Proof:

It is enough to show that $\Phi_n \circ \Psi_n = 1$. Let σ be an n-cell of X not in A, i.e. a generator of $\underline{C}_n(X, A; \Lambda)$.

$$
\begin{aligned}
\Phi_n(\Psi_n(\sigma)) &= \Phi_n(f_{\sigma *}[D^n]) \\
&= \sum_\tau \phi_n(p_{\tau *} f_{\sigma *}[D^n]) \tau \\
&= \sum_\tau \phi_n((p_\tau \circ f_\sigma)_*[D^n]) \tau
\end{aligned}
$$

Since $p_\tau \circ \sigma$ is the constant map to the base point if $\tau \neq \sigma$ we have

$$
\Phi_n(\Psi_n(\sigma)) = \phi_n((p_\sigma \circ f_\sigma)_*[D^n]) \sigma = \phi_n[S^n] \sigma = \sigma.
$$

$\square$

Now we consider $\underline{\partial}_n : \underline{C}_n(X, A) \to \underline{C}_{n-1}(X, A)$ obtained as the composite of the following maps:

$$
\underline{C}_n(X, A) \xrightarrow{\Psi_n} H_n(X^{(n)}, X^{(n-1)}) \xrightarrow{\delta_*} H_{n-1}(X^{(n-1)}, X^{(n-2)}) \xrightarrow{\Phi_{n-1}} \underline{C}_{n-1}(X, A)
$$

where the map δ_* is the connecting homomorphism coming from the triple $(X^{(n)}, X^{(n-1)}, X^{(n-2)})$. Note that $\delta_*^2 = 0$ because $\delta_* \circ \delta_* = j_* \circ (\delta_{n-1} \circ j_*) \circ \delta_n$ where δ_{n-1} and j_* are successive maps in the exact sequence of the pair $(X^{(n-1)}, X^{(n-2)})$.

$$
\cdots \xrightarrow{i_*} H_{n-1}(X^{(n-1)}) \xrightarrow{j_*} H_{n-1}(X^{(n-1)}, X^{(n-2)}) \xrightarrow{\delta_{n-1}} H_{n-2}(X^{(n-2)}) \to \cdots
$$

Theorem 2.15 (CW-Homology Theorem) *If X is a CW-complex, then $\underline{\partial}_n$: $\underline{C}_n(X, A; \Lambda) \to \underline{C}_{n-1}(X, A; \Lambda)$ satisfies $\underline{\partial}_{n-1} \circ \underline{\partial}_n = 0$ and is given by*

$$
\underline{\partial}_n(\sigma) = \sum_\tau [\tau : \sigma] \tau
$$

where $[\tau : \sigma]$ is the degree of the map $p_\tau \circ f_{\partial \sigma} : \partial D_\sigma^n \to S_\tau^{n-1}$. Moreover, there is a natural identification of the homology of the complex $(\underline{C}_*(X, A; \Lambda), \underline{\partial}_*)$ with the singular homology $H_*(X, A; \Lambda)$.

Remark 2.16 In the preceeding theorem, $f_{\partial \sigma} : \partial D_\sigma^n \to X^{(n-1)}$ is the attaching map of the n-cell σ, τ is an $(n-1)$-cell, and p_τ is the composition $X^{(n-1)} \to X^{(n-1)}/X^{(n-2)} \to S_\tau^{n-1}$. The map $p_\tau \circ f_{\partial \sigma} : \partial D_\sigma^n \to$

S_τ^{n-1} is called a **relative attaching map**. As an immediate consequence of the CW-Homology Theorem we see that the homology of the chain complex $(\underline{C}_*(X, A; \Lambda), \underline{\partial}_*)$ is independent of the choice of the CW-structure on the space X.

Proof:

Since $\underline{\partial}_n : \underline{C}_n(X, A) \to \underline{C}_{n-1}(X, A)$ is defined as $\underline{\partial}_n = \Phi_{n-1} \circ \delta_* \circ \Psi_n$ we have $\underline{\partial}_{n-1} \circ \underline{\partial}_n = \Phi_{n-2} \circ \delta_* \circ \Psi_{n-1} \circ \Phi_{n-1} \circ \delta_* \circ \Psi_n = \Phi_{n-2} \circ (\delta_* \circ \delta_*) \circ \Psi_n = 0$ since $\delta_*^2 = 0$.

Let $\sigma \in \underline{C}_n(X, A)$ be a generator, i.e. an n-cell of X not in A. $\underline{\partial}_n(\sigma)$ is defined by taking $\sigma \mapsto f_{\sigma*}[D^n_\sigma] \in H_n(X^{(n)}, X^{(n-1)})$, applying the connecting homomorphism $\delta_n : H_n(X^{(n)}, X^{(n-1)}) \to H_{n-1}(X^{(n-1)})$, sending this into $H_{n-1}(X^{(n-1)}, X^{(n-2)})$ by j_*, and then applying Φ_{n-1}. Since δ_n is a natural homomorphism, the following diagram commutes

$$
\begin{array}{ccc}
H_n(D^n_\sigma, \partial D^n_\sigma) & \xrightarrow{\ \delta_n\ } & H_{n-1}(\partial D^n_\sigma) \\
\downarrow{\scriptstyle f_{\sigma*}} & & \downarrow{\scriptstyle f_{\partial\sigma*}} \\
H_n(X^{(n)}, X^{(n-1)}) & \xrightarrow{\ \delta_n\ } & H_{n-1}(X^{(n-1)})
\end{array}
$$

and we see that $\delta_n(f_{\sigma*}[D^n_\sigma]) = f_{\partial\sigma*}[\partial D^n_\sigma]$. (See Example 2.2 and Lemma 2.11 When $n = 0$, $[\partial D^0_\sigma] = -[\sigma_{-1}] + [\sigma_1]$). Hence,

$$
\begin{aligned}
\underline{\partial}_n(\sigma) = \Phi_{n-1}(f_{\partial\sigma*}[\partial D^n_\sigma]) &= \sum_\tau \phi_{n-1}(p_{\tau*} f_{\partial\sigma*}[\partial D^n_\sigma])\tau \\
&= \sum_\tau \phi_{n-1}((p_\tau \circ f_{\partial\sigma})_*[\partial D^n_\sigma])\tau \\
&= \sum_\tau \phi_{n-1}(\deg(p_\tau \circ f_{\partial\sigma}) \cdot [S^{n-1}])\tau \\
&= \sum_\tau \deg(p_\tau \circ f_{\partial\sigma})\tau
\end{aligned}
$$

by the definition of the degree.

Now consider the exact sequence of the triple $X^{(n-2)} \subseteq X^{(n-1)} \subseteq X^{(n)}$.

$$
\begin{array}{ccccccc}
H_n(X^{(n-1)}, X^{(n-2)}) & \longrightarrow & H_n(X^{(n)}, X^{(n-2)}) & \longrightarrow & H_n(X^{(n)}, X^{(n-1)}) & \xrightarrow{\ \delta_*\ } & H_{n-1}(X^{(n-1)}, X^{(n-2)}) \\
\approx \downarrow {\scriptstyle \text{Lemma 2.11}} & & \approx \downarrow {\scriptstyle \text{Lemma 2.12}} & & \approx \downarrow {\scriptstyle \text{Lemma 2.11}} & & \approx \downarrow {\scriptstyle \text{Lemma 2.11}} \\
0 & \longrightarrow & H_n(X^{(n)}, A) & \longrightarrow & \underline{C}_n(X, A) & \xrightarrow{\ \underline{\partial}_n\ } & \underline{C}_{n-1}(X, A)
\end{array}
$$

From this we deduce that $H_n(X^{(n)}, X^{(n-2)})$ can be identified with $\ker \underline{\partial}_n$, that is

$$
H_n(X^{(n)}, X^{(n-2)}) \approx H_n(X^{(n)}, A) \approx \ker \underline{\partial}_n = \underline{Z}_n(X, A).
$$

Define $\partial_{n+1}^* : \underline{C}_{n+1}(X, A) \to \underline{Z}_n(X, A)$ to be the composite:

$$\underline{C}_{n+1}(X, A) \overset{\approx}{\to} H_{n+1}(X^{(n+1)}, X^{(n)}) \overset{\delta_*}{\to} H_n(X^{(n)}, X^{(n-2)}) \overset{\approx}{\to} \underline{Z}_n(X, A)$$

where δ_* is the connecting homomorphism in the exact sequence of the triple $X^{(n-2)} \subseteq X^{(n)} \subseteq X^{(n+1)}$.

$$\cdots \to H_{n+1}(X^{(n+1)}, X^{(n)}) \overset{\delta_*}{\to} H_n(X^{(n)}, X^{(n-2)}) \to H_n(X^{(n+1)}, X^{(n-2)}) \to \cdots$$

The last term is zero by Lemma 2.11, so we have an exact sequence

$$\underline{C}_{n+1}(X, A) \overset{\partial_{n+1}^*}{\to} \underline{Z}_n(X, A) \to H_n(X^{(n+1)}, X^{(n-2)}) \to 0$$

and

$$\underline{Z}_n(X, A)/\text{im}\, \partial_{n+1}^* \approx H_n(X^{(n+1)}, X^{(n-2)}).$$

Putting these two exact sequence together we see that we have the following commutative diagram.

$$
\begin{array}{ccccccc}
& & 0 & & & & \\
& & \downarrow & & & & \\
\underline{C}_{n+1}(X, A) & \overset{\partial_{n+1}^*}{\longrightarrow} & \underline{Z}_n(X, A) & \longrightarrow & H_n(X^{(n+1)}, X^{(n-2)}) & \longrightarrow & 0 \\
& \underset{\underline{\partial}_{n+1}}{\searrow} & \downarrow & & & & \\
& & \underline{C}_n(X, A) & & & & \\
& & \downarrow {\scriptstyle \underline{\partial}_n} & & & & \\
& & \underline{C}_{n-1}(X, A) & & & &
\end{array}
$$

From the diagram we see that $\partial_{n+1}^*(\underline{C}_{n+1}(X, A)) \subseteq \underline{Z}_n(X, A)$ is sent injectively to $\underline{\partial}_{n+1}(\underline{C}_{n+1}(X, A))$. Hence, im $\partial_{n+1}^* \approx$ im $\underline{\partial}_{n+1} = \underline{B}_n(X, A)$. Therefore,

$$
\begin{aligned}
H_n(X^{(n+1)}, X^{(n-2)}) & \approx \underline{Z}_n(X, A)/\text{im}\, \partial_{n+1}^* \\
& \approx \underline{Z}_n(X, A)/\underline{B}_n(X, A) \\
& = H_n(\underline{C}_*(X, A), \underline{\partial}).
\end{aligned}
$$

Finally, consider the exact sequence of the triple $A \subseteq X^{(n-2)} \subseteq X^{(n+1)}$.

$$H_n(X^{(n-2)}, A) \to H_n(X^{(n+1)}, A) \to H_n(X^{(n+1)}, X^{(n-2)}) \to H_{n-1}(X^{(n-2)}, A)$$

The end terms are zero by Lemma 2.12 ii). Therefore,

$$H_n(X^{(n+1)}, X^{(n-2)}) \approx H_n(X^{(n+1)}, A) \approx H_n(X, A)$$

by Lemma 2.12 i), and $H_n(X, A) \approx H_n(\underline{C}_*(X, A), \underline{\partial})$.

$\square$

Example 2.17 Since S^n is a CW-complex with one 0-cell and one n-cell, we have

$$\underline{C}_k(S^n; \Lambda) = \begin{cases} \Lambda & \text{if } k = n \text{ or } k = 0 \\ 0 & \text{otherwise} \end{cases}$$

Thus all the boundary maps in the chain complex $(\underline{C}_*(S^n), \underline{\partial}_*)$ must be zero if $n > 1$. When $n = 1$ it is easy to see that $\underline{\partial}_1$ is also zero since the boundary of an oriented 1-manifold has zero points when counted with sign. Thus,

$$H_k(S^n; \Lambda) = \begin{cases} \Lambda & \text{if } k = n \text{ or } k = 0 \\ 0 & \text{otherwise} \end{cases}$$

Note that using the CW-Homology Theorem to compute the homology of S^n is much easier than the computation done in Example 2.2 using singular homology.

Example 2.18 $\mathbb{C}P^n$ is obtained by successively attaching one cell in the even dimensions $0, 2, 4, \ldots, 2n$. Hence, $\underline{C}_k(\mathbb{C}P^n) = 0$ if k is odd and has rank 1 if k is even. For any $a \in \underline{C}_{2k}(\mathbb{C}P^n)$ we have $\underline{\partial}_k(a) = 0$ since there are no $2k - 1$ chains, and im $\underline{\partial}_{2k+1} = \{0\}$ since there are no non-zero $2k + 1$ chains. Hence,

$$H_k(\mathbb{C}P^n; \Lambda) = \begin{cases} \Lambda & \text{if } j = 0, 2, \ldots, 2n \\ 0 & \text{otherwise.} \end{cases}$$

Induced maps between CW-homology groups

A continuous map of topological pairs $f : (X, A) \to (Y, B)$ induces a Λ-module homomorphism between the singular homology groups:

$$f_* : H_*(X, A) \to H_*(Y, B).$$

For CW-homology we have to restrict to cellular maps in order to get induced maps between the CW-homology groups.

Definition 2.19 *A continuous map $f : (X, A) \to (Y, B)$ between CW-pairs is called* **cellular** *if $f(X^{(n)}) \subseteq Y^{(n)}$ for all n where $X^{(n)}$ and $Y^{(n)}$ denote the n-skeletons of X and Y respectively.*

It turns out that restricting to cellular maps isn't much of a limitation because of the following (see [30], [138] or [150]).

Theorem 2.20 (Cellular Approximation Theorem) *Let X and Y be CW-complexes, and let A be a subcomplex of X. If $f : X \to Y$ is a continuous map such that $f|_A$ is cellular, then f is homotopic rel A to a cellular map $g : X \to Y$.*

For a proof of the following theorem see Theorem IV.11.6 of [30].

Theorem 2.21 *If $g : (X, A) \to (Y, B)$ is a cellular map of CW-pairs, then there is an induced map of CW-chain complexes*

$$g_* : \underline{C}_*(X, A) \to \underline{C}_*(Y, B)$$

given by

$$g_*(\sigma) = \sum_\tau deg(g_{\tau,\sigma})\tau$$

where $g_{\tau,\sigma} : S_\sigma^n \to S_\tau^n$ is defined as the composite $g_{\tau,\sigma} = \bar{p}_\tau \circ \bar{g} \circ \bar{f}_\sigma$ from the following commutative diagram.

$$
\begin{array}{ccccccc}
D_\sigma^n & \xrightarrow{f_\sigma} & X^{(n)} & \xrightarrow{g} & Y^{(n)} & \xrightarrow{p_\tau} & S_\tau^n \\
\downarrow & & \downarrow & & \downarrow & & \downarrow = \\
S_\sigma^n & \xrightarrow{\bar{f}_\sigma} & X^{(n)}/X^{(n-1)} & \xrightarrow{\bar{g}} & Y^n/Y^{(n-1)} & \xrightarrow{\bar{p}_\tau} & S_\tau^n
\end{array}
$$

2.5   Some homotopy theory

Let $I = [0, 1]$ be the unit interval.

Definition 2.22 *Let (X, A) be a topological pair and Y a topological space. The pair (X, A) is said to have the **homotopy extension property (HEP)** with respect to Y if and only if whenever we are given continuous maps $f : X \to Y$ and $H : A \times I \to Y$ such that $f(x) = H(x, 0)$ for all $x \in A$, there exists a continuous map $F : X \times I \to Y$ extending H such that $F(x, 0) = f(x)$ for all $x \in X$. That is, the following diagram can always be completed to be commutative*

$$
\begin{array}{ccc}
A \times I \cup X \times \{0\} & \xrightarrow{H \cup \bar{f}} & Y \\
\downarrow j & \nearrow F & \\
X \times I & &
\end{array}
$$

*where $\bar{f}(x, 0) = f(x)$ and j is the inclusion map. If (X, A) has the homotopy extension property with respect to every topological space Y and $i : A \to X$ is the inclusion map, then (X, A) is called a **cofibered pair** and i is called a **cofibration**.*

Recall that if A is a subspace of a topological space X, then a continuous map $r : X \rightarrow A$ such that $r|_A = id_A$ is called a **retraction**. If such a map exists, then A is said to be a **retract** of X.

Lemma 2.23 *The inclusion $i : A \rightarrow X$ is a cofibration if and only if $A \times I \cup X \times \{0\} \subseteq X \times I$ is a retract of $X \times I$.*

Proof:

Assume that the inclusion $i : A \rightarrow X$ is a cofibration. Then choosing $Y = A \times I \cup X \times \{0\}$ and taking f and H to be inclusion maps we get a continuous map $F : X \times I \rightarrow A \times I \cup X \times \{0\}$ such that the following diagram commutes.

$$A \times I \cup X \times \{0\} \xrightarrow{\ id\ } A \times I \cup X \times \{0\}$$

$$j \downarrow \qquad \nearrow F$$

$$X \times I$$

Thus, $A \times I \cup X \times \{0\}$ is a retract of $X \times I$.

Now assume there exists a continuous map $r : X \times I \rightarrow A \times I \cup X \times \{0\}$ that is the identity on $A \times I \cup X \times \{0\}$. Given continuous maps $f : X \rightarrow Y$ and $H : A \times I \rightarrow Y$ such that $f(x) = H(x, 0)$ for all $x \in A$, we can define $F : X \times I \rightarrow Y$ by $F \overset{\text{def}}{=} (H \cup \bar{f}) \circ r$, and $F : X \times I \rightarrow Y$ is clearly a continuous map extending H such that $F(x, 0) = f(x)$ for all $x \in X$.

$\square$

Lemma 2.24 *Suppose that a topological space X is obtained from a topological space A by attaching n-cells, then the inclusion $i : A \rightarrow X$ is a cofibration.*

Proof:

Let $\{f_\sigma : D_\sigma^n \rightarrow A\}_\sigma$ be the characteristic maps of the n-cells being attached to A, let D be the disjoint union of the domains D_σ^n, and let $f : \coprod_\sigma S_\sigma^{n-1} \rightarrow A$ be the map obtained by putting together all the attaching maps $f_{\partial\sigma} : S_\sigma^{n-1} \rightarrow A$. We have, $X = A \cup_f D$.

For each σ, consider the projection $p_\sigma : D_\sigma^n \times I \rightarrow S_\sigma^{n-1} \times I \cup D_\sigma^n \times \{0\}$ where $p_\sigma(x, t)$ is the point on $S_\sigma^{n-1} \times I \cup D_\sigma^n \times \{0\}$ on the line through $(0, 2) \in D_\sigma^n \times \{2\}$ and the point (x, t). (See the diagram.)

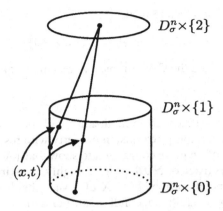

Using the projections p_σ we can define a retraction

$$r : X \times I \to A \times I \cup X \times \{0\}$$

by

$$r(x,t) = \begin{cases} (x,t) & \text{if } x \in A \\ p_\sigma(x,t) & \text{if } x \in D_\sigma^n - S_\sigma^{n-1} \end{cases}$$

which is continuous since $X = A \cup_f D$ is endowed with the quotient topology. □

Theorem 2.25 *If A is a subcomplex of a CW-complex X, then the inclusion $i : A \to X$ is a cofibration.*

Proof:

Let $r_0 : (A \cup X^{(0)}) \times I \cup X \times \{0\} \to A \times I \cup X \times \{0\}$ be the obvious retraction defined by projecting points in $(X^{(0)} - A) \times I$ into $X \times \{0\}$. Now assume that a retraction

$$r_n : (A \cup X^{(n)}) \times I \cup X \times \{0\} \to A \times I \cup X \times \{0\}$$

has been defined. The preceeding lemma shows that $A \cup X^{(n)} \subseteq A \cup X^{(n+1)}$ is a cofibration, and hence, by Lemma 2.23, there is a retraction

$$(A \cup X^{(n+1)}) \times I \to (A \cup X^{(n)}) \times I \cup (A \cup X^{(n+1)}) \times \{0\}.$$

Extending this map as the identity on $X \times \{0\}$ gives a retraction

$$(A \cup X^{(n+1)}) \times I \cup X \times \{0\} \to (A \cup X^{(n)}) \times I \cup X \times \{0\}$$

and composing with the retraction r_n gives a retraction

$$r_{n+1} : (A \cup X^{(n+1)}) \times I \cup X \times \{0\} \to A \times I \cup X \times \{0\}.$$

Hence, by induction there is a retraction

$$r : X \times I \to A \times I \cup X \times \{0\}$$

which is continuous since the CW-complex X has the weak topology.

□

Rather than showing that a given inclusion $i : A \to X$ is a cofibration by directly checking the homotopy extension property with respect to every topological space Y, it is often more convenient to show that (X, A) is a **neighborhood deformation retract** or **NDR-pair**. Roughly speaking, this means that there exists an open neighborhood $U \subseteq X$ of A such that U deforms to A with A fixed (there are some additional conditions). For a discussion of NDR-pairs based on the work of Steenrod see Section I.5 of [150]. Here, we will take a slightly different approach based on [30], [85], [141] and [142].

Recall that a subset A of a topological space U is said to be a **strong deformation retract** of U if there exists a homotopy $R : U \times I \to U$ such that

$$\begin{aligned}
R(u, 0) &= u && \text{for all } u \in U \\
R(u, 1) &\in A && \text{for all } u \in U \\
R(a, t) &= a && \text{for all } a \in A \text{ and all } t \in I.
\end{aligned}$$

Definition 2.26 *A **Strøm structure** on a topological pair (X, A) is a pair (α, h) consisting of a continuous map $\alpha : X \to I$ such that $\alpha(a) = 0$ for all $a \in A$ and a homotopy $h : X \times I \to X$ such that*

$$\begin{aligned}
h(x, 0) &= x && \text{for all } x \in X \\
h(x, t) &\in A && \text{for all } x \in X \text{ and all } t > \alpha(x) \\
h(a, t) &= a && \text{for all } a \in A \text{ and all } t \in I.
\end{aligned}$$

Note that if (X, A) admits a Strøm structure, then A is a strong deformation retract of the open neighborhood $U = \{x \in X \mid \alpha(x) < 1\}$ since $h(u, 1) \in A$ for all $u \in U$. Also, note that if A is closed in X, then the Strøm conditions imply that $A = \alpha^{-1}(0)$. This follows because if $\alpha(x) = 0$ for some $x \in X$, then $h(x, 1/n) \in A$ for all $n \in \mathbb{N}$, and hence $x = h(x, 0) \in A$ since A is closed in X.

Theorem 2.27 (Strøm [142]) *Let $i : A \to X$ be the inclusion map for a topological pair (X, A). Then $i : A \to X$ is a cofibration if and only if there exists a Strøm structure on (X, A).*

Proof:

Assume that $i : A \to X$ is a cofibration. Then by Lemma 2.23 there exists a retraction $r : X \times I \to A \times I \cup X \times \{0\}$. Let $\pi_1 : X \times I \to X$ and

$\pi_2 : X \times I \to I$ be the projections. Since I is compact we have a well defined map

$$\alpha(x) = \sup_{t \in I} |\pi_2(r(x,t)) - t|,$$

and $(\alpha, \pi_1 \circ r)$ is clearly a Strøm structure on (X, A) as long as α is continuous. To see that α is continuous, let $\beta(x,t) = |\pi_2(r(x,t)) - t|$ and $\beta_t(x) = \beta(x,t)$. It's clear that $\beta_t : X \to \mathbb{R}$ is continuous for all $t \in I$. Now,

$$\alpha^{-1}((-\infty, b]) = \{x \,|\, \beta(x,t) \leq b \text{ for all } t\} = \bigcap_{t \in I} \beta_t^{-1}((-\infty, b])$$

is an intersection of closed sets, and hence it is closed. Similarly,

$$\alpha^{-1}([a, \infty)) = \{x \,|\, \beta(x,t) \geq a \text{ for some } t\} = \pi_1(\beta^{-1}([a, \infty)))$$

is closed since the projection π_1 is closed. Since the complements of intervals of the form $(-\infty, b]$ and $[a, \infty)$ are a subbasis for the topology of $\mathbb{R}$, this shows that α is continuous.

Conversely, suppose that (α, h) is a Strøm structure on (X, A). Then we can define a retraction $r : X \times I \to A \times I \cup X \times \{0\}$ by

$$r(x,t) = \begin{cases} (h(x,t), 0) & \text{if } t \leq \alpha(x) \\ (h(x,t), t - \alpha(x)) & \text{if } t \geq \alpha(x). \end{cases}$$

It's clear that r is continuous, and since $\alpha(a) = 0$ for all $a \in A$ and $h(a,t) = a$ for all $a \in A$ and all $t \in I$, it's clear that r restricted to $A \times I \cup X \times \{0\}$ is the identity.

$\square$

Corollary 2.28 *If (X, A) is a cofibered pair, then so is $(X, \bar{A})$.*

Remark 2.29 Let X be a metric space, and let A be a closed subspace of X. Suppose that there exists an open neighborhood $U \subseteq X$ containing A such that A a strong deformation retract of U. Then it is easy to construct a Strøm structure on (X, A). Simply take an ε-neighborhood of A inside the open set U and define α to be $1/\varepsilon$ times the distance to A inside the ε-neighborhood and 1 outside of the ε-neighborhood. The deformation retraction taking U to A can then be used to construct the required homotopy $h : X \times I \to X$. The details are left to the reader.

Thus, if A is a closed subspace of a metric space X, then the inclusion map $i : A \to X$ is a cofibration if and only if there exists an open neighborhood $U \subseteq X$ of A such that A is a strong deformation retract of U.

Mapping cylinders and mapping cones

In homotopy theory, any map can be "turned into" an inclusion via the **mapping cylinder** construction. Indeed, if $f : X \to Y$ is a continuous map, then one defines the mapping cylinder M_f as

$$M_f = ((X \times I) \amalg Y) / (x, 0) \sim f(x),$$

and the space X is identified inside M_f as $X \times \{1\}$. Note that Y is also inside M_f.

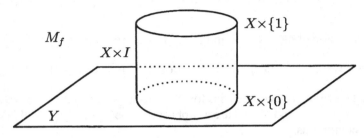

Denote these inclusions by $i : X \to M_f$ and $j : Y \to M_f$. If π is the natural projection $\pi : M_f \to Y$, then π is a homotopy inverse of the inclusion j. To see this, note that we can define a homotopy $H : M_f \times I \to M_f$ from the identity to $j \circ \pi$ that fixes Y by

$$
\begin{aligned}
H([x, t], t') &= [x, (1 - t')t] && \text{for all } x \in X \text{ and } t, t' \in I \\
H([y], t') &= [y] && \text{for all } y \in Y \text{ and } t' \in I.
\end{aligned}
$$

H is continuous since M_f is endowed with the quotient topology. Also, it's clear that (M_f, X) satisfies the conditions of Definition 2.26, and hence by Theorem 2.27, i is a cofibration. Thus, $Y \simeq M_f$, and up to homotopy equivalence of spaces, every continuous map is a cofibration. This is summarized in the following commutative diagram in which i is a cofibration and π is a homotopy equivalence.

$$
\begin{array}{ccc}
X & \xrightarrow{\ i\ } & M_f \\
& {\scriptstyle f} \searrow & \ \downarrow{\scriptstyle \simeq}{\scriptstyle \pi} \\
& & Y
\end{array}
$$

Note: For a direct proof that the mapping cylinder construction produces a cofibration (that does not rely on Theorem 2.27) see Theorem I.4.12 of [138].

In homotopy theory, one also defines the **mapping cone** C_f of a function $f : X \to Y$ as

$$C_f = M_f/(X \times \{1\}) \approx Y \cup_f CX$$

where $CX = X \times I/(X \times \{1\})$ and X is identified inside CX as $X \times \{0\}$.

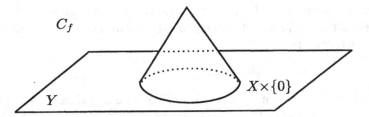

If $i : A \to X$ is the inclusion, then $C_i = X \cup CA = (X \cup (A \times I))/(A \times \{1\})$. There is a natural homeomorphism

$$X/A \xrightarrow{\cong} X \cup CA/CA$$

whose inverse when composed with the projection

$$X \cup CA \to X \cup CA/CA$$

produces a map

$$h : C_i = X \cup CA \to X \cup CA/CA \xrightarrow{\cong} X/A.$$

The following theorem gives a sufficient condition for $h : C_i \to X/A$ to be a homotopy equivalence. For the proof we will follow [30].

Theorem 2.30 *If A is a closed subset of X and the inclusion $i : A \to X$ is a cofibration, then the map $h : C_i \to X/A$ is a homotopy equivalence. In fact, there is a homotopy equivalence of pairs*

$$(C_i, CA) \simeq (X/A, *).$$

Proof:

Let $v = [A \times \{1\}]$ be the vertex of the cone CA, and identify X inside CX as $X \times \{0\}$. Let $f : A \times I \cup X \times \{0\} \to C_i$ be the collapsing map and extend f to $\bar{f} : X \times I \to C_i$ by the definition of a cofibration. Then $\bar{f}(a, 1) = v$, $\bar{f}(a, t) = (a, t)$ for all $a \in A$, and $\bar{f}(x, 0) = x$ for all $x \in X$.

Let $\bar{f}_t = \bar{f}|_{X \times \{t\}} : X \times \{t\} \to C_i$. Since $\bar{f}_1(A) = \{v\}$, the map $\bar{f}_1$ factors through X/A. That is, $\bar{f}_1 = g \circ p$, where $p : X \to X/A$ is the quotient map and $g : X/A \to C_i$ is a map that makes the following diagram commute.

By the definition of the quotient topology g is continuous. We will show that g is a homotopy inverse to h.

First we will prove that $h \circ g \simeq id_{X/A}$. Consider the homotopy $h \circ \bar{f}_t : X \to X/A$. For all t, $h \circ \bar{f}_t$ takes A to the point $[A]$, and thus it factors through X/A to give the homotopy

$$h \circ g \simeq [h \circ \bar{f}_t] \simeq [h \circ \bar{f}_0] = [p] = id_{X/A}.$$

We will now show that $g \circ h \simeq id_{C_i}$. First, note that since $\bar{f}(A \times \{1\}) = v \in C_i$, $\bar{f}$ induces a map $\bar{f}' : W \to C_i$ where $W = (X \times I)/(A \times \{1\})$.

$$
\begin{array}{ccc}
X \times I & & \\
\downarrow{\scriptstyle p'} & \searrow{\scriptstyle \bar{f}} & \\
W & \xrightarrow[\bar{f}']{} & C_i
\end{array}
$$

Moreover, there is a map $l : C_i \to W$ induced from the following diagram, where the vertical maps are quotient maps,

$$
\begin{array}{ccc}
A \times I \cup X & \xrightarrow{\text{include}} & X \times I \\
\downarrow & & \downarrow \\
C_i & \xrightarrow{\quad l \quad} & W
\end{array}
$$

that satisfies $\bar{f}' \circ l = id_{C_i}$. We also have maps $k : X/A \to W$ and $\pi : W \to X/A$ induced from the maps $x \mapsto (x, 1)$ and $(x, t) \mapsto x$ respectively.

$$
\begin{array}{ccc}
X & \xrightarrow{x \mapsto (x,1)} & X \times I \\
\downarrow & & \downarrow \\
X/A & \xrightarrow{\quad k \quad} & W
\end{array}
\qquad
\begin{array}{ccc}
X \times I & \xrightarrow{(x,t) \mapsto x} & X \\
\downarrow & & \downarrow \\
W & \xrightarrow{\quad \pi \quad} & X/A
\end{array}
$$

The above maps satisfy the following identities.

$$
\begin{aligned}
\bar{f}' \circ l &= id_{C_i} \\
\pi \circ k &= id_{X/A} \\
k \circ \pi &\simeq id_W \\
\bar{f}' \circ k &= g \quad \text{(the definition of } g\text{)} \\
\pi \circ l &= h
\end{aligned}
$$

Hence, $g \circ h = \bar{f}' \circ (k \circ \pi) \circ l \simeq \bar{f}' \circ l = id_{C_i}$, as desired.

$\square$

We have now arrived at the most important result in this section.

Corollary 2.31 *If A is a closed subset of X and the inclusion $i : A \to X$ is a cofibration, then the map $(X, A) \to (X/A, *)$ induces isomorphisms*

$$H_k(X, A; \Lambda) \approx H_k(X/A, *; \Lambda)$$

and

$$H^k(X/A, *; \Lambda) \approx H^k(X, A; \Lambda)$$

for all $k \in \mathbb{Z}_+$.

Proof:

We use the same argument as in Lemma 2.11:

$$H_k(X/A, *) \approx H_k(C_i, CA) \approx H_k(X \cup A \times [0, 1/2], A \times [0, 1/2])$$

by homotopy, and the last group is isomorphic to $H_k(X, A)$ by excision. The proof for cohomology is similar.

$\square$

Note that the proof of the preceeding theorem and corollary relied heavily on the assumption that the inclusion $i : A \to X$ is a cofibration. The following example shows that if the inclusion $i : A \to X$ is not a cofibration, then it is possible for $H_k(X, A) \not\approx H_k(X/A, *)$ for some $k \in \mathbb{Z}_+$.

Example 2.32 (The wedge of the cone on the Hawaiian earring [40]) The Hawaiian earring E is the compact subset of the plane consisting of the union of countably many circles C_j, where C_j is the circle of radius $1/j$ centered at $(-1/j, 0)$ for all $j \in \mathbb{N}$.

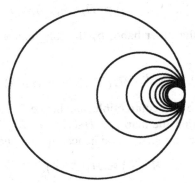

The Hawaiian Earring

Note that the topology on the Hawaiian earring is the subspace topology it inherits as a subspace of $\mathbb{R}^2$, and this subspace topology is not the same as the topology on the CW-complex consisting of the wedge of countably many circles. In fact, it is possible to show that $H_1(E)$ is not a free Abelian group (see for instance Theorem 2.5 of [33]).

Let X be the topological space consisting of the wedge of two copies of the cone on E, where the basepoint is taken to be the point $c_0 \in CE$ corresponding to the point where the circles are all tangent.

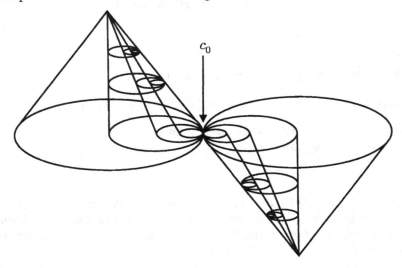

$$X = CE \vee_{c_0} CE$$

If we let $A \approx CE$ be one of the cones inside X, then $X/A \simeq CE \simeq v_0$, where v_0 is the vertex of CE. Therefore, $(X/A, v_0) \simeq (v_0, v_0)$, and hence,

$$H_k(X/A, v_0) \approx H_k(v_0, v_0) = 0$$

for all $k \in \mathbb{Z}_+$. On the other hand, by the long exact sequence of the pair (X, A) we have

$$\cdots \longrightarrow H_1(A) \overset{i_*}{\longrightarrow} H_1(X) \overset{j_*}{\longrightarrow} H_1(X, A) \overset{\delta_1}{\longrightarrow} H_0(A) \overset{i_*}{\longrightarrow} H_0(X) \longrightarrow \cdots$$

where $H_1(A) = 0$ since A is contractible, and hence $j_* : H_1(X) \to H_1(X, A)$ is an injection. Moreover, the map $i_* : H_0(A) \to H_0(X)$ is an isomorphism since both X and A are connected, and hence, $\delta_1 = 0$ and

$$H_1(X) \approx H_1(X, A).$$

Now, it can be shown that $H_1(X)$ is uncountable (see for instance Theorem 2.6 of [33] or [155]). In particular, $H_1(X, A)$ is not trivial, and hence,

$$H_1(X, A) \not\approx H_1(X/A, v_0).$$

Remark 2.33 The homotopy extension property (HEP) has a counterpoint called the **homotopy lifting property (HLP)**. A map $\pi : Y \to B$ has the homotopy lifting property with respect to the space X if and only if given any map $f : X \to Y$ and a homotopy g_t of $\pi \circ f$, there exists a homotopy f_t of f such that $\pi \circ f_t = g_t$. That is, the following diagram can always be completed to be commutative

$$
\begin{array}{ccc}
X \times \{0\} & \xrightarrow{\ \bar{f}\ } & Y \\
{\scriptstyle j}\big\downarrow & \nearrow & \big\downarrow{\scriptstyle \pi} \\
X \times I & \xrightarrow{\ g\ } & B
\end{array}
$$

where $\bar{f}(x,0) = f(x)$ and $g_t(x) = g(x,t)$. If the map $\pi : Y \to B$ has the homotopy lifting property with respect to $X = D^n$ for all $n \in \mathbb{Z}_+$, then π is called a **Serre fibration** (or simply a **fibration**). If the map $\pi : Y \to B$ has the homotopy lifting property with respect to every space X, then π is called a **Hurewicz fibration**.

It is easy to see that, up to homotopy equivalence of spaces, every map is a fibration. For any topological space B let $P(B)$ be the space of paths in B with the compact open topology. Given a map $p : Y \to B$ consider $\widetilde{Y} = \{(y, w) \in Y \times P(B) \mid w(0) = p(y)\}$. Define $\pi : \widetilde{Y} \to B$ by $\pi(y, w) = w(1)$. There is an inclusion map $i : Y \to \widetilde{Y}$ given by $i(y) = (y, w_y)$, where w_y is the constant path $w_y(t) = p(y)$ for all t. Clearly, $p = \pi \circ i$ and $i : Y \to \widetilde{Y}$ is a homotopy equivalence.

$$
\begin{array}{ccc}
\widetilde{Y} & \xleftarrow[\simeq]{\ i\ } & Y \\
{\scriptstyle \pi}\big\downarrow & \swarrow{\scriptstyle p} & \\
B & &
\end{array}
$$

Moreover, it is easy to show that $\pi : \widetilde{Y} \to B$ is a fibration. See Section VII.6 of [30] for more details.

Problems

1. A **chain complex** is a graded abelian group $C_* = \{C_k\}$ together with a sequence of homomorphisms $\partial : C_k \to C_{k-1}$ such that $\partial^2 : C_k \to C_{k-2}$ is zero. The **homology** of the chain complex (C_*, ∂) is defined to be

$$
H_k(C_*) = \frac{\ker \partial : C_k \to C_{k-1}}{\operatorname{im} \partial : C_{k+1} \to C_k}.
$$

If A_* and B_* are chain complexes, then a **chain map** $f : A_* \to B_*$ is a homomorphism that preserves the grading and satisfies $f \circ \partial = \partial \circ f$. Assume that $i : A_* \to B_*$ and $j : B_* \to C_*$ are chain maps and show that the short exact sequence

$$0 \longrightarrow A_* \xrightarrow{\ i\ } B_* \xrightarrow{\ j\ } C_* \longrightarrow 0$$

induces a long exact sequence

$$\cdots \xrightarrow{\delta_{k+1}} H_k(A_*) \xrightarrow{\ i_*\ } H_k(B_*) \xrightarrow{\ j_*\ } H_k(C_*) \xrightarrow{\delta_k} H_{k-1}(A_*) \xrightarrow{\ i_*\ } \cdots$$

where $\delta_*[c] = [i^{-1} \circ \partial \circ j^{-1}(c)]$ is the **connecting homomorphism**.

2. Prove **The 5-Lemma**: Suppose that the following diagram consisting of abelian groups and homomorphisms is commutative and has exact rows. If f_1, f_2, f_4, and f_5 are isomorphisms, then f_3 is also an isomorphism.

$$
\begin{array}{ccccccccc}
A_1 & \longrightarrow & A_2 & \longrightarrow & A_3 & \longrightarrow & A_4 & \longrightarrow & A_5 \\
\approx \downarrow f_1 & & \approx \downarrow f_2 & & \downarrow f_3 & & \approx \downarrow f_4 & & \approx \downarrow f_5 \\
B_1 & \longrightarrow & B_2 & \longrightarrow & B_3 & \longrightarrow & B_4 & \longrightarrow & B_5
\end{array}
$$

3. Show that if A, B are open in $A \cup B \subset X$, then there is a long exact sequence

$$\rightarrow H_k(X, A \cap B) \rightarrow H_k(X, A) \oplus H_k(X, B) \rightarrow H_k(X, A \cup B) \rightarrow H_{k-1}(X, A \cap B) -$$

where the map to the direct sum is induced by the sum of the inclusions and the map from the direct sum is the difference of those induced by the inclusions. This sequence is known as the **Mayer-Vietoris sequence**.

4. Use the Mayer-Vietoris sequence to derive the homology groups of S^n for all $n \in \mathbb{Z}_+$.

5. Suppose that (X, A) is a topological pair such that A is a deformation retract of X. Show that $H_*(X, A) = 0$.

6. Show that there is a homeomorphism

$$\mathbb{R}P^n \approx \mathbb{R}P^{n-1} \cup_f D^n$$

where $f : S^{n-1} \to \mathbb{R}P^{n-1}$ is the map assigning to each point $p \in S^{n-1} \subset \mathbb{R}^n$ the line through p and the origin. Similarly, show that there is a homeomorphism

$$\mathbb{C}P^n \approx \mathbb{C}P^{n-1} \cup_f D^{2n}$$

where $f : S^{2n-1} \to \mathbb{C}P^{n-1}$ is the map assigning to each point $p \in S^{2n-1} \subset \mathbb{C}^n$ the complex line through p and the origin.

7. Consider the 2-dimensional torus T^2 as the space obtained by identifying opposite sides of the unit square in $\mathbb{R}^2$. Show that T^2 has a CW-structure, and use this CW-structure to compute the homology of T^2.

8. Consider the 2-dimensional Klein bottle K^2 as the space obtained by identifying opposite sides of the unit square in $\mathbb{R}^2$, but with a "flip" along one of the sides. Show that K^2 has a CW-structure, and use this CW-structure to compute the homology of K^2.

9. Show that $\mathbb{R}P^2 \approx S^2/(x \sim -x)$. Use this description of $\mathbb{R}P^2$ to define a CW-complex structure on $\mathbb{R}P^2$, and use Theorem 2.15 to to compute its homology.

10. Use the CW-structure of $\mathbb{R}P^n$ to compute its homology.

11. Let $\mathbb{H}P^n$ be the quaternionic projective space, i.e. the space of all lines through 0 in $\mathbb{H}^{n+1}$ (where $\mathbb{H}$ is the space of quaternions). Show that $\mathbb{H}P^n$ has the structure of a CW-complex with one cell in every 4th dimension from 0 to $4n$. Deduce the homology of $\mathbb{H}^n$ from this CW-structure.

12. Let X be a CW-complex, and let $p : \tilde{X} \to X$ be a covering map. Show that $\tilde{X}$ has a CW-structure given by $\tilde{X}^{(n)} = p^{-1}(X^{(n)})$, and the restriction of p to each cell of $\tilde{X}$ is a relative homeomorphism onto a cell of X.

13. If (X, x_0) and (Y, y_0) are two based topological spaces, one denotes by $\text{Map}(X, Y)_0$ the space of continuous maps of the pairs $(X, x_0) \to (Y, y_0)$, with the compact open topology, and by $[X, Y]_0$ the set of homotopy classes of elements of $\text{Map}(X, Y)_0$. Show that for $n \geq 1$, $\pi_n(X, x_0) = [S^n, X]_0$ is a group, and for $n \geq 2$, $\pi_n(X, x_0)$ is abelian.

14. The group $\pi_n(X, x_0)$ from the previous problem is called the nth **homotopy group** of (X, x_0). The natural map

$$\rho : \pi_n(X, x_0) \to H_n(X; \mathbb{Z})$$

given by $\rho([\alpha]) = \alpha_*[S^n]$, where $[S^n] \in H_n(S^n; \mathbb{Z})$ is the generator of $H_n(S^n; \mathbb{Z}) \approx \mathbb{Z}$, is called the **Hurewicz homomorphism**.

   a. Show that is X is path connected, then $\rho : \pi_1(X, x_0) \to H_1(X, x_0)$ is onto and its kernel is the commutator subgroup of $\pi_1(X, x_0)$.

   b. Show that $\rho : \pi_n(S^n, x_0) \to H_n(S^n; \mathbb{Z}) \approx \mathbb{Z}$ is an isomorphism. (This is the degree of a map from S^n to S^n.)

15. Prove the **Hurewicz Theorem**: If (X, x_0) is a path connected space such that $\pi_i(X, x_0) = 0$ for all $i \leq n - 1$ where $n \geq 2$, then the Hurewicz homomorphism $\rho : \pi_n(X, x_0) \to H_n(X; \mathbb{Z})$ is an isomorphism.

16. Prove the **Whitehead Theorem**: Let X and Y be connected CW-complexes with basepoints $x_0 \in X^{(0)}$ and $y_0 \in Y^{(0)}$ respectively. If $f : (X, x_0) \rightarrow (Y, y_0)$ is a continuous map that induces isomorphisms $f_* : \pi_n(X, x_0) \rightarrow \pi_n(Y, y_0)$ for all n, then $f : X \rightarrow Y$ is a homotopy equivalence.

17. Let (X, x_0) and (Y, y_0) be pointed spaces. Let $X \vee Y = (X \times \{y_0\}) \cup (\{x_0\} \times Y)$ denote the **wedge product** and $X \wedge Y = X \times Y/(X \vee Y)$ denote the **smash product**. Show that $\mathrm{Map}(X \wedge Y, Z)_0 \approx \mathrm{Map}(X, \mathrm{Map}(Y, Z)_0)_0$.

18. Show that $S^p \wedge S^q \approx S^{p+q}$.

19. Show that $S^2/S^0 \simeq S^1 \vee S^2$. Also, show that $S^3/S^1 \simeq S^2 \vee S^3$.

20. Show that $S^n/S^{n-2} \simeq S^{n-1} \vee S^n$ for all $n \geq 2$. More generally, show that $S^n/S^m \simeq S^{m+1} \vee S^n$ for all $0 \leq m < n - 1$.

21. Show that $S^2 \times S^4 - \{pt.\} \simeq S^2 \vee S^4$.

22. Show that $S^2 \times S^4 \not\simeq \mathbb{C}P^3$.

23. Show that S^{n-1} is not a retract of the closed disk D^n.

24. Show that if X is Hausdorff and $A \subset X$ is a retract of X, then A is closed in X.

25. Show that $A \subset X$ is a retract of X if and only if for every space Z, any continuous map $f : A \rightarrow Z$ is extendable over X.

26. Suppose that A is a strong deformation retract of X and $f : A \rightarrow Z$ is continuous. Show that Z is a strong deformation retract of $X \cup_f Z$.

27. Show that the composite of cofibrations is a cofibration.

28. Show that if A is a subspace of a Hausdorff space X and the inclusion $i : A \rightarrow X$ is a cofibration, then A is closed in X.

29. Suppose that A is a closed subset of X and the inclusion $i : A \rightarrow X$ is a cofibration. Show that if A is contractible, then the identification map $\pi : X \rightarrow X/A$ is a homotopy equivalence.

30. Show that a cofibration is a homeomorphism onto its image.

Chapter 3

Basic Morse Theory

The main goal of this chapter is to show how to construct a CW-complex that is homotopy equivalent to a given smooth manifold M using some special functions on M called "Morse" functions (Theorem 3.28). The CW-homology of the resulting CW-complex is isomorphic to the singular homology of M by Theorem 2.15, and hence it is independent of the choice of the Morse function used to build the CW-complex. As a consequence we derive the Morse inequalities. The last section of this chapter is an introduction to Morse-Bott functions.

3.1 Morse functions

If p is a point on a smooth manifold M of dimension m, we denote by $T_p M$ the tangent space of M at p. Recall that a tangent vector $V \in T_p M$ is an equivalence class of curves $\gamma : (-\varepsilon, \varepsilon) \to M$ for some $\varepsilon > 0$ with $\gamma(0) = p$. Two curves are equivalent if they have the same velocity at 0, i.e.

$$V = \frac{d\gamma}{dt}\bigg|_{t=0}.$$

We will also write $V = [\gamma]$.

A **critical point** of a smooth function $f : M \to \mathbb{R}$ is a point p at which the differential

$$df_p : T_p M \to T_{f(p)}\mathbb{R} \approx \mathbb{R}$$

vanishes. Recall that if $V = [\gamma] \in T_p M$, then

$$df_p(V) = \frac{d}{dt}(\gamma(t))\bigg|_{t=0} \quad \text{or} \quad df_p([\gamma]) = [f \circ \gamma].$$

Also, one often finds the following notation: $df_p(V) = (V \cdot f)(p)$ or $V_p \cdot f$.
If p is a critical point, then $V \cdot f = 0$ for all tangent vectors $V \in T_pM$. If
$\phi : U \to \mathbb{R}^m$ is a local chart $\phi(x) = (x_1, \ldots, x_m)$ around p, then

$$\frac{\partial}{\partial x_j}(f \circ \phi^{-1})(\phi(p)) = 0$$

for all $j = 1, \ldots, m$.

The **Hessian** $H_p(f)$ of a smooth function $f : M \to \mathbb{R}$ at a critical point p is
the symmetric bilinear map

$$H_p(f) : T_pM \times T_pM \to \mathbb{R}$$

defined as follows. For any tangent vectors $V, W \in T_pM$ choose extensions
$\tilde{V}$ and $\tilde{W}$ to vector fields on an open neighborhood of p and set

$$\begin{aligned} H_p(f)(V, W) &= (\tilde{V} \cdot (\tilde{W} \cdot f))(p) \\ &= V_p \cdot (\tilde{W} \cdot f) \end{aligned}$$

By definition the expression above is independent of the extension $\tilde{V}$ of V.
Since

$$\tilde{V} \cdot (\tilde{W} \cdot f)(p) - \tilde{W} \cdot (\tilde{V} \cdot f)(p) = ([\tilde{V}, \tilde{W}] \cdot f)(p) = 0$$

the above expression for $H_p(f)$ is symmetric and hence is independent of the
extension $\tilde{W}$ of W as well. If $\phi(x) = (x_1, \ldots, x_m)$ is a local coordinate
system near p, then $\frac{\partial}{\partial x_1}\big|_p, \ldots, \frac{\partial}{\partial x_m}\big|_p$ is a basis for T_pM, and the matrix of
$H_p(f)$ with respect to this basis is expressed by the $m \times m$ matrix of second
partial derivatives:

$$M_p(f) = \left(\frac{\partial^2(f \circ \phi^{-1})}{\partial x_i \partial x_j} \phi(p) \right).$$

Since this matrix is symmetric, it is diagonalizable with real eigenvalues. Note
that the signs of these eigenvalues are uniquely determined by $H_p(f)$, but the
magnitudes of the eigenvalues depends on the choice of coordinate chart.

Definition 3.1 *Let p be a critical point of a smooth function $f : M \to \mathbb{R}$.*

1. *The dimension of the subspace of T_pM on which $H_p(f)$ is negative definite
   is called the **index** of p, i.e. the number of negative eigenvalues of $M_p(f)$,
   and is denote by λ_p.*

2. *The critical point p is said to be **non-degenerate** if and only if the Hessian
   $H_p(f)$ is non-degenerate, i.e. the determinant of $M_p(f)$ is non-zero.*

3. A ***Morse function*** *on a smooth manifold is a smooth function whose critical points are all non-degenerate.*

Our first observation is that non-degenerate critical points do not accumulate. More precisely, we have the following.

Lemma 3.2 *Non-degenerate critical points are isolated.*

Proof:

Let $p \in M$ be a non-degenerate critical point of $f : M \to \mathbb{R}$, and let $\phi : U \to \mathbb{R}^m$ be a chart around p such that $\phi(p) = 0$. Consider the C^∞-map $g : \phi(U) \subseteq \mathbb{R}^m \to \mathbb{R}^m$ given by

$$g(x) = \left(\frac{\partial}{\partial x_1}(f \circ \phi^{-1})(x), \ldots, \frac{\partial}{\partial x_n}(f \circ \phi^{-1})(x) \right).$$

We have $g(0) = 0$ and $dg_0 = M_0(f)$ is non-singular. By the Inverse Function Theorem (see for instance Theorem 6.1.2 of [94]), g is a diffeomorphism of some neighborhood U_0 of 0 to another neighborhood U_0' of 0. In particular, g is injective on U_0, i.e. for all $x \in U_0 \backslash \{0\}$, $g(x) \neq g(0) = 0$. Therefore, x is not a critical point of f.

$\square$

Corollary 3.3 *A Morse function on a finite dimensional compact smooth manifold has a finite number of critical points.*

Examples of Morse functions

Example 3.4 (The height function on the n-sphere) Let

$$S^n = \{(x_1, \ldots, x_{n+1}) \in \mathbb{R}^{n+1} \mid x_1^2 + \cdots + x_{n+1}^2 = 1\}$$

be the n-sphere and define $f : S^n \to \mathbb{R}$ by $f(x_1, \ldots, x_{n+1}) = x_{n+1}$. The function f is a smooth Morse function on S^n with only two critical points, the north pole $N = (0, \ldots, 0, +1)$ (the maximum) and the south pole $S = (0, \ldots, 0, -1)$ (the minimum).

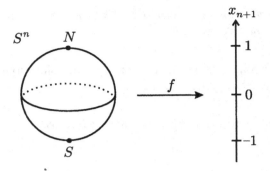

The critical points N and S are both non-degenerate and are of index $\lambda_N = n$ and $\lambda_S = 0$ respectively.

Example 3.5 (A function with degenerate critical points) The function $g = f^2 : S^n \to [0, 1]$ is not a Morse function because it has infinitely many critical points. The north pole N, the south pole S, and any point on the equator $E = \{(x_1, \ldots, x_{n+1}) \in S^n |\ x_{n+1} = 0\}$ is a critical point.

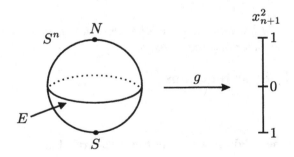

Example 3.6 (The standard height function on the torus) Consider the torus T^2 resting vertically on the plane $z = 0$ in $\mathbb{R}^3$. The height function $f : T^2 \to \mathbb{R}$ is a Morse function.

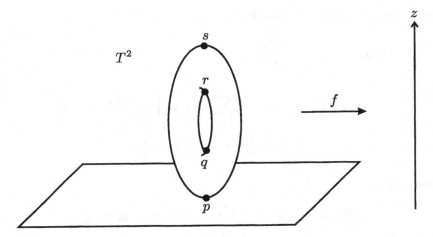

There are 4 critical points: the minimum p, the saddle points q and r, and the maximum s. These are all non-degenerate and have indices 0, 1, 1, and 2 respectively.

Example 3.7 (Bott's perfect Morse functions) Consider

$$S^{2n+1} = \{(z_0, \ldots, z_n) \in \mathbb{C}^{n+1} | \sum_{k=0}^{n} |z_k|^2 = 1\}$$

and define $f : S^{2n+1} \to \mathbb{R}$ by

$$f(z) = \sum_{k=1}^{n} k|z_k|^2$$

where $z = (z_0, \ldots, z_n)$. This function is invariant under the natural action of S^1 on S^{2n+1} given by $s \cdot z = (sz_0, sz_1, \ldots, sz_n)$. Hence, it descends to the quotient $S^{2n+1}/S^1 = \mathbb{C}P^n$. We will still denote by f the induced function $f : \mathbb{C}P^n \to \mathbb{R}$.

The projective space $\mathbb{C}P^n$ is covered by the $n+1$ open sets

$$U_j = \{[z_0, \ldots, z_n] | z_j \neq 0\}$$

which are the domains of the charts $\phi_j : U_j \to \mathbb{R}^{2n}$ given by

$$\phi_j([z_0, \ldots, z_n]) = (x_0, \ldots, \widehat{x_j}, \ldots, x_n, y_0, \ldots, \widehat{y_j}, \ldots, y_n)$$

where $|z_j|\dfrac{z_k}{z_j} = x_k + iy_k$ and $\widehat{x_j}$ and $\widehat{y_j}$ denote deleted coordinates. Clearly we have $|z_k|^2 = x_k^2 + y_k^2$ for $k \neq j$ and

$$|z_j|^2 = 1 - \left(\sum_{k \neq j} |z_k|^2\right) = 1 - \left(\sum_{k \neq j} x_k^2 + y_k^2\right).$$

Hence, in U_j we have the following:

$$f(z_0, \ldots, z_n) = \sum_{k=1}^{n} k|z_k|^2$$

$$= j\left(1 - \sum_{k \neq j} x_k^2 + y_k^2\right) + \sum_{k \neq j} k(x_k^2 + y_k^2)$$

$$= j + \sum_{k \neq j}(k - j)(x_k^2 + y_k^2).$$

Therefore,

$$(df)(z_0, \ldots, z_n) = 2\sum_{k \neq j}(k - j)(x_k e_k + y_k e_{k+n})$$

where $e_k = (0, \ldots, 0, 1, 0, \ldots, 0)$ with the 1 in the kth position. Thus, $df = 0$ if and only if $x_0 = \cdots = x_n = y_0 = \cdots = y_n = 0$. In U_j there is exactly one critical point, namely $p_j = [0, \ldots, 0, 1, 0, \ldots, 0]$ where the 1 is in the jth position. The function $f : \mathbb{C}P^n \to \mathbb{R}$ has $n + 1$ critical points $p_j = [e_j]$ for $j = 0, \ldots, n$, and the Hessian matrix $M_{p_j}(f)$ is the $2n \times 2n$ diagonal matrix

$$M_{p_j}(f) = \begin{pmatrix} 0 - j & & & & & & & \\ & 1 - j & & & & & 0 & \\ & & \ddots & & & & & \\ & & & n - j & & & & \\ & & & & 0 - j & & & \\ & 0 & & & & 1 - j & & \\ & & & & & & \ddots & \\ & & & & & & & n - j \end{pmatrix}$$

in which $j - j$ has been deleted. Clearly, p_j has index $2j$ for all $j = 0, \ldots, n$.

Existence of Morse functions

In this section we show that Morse functions are abundant.

Let M be a smooth manifold of dimension m. We can always embed M into $\mathbb{R}^r$ for some $r > m$ (see for instance Theorem II.10.8 of [30]). Denote by $(x_1, \ldots, x_r)$ the coordinates of a point $x \in M$. We have the following result.

Theorem 3.8 *For almost all* $a = (a_1, \ldots, a_r) \in \mathbb{R}^r$ *(for the Lebesgue measure on* $\mathbb{R}^r$*), the function* $f : M \to \mathbb{R}$ *given by* $f(x) = a_1 x_1 + a_2 x_2 + \cdots + a_r x_r$ *is a Morse function.*

To prove this we will need the following.

Lemma 3.9 *Let* $f : U \to \mathbb{R}$ *be a smooth function on an open set* $U \subseteq \mathbb{R}^m$. *For almost all* $a = (a_1, a_2, \ldots, a_m) \in \mathbb{R}^n$, *the function*

$$f_a(x) = f(x) - \sum_{j=1}^{m} x_j a_j$$

is a Morse function.

Proof:
     Consider the smooth function $g : \mathbb{R}^m \to \mathbb{R}^m$ given by

$$g(x) = \left(\frac{\partial f}{\partial x_1}(x), \ldots, \frac{\partial f}{\partial x_m}(x) \right).$$

Let $a \in \mathbb{R}^m$ be a regular value of g. By Sard's Theorem (see for instance Theorem II.6.2 of [30]), the set of regular values $a \in \mathbb{R}^m$ is dense in $\mathbb{R}^m$. Consider the function

$$f_a(x) = f(x) - \sum_{j=1}^{m} x_j a_j.$$

If $p \in \mathbb{R}^m$ is a critical point of f_a, then

$$(df_a)_p = g(p) - a = 0.$$

Since a is a regular value of g, dg_p is surjective, and hence it is invertible. The Hessian $H_p(f_a)$ is precisely dg_p, and thus p is non-degenerate.

$\square$

Remark 3.10 We may choose the regular value a as close to $0 \in \mathbb{R}^n$ as we wish. Hence, the Morse function f_a may be made as close to f as we wish. For a global version of the preceeding lemma see Theorem 5.27.

Proof of Theorem 3.8:
     Let $(x_1, \ldots, x_r)$ be the coordinates of a point $x \in M \subseteq \mathbb{R}^r$. There is a neighborhood U of x on which some of the x_j's, say $(x_{j_1}, \ldots, x_{j_m})$, form local coordinates on M. Indeed, since $T_x M \hookrightarrow T_x \mathbb{R}^r$ is injective, its dual $T_x^* \mathbb{R}^r \to T_x^* M$ is surjective. Therefore, for some neighborhood U of x, $T^* M|_U$ is spanned by some linearly independent set $dx_{j_1}, \ldots, dx_{j_m}$. Consequently, $(x_{j_1}, \ldots, x_{j_m})$ are linearly independent and hence form a coordinate system on U.

Now cover M by such domains U_j, and let $(x_1, \ldots, x_m)$ be a coordinate system on U_j. Fix $(a_{m+1}, \ldots, a_r) \in \mathbb{R}^{r-m}$ and consider the function $f(x) = x_{m+1}a_{m+1} + \cdots + x_r a_r$. By Lemma 3.9, the function

$$f_a(x) = f(x) - \sum_{j=1}^{m} a_j x_j$$

is a Morse function for almost all $a = (a_1, \ldots, a_m) \in \mathbb{R}^m$. Hence, for almost all $a = (a_1, \ldots, a_r) \in \mathbb{R}^r$, the function $f_a(x)$ is a Morse function on U_j. Let $A_j = \{a \in \mathbb{R}^r | f_a$ is not a Morse function on $U_j\}$. Each A_j has measure zero, and hence $A = \bigcup_j A_j$ has measure zero as well. If $a \in \mathbb{R}^r - A$, then f_a is a Morse function on M.

$\square$

The Morse Lemma

In this section we prove an important result which says that near a nondegenerate critical point a function assumes a canonical form in suitable local coordinates. This lemma was first proved by M. Morse in 1925 [105].

Lemma 3.11 (Morse Lemma) *Let $p \in M$ be a nondegenerate critical point of a smooth function $f : M \to \mathbb{R}$ of index k. There exists a smooth chart $\phi : U \to \mathbb{R}^m$, where U is an open neighborhood of p, with $\phi(p) = 0$ such that if $\phi(x) = (x_1, \ldots, x_m)$ for $x \in U$, then*

$$(f \circ \phi^{-1})(x_1, \ldots, x_m) = f(p) - x_1^2 - x_2^2 - \cdots - x_k^2 + x_{k+1}^2 + x_{k+2}^2 + \cdots + x_m^2.$$

To prove this we will need the following.

Proposition 3.12 *Let $f : U \to \mathbb{R}$ be a C^∞ function on a convex neighborhood U of $0 \in \mathbb{R}^m$ such that $f(0) = 0$. Then there exist functions g_i with $g_i(0) = \frac{\partial f}{\partial x_i}(0)$ and $f(x_1, \ldots, x_m) = \sum_{i=1}^{m} x_i g_i(x_1, \ldots, x_m)$.*

Proof:
We have,

$$\frac{d}{dt} f(tx) = \sum_{i=1}^{m} x_i \frac{\partial f}{\partial x_i}(tx_i).$$

Hence, setting

$$g_i(x_1, \ldots, x_m) = \int_0^1 \frac{\partial f}{\partial x_i}(tx) \, dt$$

we get

$$f(x) = f(x) - f(0) = \int_0^1 \frac{d}{dt} f(tx)\, dt$$

$$= \int_0^1 \sum_{i=1}^m x_i \frac{\partial f}{\partial x_i}(tx)\, dt$$

$$= \sum_{i=1}^m x_i g_i(x_1, \ldots, x_m)$$

and $g_i(0) = \int_0^1 \frac{\partial f}{\partial x_i}(0)\, dt = \frac{\partial f}{\partial x_i}(0)$.

$\square$

Remark 3.13 If $g_i(0) = 0$ then we can apply the preceeding proposition to g_i, and we have

$$g_i(x_1, \ldots, x_m) = \sum_{j=1}^m x_j h_{ij}(x_1, \ldots, x_m)$$

where the h_{ij} are C^∞ functions with $h_{ij}(0) = \frac{\partial g_i}{\partial x_j}(0) = \frac{\partial^2 f}{\partial x_i \partial x_j}(0)$. We have

$$f(x_1, \ldots, x_m) = \sum_{i=1}^m x_i \left(\sum_{j=1}^m x_j h_{ij}(x_1, \ldots, x_m) \right)$$

$$= \sum_{i,j} x_i x_j h_{ij}(x_1, \ldots, x_m)$$

$$= {}^t x S_x x$$

where $x = \begin{pmatrix} x_1 \\ \vdots \\ x_m \end{pmatrix}$ and $S_x = (s_{ij}(x))$ is the symmetric matrix with entries

$$s_{ij}(x) = \frac{1}{2}(h_{ij}(x) + h_{ji}(x)).$$

The expression $f(x) = {}^t x S_x x$ is a Taylor formula of order 2 for the function $f : U \to \mathbb{R}$.

Proof of Lemma 3.11:

The following proof of Lemma 3.11 is due to Palais [111] (see also [72]), and it is based on the celebrated Moser "path method" [107].

By replacing f by $f - f(p)$ and by choosing a suitable coordinate chart on M we may assume that the function f is defined on a convex neighborhood U_0 of $0 \in \mathbb{R}^m$ where $f(0) = 0$, $df(0) = 0$, and the matrix of the Hessian at $0 \in \mathbb{R}^m$,

$$M_0(f) = A = \left(\frac{\partial^2 f}{\partial x_i \partial x_j}(0) \right),$$

is a diagonal matrix with the first k diagonal entries equal to -1 and the rest equal to $+1$.

The matrix A induces a function $\tilde{A} : \mathbb{R}^m \to \mathbb{R}$ given by

$$\tilde{A}(x) = {}^t x A x = <Ax, x> = \sum_{j=1}^{m} \delta_j x_j^2$$

where $\delta_j = \frac{\partial^2 f}{\partial x_j^2}\Big|_0 = \pm 1$ for all $j = 1, \ldots, m$. We want to prove that there are neighborhoods U and U' of 0 with $U \subseteq U_0$ and a diffeomorphism $\varphi : U \to U'$ such that

$$f \circ \varphi = \tilde{A}. \tag{3.1}$$

The idea of the path method is to interpolate f and $\tilde{A}$ by a path such as,

$$f_t = \tilde{A} + t(f - \tilde{A}), \tag{3.2}$$

and to look for a smooth family φ_t of diffeomorphisms such that

$$f_t \circ \varphi_t = f_0 = \tilde{A}. \tag{3.3}$$

Then $\varphi = \varphi_1$ will satisfy $f \circ \varphi = \tilde{A}$.

We get φ_t as the solution of the differential equations

$$\frac{d\varphi_t}{dt}(x) = \xi_t(\varphi_t(x)); \qquad \varphi_0(x) = x$$

where the smooth family ξ_t is the tangent along the curves $t \mapsto \varphi_t(x)$. Taking the partial derivative with respect to t of both sides of (3.3) gives

$$(\dot{f}_t \circ \varphi_t + (\xi_t \cdot f_t) \circ \varphi_t)(x) = 0 \tag{3.4}$$

for all $x \in U$ where $\dot{f}_t$ denotes $\frac{\partial f_t}{\partial t}$. Thus,

$$(\dot{f}_t + \xi_t \cdot f_t)(y) = 0 \tag{3.5}$$

for all $y \in U'$. But $\dot{f}_t = f - \tilde{A}$ by (3.2), and therefore (3.5) becomes

$$df_t(\xi_t) = g \tag{3.6}$$

where $g = \tilde{A} - f$.

Since $g(0) = \tilde{A}(0) - f(0) = 0$ we have $dg(0) = d\tilde{A}(0) - df(0) = 0$, and Proposition 3.12 (applied twice as in Remark 3.13) gives

$$g(x) = < G_x x, x >$$

where G_x is a symmetric matrix depending smoothly on x. The proof of Proposition 3.12 can be modified as follows to show that

$$df_t(x)(\xi_t) = < B_x^t \xi_t, x > \tag{3.7}$$

where B_x^t is an $m \times m$ matrix with entries

$$(B_x^t)_{ij} = \int_0^1 \frac{\partial^2 f_t}{\partial x_i \partial x_j}(sx) \, ds.$$

We have,

$$
\begin{aligned}
\frac{d}{ds}(df_t(sx)(\xi_t)) &= \frac{d}{ds}\left(\sum_j \xi_t^j \frac{\partial f_t}{\partial x_j}(sx) \right) \\
&= \sum_{i,j} \xi_t^j \frac{\partial}{\partial x_i}\left(\frac{\partial f_t}{\partial x_j}(sx) \right) \cdot \frac{d}{ds} sx_i \\
&= \sum_{i,j} x_i \xi_t^j \frac{\partial^2 f_t}{\partial x_i \partial x_j}(sx).
\end{aligned}
$$

Hence, since $df_t(0) = 0$ by (3.6), we get

$$
\begin{aligned}
df_t(x)(\xi_t) &= df_t(x)(\xi_t) - df_t(0)(\xi_t) \\
&= \int_0^1 \frac{d}{ds} df_t(sx)(\xi_t) \, ds \\
&= \int_0^1 \sum_{i,j} x_i \xi_t^j \frac{\partial^2 f_t}{\partial x_i \partial x_j}(sx) \, ds \\
&= \sum_{i,j} x_i \xi_t^j (B_x^t)_{ij}
\end{aligned}
$$

and formula (3.7) follows.

Now observe that by (3.2)

$$B_0^t = \left(\frac{\partial^2 f_t}{\partial x_i \partial x_j}(0) \right)$$

is a diagonal matrix whose $(j, j)^{\text{th}}$ entry is

$$2\delta_j + t(\frac{\partial^2 f}{\partial x_j^2}\bigg|_0 - 2\delta_j) = (2 - t)\delta_j$$

for all $j = 1, \ldots, m$. Hence, B_0^t is non-degenerate for all $0 \leq t \leq 1$, and there exists a neighborhood $\tilde{U}$ of $0 \in \mathbb{R}^m$ such that B_x^t is also non-degenerate for all t. For $x \in \tilde{U}$, we have a unique solution ξ_t of

$$< B_x^t \xi_t, x > = < G_x x, x >$$

and this solution depends smoothly on both x and t. That is, we have a smooth solution to (3.6) defined on $\tilde{U}$. Clearly, $\xi_t(0) = 0$ since B_0^t is non-degenerate. Hence, by shrinking $\tilde{U}$ we can integrate ξ_t and get a smooth family of diffeomorphisms φ_t from a smaller neighborhood U of 0 to another neighborhood U' of 0 which satisfies $f_t \circ \varphi_t = f_0 = \tilde{A}$.

$\square$

Remark 3.14 Palais' method also works on Banach spaces [111]. Moreover, the above proof yields a Morse lemma with "parameters". If the quadratic form $\tilde{A}$ depends on "parameters" (for instance if $\tilde{A}$ is parameterized by points in some submanifold), then the solution ξ_t depends smoothly on the parameters. This method will yield the Morse-Bott Lemma (Lemma 3.51).

Remark 3.15 The "traditional" proof of the Morse lemma can be found in Section 6.1 of [77] or Section 2 of [100]. The basic ingredients are Proposition 3.12 and the following fact.

Proposition 3.16 *Let $A = diag(a_1, a_2, \ldots, a_m)$ be a diagonal matrix with diagonal entries $a_j = \pm 1$ for all $j = 1, \ldots, m$. Then there is a neighborhood U of A in the vector space of symmetric matrices ($\approx \mathbb{R}^{m(m+1)/2}$) and a C^∞ map $P : U \to GL_m(\mathbb{R})$ that satisfies*

1. $P(A) = I_{m \times m}$

2. *If $P(S) = Q$, then ${}^t QSQ = A$.*

Proof:
   We proceed by induction on the dimension m. First, suppose $m = 1$ and $A = (a)$ with $a = \pm 1$. If $S = (s)$ is any 1×1 matrix sufficiently close to A, then s will be non-zero with the same sign as a, and we define

$$P(S) = Q = \left(\frac{1}{\sqrt{|s|}}\right).$$

Now let $A = \text{diag}(a_1, a_2, \ldots, a_m)$ where $a_j = \pm 1$ for all $j = 1, \ldots, m$, and assume for the purpose of induction that there is a neighborhood U_1 of $A_1 = \text{diag}(a_2, \ldots, a_m)$ in the vector space of $(m-1) \times (m-1)$ symmetric matrices such that for every $S_1 \in U_1$ there exists a smooth map $P_1 : U_1 \to GL_{m-1}(\mathbb{R})$ such that $P_1(A_1) = I_{(m-1) \times (m-1)}$ and $^tQ_1S_1Q_1 = A_1$ where $Q_1 = P_1(S_1) \in GL_{m-1}(\mathbb{R})$.

Let $S = (s_{ij})$ be a symmetric $m \times m$ symmetric matrix near enough to $A = \text{diag}(a_1, a_2, \ldots, a_m)$ so that s_{11} is non-zero and has the same sign as a_1. The symmetric matrix S determines a symmetric bilinear form $B : \mathbb{R}^m \times \mathbb{R}^m \to \mathbb{R}$ given by $B(x, y) = {}^txSy$ for all $x, y \in \mathbb{R}^m$. Following the first step in the Gram-Schmidt orthogonalization process, we change the standard basis $e_1, \ldots, e_m$ of $\mathbb{R}^m$ to a basis $v_1, \ldots, v_m$ where

$$v_1 = \frac{e_1}{\sqrt{|s_{11}|}}$$

and

$$v_j = e_j - B(v_1, v_1)B(v_1, e_j)\, v_1 = e_j - \frac{s_{1j}}{s_{11}} e_1$$

for all $j = 2, \ldots, m$. The corresponding change of basis matrix $C \in GL_m(\mathbb{R})$ is given by

$$C = \begin{pmatrix} \dfrac{1}{\sqrt{|s_{11}|}} & -\dfrac{s_{12}}{s_{11}} & \cdots & -\dfrac{s_{1m}}{s_{11}} \\ 0 & & & \\ \vdots & & I & \\ 0 & & & \end{pmatrix}$$

where I denotes the $(m-1) \times (m-1)$ identity matrix. The new basis satisfies $B(v_1, v_j) = 0$ for all $j = 2, \ldots, m$, and thus it is easy to see that

$$^tCSC = \begin{pmatrix} a_1 & 0 & \cdots & 0 \\ 0 & & & \\ \vdots & & S_1 & \\ 0 & & & \end{pmatrix}$$

where S_1 is an $(m-1) \times (m-1)$ symmetric matrix depending smoothly on S. If S is sufficiently close to A, then $S_1 \in U_1$, and we can apply the induction hypothesis to conclude that there exists some $Q_1 \in GL_{m-1}(\mathbb{R})$ depending smoothly on S_1 such that $^tQ_1S_1Q_1 = A_1 = \text{diag}(a_2, \ldots, a_m)$. Define $P(S) = Q = CR$ where

$$R = \begin{pmatrix} 1 & 0 & \cdots & 0 \\ 0 & & & \\ \vdots & & Q_1 & \\ 0 & & & \end{pmatrix}.$$

Then we have ${}^tQSQ = A$ where $P(S) = Q \in GL_m(\mathbb{R})$ depends smoothly on S and $P(A) = I_{m \times m}$.

$\square$

3.2   The gradient flow of a Morse function

Recall that an **inner product** on a vector space V over the field $\mathbb{R}$ is a bilinear function $< \cdot, \cdot >: V \times V \to \mathbb{R}$ which is **symmetric**, that is $< v, w >=< w, v >$ for all $v, w \in V$, and **non-degenerate**, i.e. if $v \neq 0$, then there exists some $w \neq 0$ such that $< v, w > \neq 0$. An inner product is said to be **positive definite** if and only if $< v, v >$ is strictly greater than zero for all $v \in V$. The tangent bundle T_*M over a smooth manifold M is a smooth vector bundle whose fiber at each point $x \in M$ is the tangent space T_xM, and a Riemannian metric g on T_*M is a smooth function that assigns to each $x \in M$ a positive definite inner product $< \cdot, \cdot >_x$ on T_xM. For more details concerning inner products and Riemannian metric see for instance Chapter 9 of [139].

A non-degenerate inner product on a vector space V induces an isomorphism between V and its dual V^*, and hence, a Riemannian metric g on a smooth manifold M defines an isomorphism $\tilde{g} : T_*M \to T^*M$ between the tangent and cotangent bundles. For any vector field W, $\tilde{g}(W)$ is the unique 1-form such that for any vector field V

$$\tilde{g}(W)(V) = g(W, V).$$

Definition 3.17 *If $f : M \to \mathbb{R}$ is a smooth function on a Riemannian manifold (M, g), then the* **gradient vector field** *of f with respect to the metric g is the unique smooth vector field ∇f such that*

$$g(\nabla f, V) = df(V) = V \cdot f$$

for all smooth vector fields V on M, i.e. $\nabla f = \tilde{g}^{-1}(df)$. In particular,

$$(\nabla f) \cdot f = g(\nabla f, \nabla f) = ||\nabla f||^2.$$

Let $\varphi_t : M \to M$ be the local 1-parameter group of diffeomorphisms generated by $-\nabla f$ (the negative gradient), i.e.

$$\frac{d}{dt}\varphi_t(x) = -(\nabla f)(\varphi_t(x))$$
$$\varphi_0(x) = x.$$

(See for instance [77] Section 6.2 or [91] Section I.6.) The integral curve $\gamma_x : (a, b) \to M$ given by
$$\gamma_x(t) = \varphi_t(x)$$

is called a gradient **flow line**.

We have the following easy but important facts.

Proposition 3.18 *Every smooth function* $f : M \to \mathbb{R}$ *on a finite dimensional smooth Riemannian manifold* (M, g) *decreases along its gradient flow lines.*

Proof:

$$
\begin{aligned}
\frac{d}{dt} f(\gamma_x(t)) &= \frac{d}{dt}(f \circ \varphi_t(x)) \\
&= df_{\varphi_t(x)} \circ \frac{d}{dt}\varphi_t(x) \\
&= df_{\varphi_t(x)}(-(\nabla f)(\varphi_t(x))) \\
&= -\|(\nabla f)(\varphi_t(x))\|^2 \le 0
\end{aligned}
$$

$\square$

Proposition 3.19 *Let* $f : M \to \mathbb{R}$ *be a Morse function on a finite dimensional compact smooth Riemannian manifold* (M, g). *Then every gradient flow line of* f *begins and ends at a critical point, i.e. for any* $x \in M$, $\lim_{t \to +\infty} \gamma_x(t)$ *and* $\lim_{t \to -\infty} \gamma_x(t)$ *exist, and they are both critical points of* f.

Proof:

Let $x \in M$ and let $\gamma_x(t)$ be the gradient flow line through x. Since M is compact, $\gamma_x(t)$ is defined for all $t \in \mathbb{R}$ (see for instance Section 6.2 of [77] or Corollary I.6.2 of [91]), and the image of $f \circ \gamma_x : \mathbb{R} \to \mathbb{R}$ is a bounded subset of $\mathbb{R}$. Hence by Proposition 3.18 we must have

$$
\lim_{t \to \pm\infty} \frac{d}{dt} f(\gamma_x(t)) = \lim_{t \to \pm\infty} -\|(\nabla f)(\varphi_t(x))\|^2 = 0.
$$

Let $t_n \in \mathbb{R}$ be a sequence with $\lim_{n \to \infty} t_n = -\infty$. $\{\gamma_x(t_n)\} \subseteq M$ is an infinite set of points in a compact manifold, and so it has an accumulation point q. The point q is a critical point of f since $\|(\nabla f)(\gamma_x(t_n))\| \to 0$ as $n \to \infty$, and by Lemma 3.2 we can pick a closed neighborhood U of q where q is the only critical point in U. If $\lim_{t \to -\infty} \gamma_x(t) \ne q$, then there is an open neighborhood $V \subset U$ of q and a sequence $\tilde{t}_n \in \mathbb{R}$ with $\lim_{n \to \infty} \tilde{t}_n = -\infty$ and $\gamma_x(\tilde{t}_n) \in U - V$. Thus, the sequence $\{\gamma_x(\tilde{t}_n)\}$ has an accumulation point in the compact set $U - V$ which, as above, must be a critical point of f. This contradicts the choice of U, and therefore, $\lim_{t \to -\infty} \gamma_x(t) = q$. A similar argument shows that $\lim_{t \to +\infty} \gamma_x(t) = p \in M$ for some critical point p.

$\square$

Proposition 3.18 yields the first significant result in Morse theory.

Theorem 3.20 *Let* $f : M \to \mathbb{R}$ *be a smooth function on a finite dimensional smooth manifold with boundary. For all* $a \in \mathbb{R}$, *let*

$$M^a = f^{-1}((-\infty, a]) = \{x \in M | \ f(x) \le a\}.$$

Let $a < b$ *and assume that* $f^{-1}([a, b])$ *is compact and contains no critical points of* f. *Then* M^a *is diffeomorphic to* M^b, *and* M^a *is a deformation retract of* M^b. *Moreover, there is a smooth diffeomorphism* $F : f^{-1}(a) \times [a, b] \to f^{-1}([a, b])$ *such that the diagram*

$$
\begin{array}{ccc}
f^{-1}(a) \times [a, b] & \xrightarrow{\ F\ } & f^{-1}([a, b]) \\
& \searrow{\scriptstyle \pi_2} & \ \downarrow{\scriptstyle f} \\
& & [a, b]
\end{array}
$$

commutes. In particular, all the level surfaces of f *between* a *and* b *are diffeomorphic.*

Proof:

If $f^{-1}([a, b])$ is contained in the interior of M, then there exists an open neighborhood U of $f^{-1}([a, b])$ in M such that $\overline{U}$ is compact. If $f^{-1}([a, b])$ is not contained in the interior of M, then we can add an external collar to M and find an open neighborhood U of $f^{-1}([a, b])$ in M union the external collar such that $\overline{U}$ is compact. (See Section 4.6 of [77] or Section I.7 of [91] for an explanation of collars on manifolds with boundary. Also, see Section 3 of [102].)

Since f has no critical points on $f^{-1}([a, b])$, the gradient vector field ∇f does not vanish there. Let ρ be a smooth function which is equal to $1/\|\nabla f\|^2$ on $f^{-1}([a, b])$ and vanishes outside of U. The vector field $X = \rho \nabla f$ has compact support and thus defines a global 1-parameter group of diffeomorphisms $\Psi_t : U \to U$. For each $x \in U$ consider the curve $c : \mathbb{R} \to \mathbb{R}$ defined by $c(t) = f(\Psi_t(x))$. We have,

$$
\begin{aligned}
\frac{d}{dt}c(t) &= df_{\Psi_t(x)} \circ \frac{d}{dt}\Psi_t(x) \\
&= df(X)(\Psi_t(x)) \\
&= g(\nabla f, X)(\Psi_t(x)).
\end{aligned}
$$

Therefore, for $\Psi_t(x) \in f^{-1}[a, b]$,

$$
\frac{d}{dt}c(t) = g\left(\nabla f, \frac{\nabla f}{\|\nabla f\|^2}\right)(\Psi_t(x)) = 1,
$$

and

$$c(t) - c(0) = \int_0^t \frac{d}{ds} c(s) \, ds = t,$$

i.e.

$$f(\Psi_t(x)) = c(t) = c(0) + t = f(\Psi_0(x)) + t = f(x) + t.$$

Suppose now that $f(x) = a$, then $c(t) = a + t$, and hence,

$$c(b - a) = a + (b - a) = b.$$

So, if $x \in f^{-1}(a)$ and $y = \Psi_{b-a}(x)$, then $f(y) = b$.

Consider the diffeomorphism $\Psi_{b-a} : M \to M$. We just saw that any $x \in M$ with $f(x) = a$ is taken to y with $f(y) = b$, and clearly if $f(x) < a$, then for $y = \Psi_{b-a}(x)$ we have $f(y) < b$. Therefore, Ψ_{b-a} maps M^a into M^b and its inverse $\Psi_{a-b} = \Psi_{b-a}^{-1}$ maps M^b into M^a. Hence, M^a and M^b are diffeomorphic.

Now consider the 1-parameter family of maps $r_t : M^b \to M^b$ given by

$$r_t(x) = \begin{cases} x & \text{if } f(x) \le a \\ \Psi_{t(a-f(x))}(x) & \text{if } a \le f(x) \le b. \end{cases}$$

This family is continuous since for $a = f(x)$ we have $\Psi_{t(a-f(x))}(x) = \Psi_0(x) = x$. Since r_0 is the identity map and r_1 is a retraction from M^b to M^a, we see that M^a is a deformation retract of M^b.

For the second part of the theorem, define $F : f^{-1}(a) \times [a, b] \to f^{-1}([a, b])$ by $F(x, t) = \Psi_{t-a}(x)$. The above computation shows that for any $x \in f^{-1}([a, b])$ we have $\Psi_{a-f(x)}(x) \in f^{-1}(a)$. Thus, $F(\Psi_{a-f(x)}(x), f(x)) = x$, and F is surjective. Since f is increasing along its gradient flow lines, F is also increasing along the flow lines of X. Hence, F is injective, and F is an immersion because gradient lines are transverse to level sets. Therefore, F is a diffeomorphism.

$\square$

Corollary 3.21 *Let M be a compact smooth manifold with boundary $\partial M = A \amalg B$, i.e. $A \cap B = \emptyset$. Suppose there exists a C^∞ function $f : M \to [0, 1]$ with no critical points such that $f(A) = 0$ and $f(B) = 1$. Then M is diffeomorphic to $A \times [0, 1] \approx B \times [0, 1]$.*

Proof:

This follows from the second part of the theorem since $f^{-1}([0, 1]) = M$.

$\square$

Corollary 3.22 (Reeb) *If M is a compact smooth manifold without boundary of dimension m admitting a Morse function $f : M \to \mathbb{R}$ with only 2 critical points, then M is homeomorphic to the m-sphere S^m.*

Proof:

Let $f : M \to \mathbb{R}$ be a smooth function with exactly 2 critical points. Since M is compact, f attains a maximum at some point $p_+ \in M$ and a minimum at some point $p_- \in M$. Let $f(p_+) = z_+$ and $f(p_-) = z_-$. The Morse Lemma (Lemma 3.11) implies that there exists an open neighborhood U_+ of z_+ and coordinates $(u_1, \ldots, u_m)$ on which $f(x) = z_+ - (u_1^2 + \cdots + u_m^2)$. Hence for some $b < z_+$ but close to z_+, the set $D_+ = f^{-1}([b, z_+]) = \{(u_1, \ldots, u_m) | u_1^2 + \cdots + u_m^2 \leq z_+ - b\}$ is diffeomorphic to the closed m-disk D^m. Similarly, for some $a > z_-$ but close to z_- the set $D_- = f^{-1}([z_-, a])$ is also diffeomorphic to D^m. By Corollary 3.21, $f^{-1}([a, b])$ is diffeomorphic to $S^{m-1} \times [0, 1]$. To get the homeomorphism with S^m we put together $D_-, S^{m-1} \times [0, 1]$, and D_+.

Let $q_\pm$ be the north and south pole of S^m respectively, and let $B_\pm$ be disjoint neighborhoods of $q_\pm \in S^m$ diffeomorphic to D^m so that $C = S^m - \text{Int}(B_+ \cup B_-) \approx S^{m-1} \times [0, 1]$ and $\partial C = \partial B_+ \cup \partial B_-$. Let $h_+ : D_+ \to B_+ \approx D^m$ be the diffeomorphism given by the Morse Lemma. Extend $h_+|_{\partial D_+} : \partial D_+ \to \partial B_+$ to a diffeomorphism from $\partial D_+ \times [0, 1] \to \partial B_+ \times [0, 1]$. Since $\partial D_+ \times [0, 1]$ is diffeomorphic to $f^{-1}([a, b])$, this gives an extension of h_+ to a homeomorphism

$$h : D_+ \cup f^{-1}([a, b]) \to B_+ \cup C.$$

Let g_0 be the restriction of h to ∂D_-,

$$g_0 : \partial D_- \approx S^{m-1} \to \partial B_- \approx S^{m-1}$$

and extend g_0 radially to a homeomorphism $g : D_- \approx D^m \to B_- \approx D^m$, i.e.

$$g(x) = \begin{cases} 0 & \text{if } x = 0 \\ \|x\| g_0\left(\frac{x}{\|x\|}\right) & \text{if } x \neq 0. \end{cases}$$

Putting together h and g we get a homeomorphism from M to S^m.

$\square$

Remark 3.23 The homeomorphism in Corollary 3.22 may fail to be a diffeomorphism. We refer the reader to Chapter X of [91] for more details. In particular, see Section X.7 of [91] for a good historical introduction to the classification of smooth structures on S^m.

Remark 3.24 Note that we proved the Morse Lemma (Lemma 3.11) before we introduced a Riemannian metric g on M. The Morse Lemma gives local coordinates near a non-degenerate critical point in which the function f has the form

$$f(x_1, \ldots, x_m) = f(p) - x_1^2 - x_2^2 - \cdots - x_k^2 + x_{k+1}^2 + x_{k+2}^2 + \cdots + x_m^2.$$

However, the Morse Lemma says nothing about the Riemannian metric g. In particular, we cannot in general compute (even locally) ∇f from the above expression because ∇f depends on the choice of the Riemannian metric g. See Lemma 4.3 for an explicit formula for ∇f in terms of the Riemannian metric g in local coordinates.

3.3 The CW-complex associated to a Morse function

The results in this section are based on Sections 1 and 3 of [100].

Theorem 3.20 can be illustrated by the following picture where $M = T^2$ is the torus sitting vertically on the plane $z = 0$, and the Morse function $f : T^2 \to [0, 1]$ is the height function.

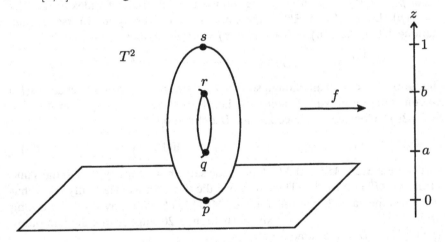

We see that $f^{-1}(0)$ is the point p, and $M^t = f^{-1}((-\infty, t])$ is homeomorphic to a disk for all $0 < t < f(q) = a$. Hence, M^t has the same homotopy type for all $0 \le t < a$, but the homotopy type of M^t for $a \le t$ is different. For all $0 \le t < a$ we have $M^t \simeq *$, and $M^t \simeq S^1$ for all $a \le t < b$. As t increases the homotopy type of M^t changes again when $t = b$ and when $t = 1$.

The next theorem describes what happens to M^t when t crosses a critical value.

Theorem 3.25 *Let $f : M \to \mathbb{R}$ be a smooth function. Suppose that for $a < b$, $f^{-1}([a, b])$ is compact and inside $f^{-1}([a, b])$ there is exactly one critical point. Assume that this critical point is non-degenerate and of index k. Then M^b has the homotopy type of M^a with one k-cell attached. In fact, there exists a set $e^k \subseteq M^b$ diffeomorphic to the closed k-disk $D^k = \{x \in \mathbb{R}^k | |x| < 1\}$ such that $M^a \cup e^k \subseteq M^b$ is a deformation retract of M^b.*

Main Idea: We will introduce a function $F : M \to \mathbb{R}$ that agrees with f except that $F < f$ in a small neighborhood of the critical point p. Thus, for a small $\varepsilon > 0$ and $c = f(p)$ the set $F^{-1}((-\infty, c - \varepsilon])$ consists of $M^{c-\varepsilon}$ unioned with a set H around the critical point. The functions F and f have the same critical points, and hence Theorem 3.20 implies that $M^{c-\varepsilon} \cup H$ is a deformation retract of $M^{c+\varepsilon}$ (which is in turn a deformation retract of M^b by Theorem 3.20). A direct argument then shows that for a suitable k-cell $e^k \subseteq H$ the set $M^{c-\varepsilon} \cup e^k$ is a deformation retract of $M^{c-\varepsilon} \cup H$. Appealing to Theorem 3.20 again we see that $M^a \cup e^k$ is a deformation retract of M^b where we have used the same notation e^k to denote the original k-cell with an added collar.

Proof of Theorem 3.25:

Let $p \in \text{int}(f^{-1}([a, b]))$ be the unique critical point of index k, and let $c = f(p)$. Let $\phi : U \to \mathbb{R}^m$ be the smooth chart given by the Morse Lemma (Lemma 3.11), i.e. $\phi(p) = 0$ and if $\phi(x) = (x_1, \ldots, x_m)$, then

$$(f \circ \phi^{-1})(x_1, \ldots, x_m) = c - x_1^2 - \cdots - x_k^2 + x_{k+1}^2 + \cdots + x_m^2.$$

Choose $0 < \varepsilon < 1$ small enough so that $f([c - \varepsilon, c + \varepsilon])$ is compact and $\varphi(U)$ contains the closed ball of radius 2ε, i.e. $\{(x_1, \ldots, x_m) \in \phi(U) | x_1^2 + \cdots + x_m^2 \leq 2\varepsilon\}$. Consider the k-cell inside U defined by

$$e^k = \phi^{-1} \left(\{(x_1, \ldots, x_m) | x_1^2 + \cdots + x_k^2 \leq \varepsilon \text{ and } x_{k+1} = \cdots = x_m = 0\} \right).$$

By Theorem 3.20, M^a and $M^{c-\varepsilon}$ have the same homotopy type, and the same is true for M^b and $M^{c+\varepsilon}$. Thus, to prove the theorem we need only show that $M^{c+\varepsilon}$ has the same homotopy type as $M^{c-\varepsilon} \cup e^k$. We may do this working inside $f^{-1}([c - \varepsilon, c + \varepsilon]) \cap U$ since Theorem 3.20 implies that the homotopy type of $M^t - U$ is the same for all $a \leq t \leq b$.

Denote the local coordinates of $x \in U$ by $\phi(x) = (x_1, \ldots, x_m)$. Then,

$$\begin{aligned} M^{c+\varepsilon} \cap U &= \{x \in U | \eta^2 - \xi^2 \leq \varepsilon\} &&\text{and} \\ M^{c-\varepsilon} \cap U &= \{x \in U | \xi^2 - \eta^2 \geq \varepsilon\} \end{aligned}$$

where

$$\begin{aligned} \xi^2 &= x_1^2 + \cdots + x_k^2 &&\text{and} \\ \eta^2 &= x_{k+1}^2 + \cdots + x_m^2. \end{aligned}$$

With this notation $f(x_1, \ldots, x_m) = c - \xi^2 + \eta^2$, and we have

$$e^k = \{x \in U | \xi^2 \leq \varepsilon \text{ and } \eta^2 = 0\}.$$

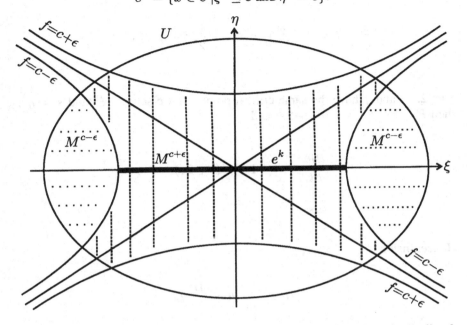

In this picture the horizontally lined area is $M^{c-\varepsilon} \cap U$ and the vertically lined area is $(M^{c+\varepsilon} - M^{c-\varepsilon}) \cap U$. The cell e^k is the dark horizontal line in the middle that is attached to $M^{c-\varepsilon}$, i.e. $\partial e^k \subset M^{c-\varepsilon}$.

Now let $\mu : \mathbb{R} \to \mathbb{R}$ be a smooth function such that $\mu(0) \geq \varepsilon$, $\mu(r) = 0$ for $r \geq 2\varepsilon$, and $-1 < \mu'(r) \leq 0$. Define a new function $F : M \to \mathbb{R}$ by

$$F = \begin{cases} f & \text{outside } U \\ f - \mu(\xi^2 + 2\eta^2) & \text{inside } U. \end{cases}$$

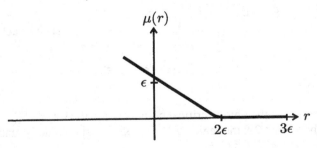

Observe that $F(p) = f(p) - \mu(0) = c - \mu(0) < c - \varepsilon$, and $F(x) \leq f(x)$ for all $x \in M$. Moreover, we have the following.

Fact 1: $F^{-1}((-\infty, c+\varepsilon]) = M^{c+\varepsilon}$. Indeed, outside the ellipsoid e defined by $\xi^2 + 2\eta^2 \leq 2\varepsilon$ we have $f = F$, and inside e we have

$$F \leq f = c - \xi^2 + \eta^2 \leq c + \frac{1}{2}\xi^2 + \eta^2 \leq c + \varepsilon.$$

Fact 2: F and f have the same critical points. If we set $t = \xi^2$ and $s = \eta^2$, then $F = c - t + s - \mu(t + 2s)$ and

$$\frac{\partial F}{\partial t} = -1 - \mu'(t + 2s) < 0$$
$$\frac{\partial F}{\partial s} = 1 - 2\mu' \geq 1.$$

Hence, inside U

$$DF = \frac{\partial F}{\partial t}dt + \frac{\partial F}{\partial s}ds$$

vanishes only when $dt = ds = 0$. This shows that F and f have the same critical point inside U, namely p, and outside of U we have $F = f$.

Fact 3: $F^{-1}([c - \varepsilon, c + \varepsilon]) \subseteq f^{-1}([c - \varepsilon, c + \varepsilon])$. Indeed, since $F \leq f$ we have $f^{-1}((-\infty, c - \varepsilon]) \subseteq F^{-1}((-\infty, c - \varepsilon])$. Therefore,

$$
\begin{aligned}
F^{-1}([c - \varepsilon, c + \varepsilon]) &= F^{-1}((-\infty, c + \varepsilon]) - F^{-1}((-\infty, c - \varepsilon]) \\
&\subseteq M^{c+\varepsilon} - F^{-1}((-\infty, c - \varepsilon]) \text{ by Fact 1} \\
&\subseteq M^{c+\varepsilon} - f^{-1}((-\infty, c - \varepsilon]) \\
&= f^{-1}([c - \varepsilon, c + \varepsilon]).
\end{aligned}
$$

Since $F(p) < c - \varepsilon$, the unique critical point is not in the set $F^{-1}([c-\varepsilon, c+\varepsilon])$. Hence by Theorem 3.20, the region $F^{-1}((-\infty, c - \varepsilon])$ is a deformation retract of $F^{-1}((-\infty, c + \varepsilon]) = M^{c+\varepsilon}$.

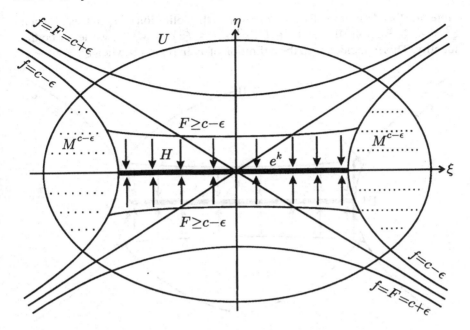

<u>Fact 4</u>: If H is the closure of $F^{-1}((-\infty, c - \varepsilon]) - M^{c-\varepsilon}$, then $M^{c-\varepsilon} \cup e^k$ is a deformation retract of $M^{c-\varepsilon} \cup H$. Since we have already shown that $F^{-1}((-\infty, c - \varepsilon]) = M^{c-\varepsilon} \cup H$ is a deformation retract of $M^{c+\varepsilon}$.

We construct the deformation retraction $r_t : M^{c-\varepsilon} \cup H \to M^{c-\varepsilon} \cup H$ as the identity outside of U. Inside U the retraction r_t is defined as follows. In region I, i.e. the set $\{x \in U \,|\, \xi^2 \leq \varepsilon\}$, r_t is defined by

$$r_t(x_1, \ldots, x_m) = (x_1, \ldots, x_k, tx_{k+1}, \ldots, tx_k)$$

for $0 \leq t \leq 1$. Here r_1 is the identity and r_0 maps region I to e^k. Also, r_t maps $F^{-1}((-\infty, c - \varepsilon])$ into itself since $\frac{\partial F}{\partial s} > 0$.

In region II, i.e. $\{x \in U \,|\, \varepsilon \leq \xi^2 \leq \eta^2 + \varepsilon\}$, r_t is defined by

$$r_t(x_1, \ldots, x_m) = (x_1, \ldots, x_k, \lambda_t x_{k+1}, \ldots, \lambda_t x_m)$$

where $\lambda_t \in [0, 1]$ is defined by

$$\lambda_t = t + (1 - t)\sqrt{(\xi^2 - \varepsilon)/\eta^2}.$$

Thus r_1 is the identity, and r_0 maps region II into $M^{c-\varepsilon}$ since

$$\begin{aligned} f(r_0(x_1, \ldots, x_m)) &= f(x_1, \ldots, x_k, \sqrt{(\xi^2 - \varepsilon)/\eta^2}x_{k+1}, \ldots, \sqrt{(\xi^2 - \varepsilon)/\eta^2}x_m) \\ &= c - \varepsilon. \end{aligned}$$

Note that this definition of r_t coincides with the definition of r_t in Case I when $\xi^2 = \varepsilon$. In Region III, i.e. $\{x \in U \mid \eta^2 + \varepsilon \le \xi^2\} = M^{c-\varepsilon}$, we let r_t be the identity. This coincides with the definition of r_t in region II when $\xi^2 = \eta^2 + \varepsilon$.

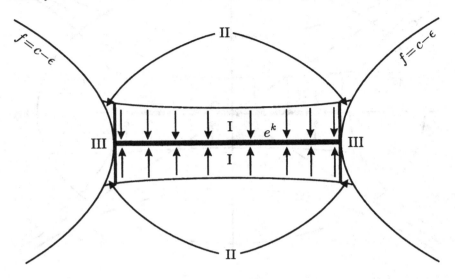

The above shows that $M^{c-\varepsilon} \cup e^k$ is a deformation retract of M^b. Theorem 3.20 that implies $f^{-1}(c-\varepsilon) \times [a, c-\varepsilon]$ is diffeomorphic to $f^{-1}([a, c-\varepsilon])$. Hence, by taking the union of the collar given by the image of $\partial e^k \times [a, c-\varepsilon]$ under this diffeomorphism with e^k we see that $M^a \cup e^k$ is a deformation retract of M^b (where we have used the same notation to denote e^k with the collar added).

$\square$

Remark 3.26 If there are j critical points, $p_1, \ldots, p_j$ in the level set $f^{-1}(c)$ with indices $\lambda_{p_1}, \ldots, \lambda_{p_j}$ respectively, then a similar proof shows that

$$M^{c+\varepsilon} \simeq M^{c-\varepsilon} \cup_{f_1} D^{\lambda_{p_1}} \cup_{f_2} \cdots \cup_{f_j} D^{\lambda_{p_j}}$$

for some collection of attaching maps $f_1, \ldots, f_j$.

Remark 3.27 An easy modification of the above proof shows that M^c is a deformation retract of $M^{c+\varepsilon}$. Indeed, $F^{-1}((-\infty, c])$ is a deformation retract of $F^{-1}((-\infty, c+\varepsilon]) = M^{c+\varepsilon}$ by Theorem 3.20, and M^c is a deformation retract of $F^{-1}((-\infty, c])$. The following diagram illustrates a deformation retraction of $F^{-1}((-\infty, c])$ to M^c.

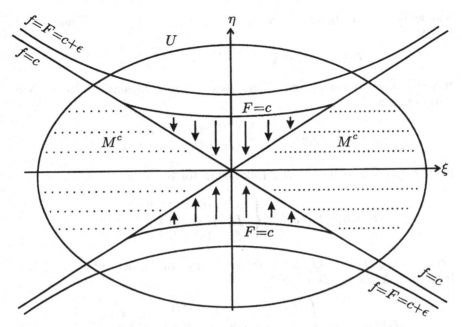

Combining the above with Theorem 3.25 we see that $M^{c-\varepsilon} \cup e^k$ has the same homotopy type as M^c. In fact, $M^{c-\varepsilon} \cup e^k$ is a deformation retract of M^c.

We now arrive at the main result of this chapter.

Theorem 3.28 *Let $f : M \to \mathbb{R}$ be a Morse function on a smooth manifold M. Suppose that M^t is compact for all $t \in \mathbb{R}$. Then M has the homotopy type of a CW-complex X with one cell of dimension k for each critical point of index k.*

The proof of this theorem will be based on the following two lemmas concerning attaching cells to a topological space.

Lemma 3.29 (J.H.C. Whitehead [152]) *Let X be a topological space, and suppose that $f_0 : S^{k-1} \to X$ and $f_1 : S^{k-1} \to X$ are homotopic. Then the identity map of X extends to a homotopy equivalence*

$$h : X \cup_{f_0} D^k \to X \cup_{f_1} D^k.$$

Proof:
Denote the characteristic maps by $f_0 : D^k \to X \cup_{f_0} D^k$ and $f_1 : D^k \to X \cup_{f_1} D^k$, and let $f_t : [0,1] \times S^{k-1} \to X$ be a homotopy from $f_0 : S^{k-1} \to X$ to $f_1 : S^{k-1} \to X$. Define $h_0 : X \cup_{f_0} D^k \to X \cup_{f_1} D^k$ by $h_0(x) = x$ if $x \in X$ and for all $u \in S^{k-1}$

$$h_0(f_0(ru)) = \begin{cases} f_1(2ru) & \text{if } 0 \le 2r \le 1 \\ f_{2-2r}(u) & \text{if } 1 \le 2r \le 2, \end{cases}$$

and define $h_1 : X \cup_{f_1} D^k \to X \cup_{f_0} D^k$ by $h_1(x) = x$ if $x \in X$ and for all $u \in S^{k-1}$

$$h_1(f_1(ru)) = \begin{cases} f_0(2ru) & \text{if } 0 \le 2r \le 1 \\ f_{2r-1}(u) & \text{if } 1 \le 2r \le 2. \end{cases}$$

It is easy to verify that h_0 and h_1 are single valued and hence continuous. We have for all $u \in S^{k-1}$

$$(h_0 \circ h_1)(f_0(ru)) = \begin{cases} h_1(f_1(2ru)) & \text{if } 0 \le 2r \le 1 \\ h_1(f_{2-2r}(u)) & \text{if } 1 \le 2r \le 2. \end{cases}$$

Since $h_1(x) = x$ for all $x \in X$ it follows that for all $u \in S^{k-1}$

$$(h_0 \circ h_1)(f_0(ru)) = \begin{cases} f_0(4ru) & \text{if } 0 \le 4r \le 1 \\ f_{4r-1}(u) & \text{if } 1 \le 4r \le 2 \\ f_{2-2r}(u) & \text{if } 1 \le 2r \le 2. \end{cases}$$

Let $\xi_t : X \cup_{f_0} D^k \to X \cup_{f_1} D^k$ be the homotopy which is defined by $\xi_t(x) = x$ for all $x \in X$ and for all $u \in S^{k-1}$

$$\xi_t(f_0(ru)) = \begin{cases} f_0((4-3t)ru) & \text{if } 0 \le r \le \frac{1}{4-3t} \\ f_{(4-3t)r-1}(u) & \text{if } \frac{1}{4-3t} \le r \le \frac{2-t}{4-3t} \\ f_{\frac{1}{2}(4-3t)(1-r)}(u) & \text{if } \frac{2-t}{4-3t} \le r \le 1. \end{cases}$$

It is easy to verify that ξ_t is single valued and hence continuous, $\xi_0 = h_1 \circ h_0$, and $\xi_1 = 1$. A homotopy $\eta_t : X \cup_{f_1} D^k \to X \cup_{f_0} D^k$ such that $\eta_0 = h_0 \circ h_1$ and $\eta_1 = 1$ is defined by replacing f_0 with f_1 and g_λ with $g_{1-\lambda}$ in the above expression for ξ_t where $\lambda = (4-3t)r - 1$ or $(4-3t)(1-r)/2$.

$\square$

Lemma 3.30 (P. Hilton [100]) *Let X be a topological space and let*

$$f : S^{k-1} \to X$$

be an attaching map. Any homotopy equivalence $h : X \to Y$ extends to a homotopy equivalence

$$H : X \cup_f D^k \to Y \cup_{hof} D^k.$$

Proof:
    Define $H : X \cup_f D^k \to Y \cup_{hof} D^k$ by

$$H(x) = \begin{cases} h(x) & \text{if } x \in X \\ x & \text{if } x \in D^k. \end{cases}$$

Let $g : Y \to X$ be a homotopy inverse of h and define

$$G : Y \cup_{hof} D^k \to X \cup_{gohof} D^k$$

by

$$G(y) = \begin{cases} g(y) & \text{if } x \in Y \\ y & \text{if } x \in D^k. \end{cases}$$

Since $g \circ h \circ f$ is a homotopic to f, it follows from Lemma 3.29 that there is a homotopy equivalence

$$F : X \cup_{gohof} D^k \to X \cup_f D^k.$$

We will first prove that the composition

$$F \circ G \circ H : X \cup_f D^k \to X \cup_f D^k$$

is homotopic to the identity. Let h_t be a homotopy between $g \circ h$ and the identity. Using the specific definitions of F, G, and H, we see that

$$\begin{aligned} (F \circ G \circ H)(x) &= (g \circ h)(x) & \text{for } x \in X \\ (F \circ G \circ H)(tu) &= 2tu & \text{for } 0 \le t \le \tfrac{1}{2}, \quad u \in \partial D^k \\ (F \circ G \circ H)(tu) &= (h_{2-2t} \circ f)(u) & \text{for } \tfrac{1}{2} \le t \le 1, \quad u \in \partial D^k. \end{aligned}$$

The required homotopy $q_\tau : X \cup_f D^k \to X \cup_f D^k$ is now defined by the formula

$$\begin{aligned} q_\tau(x) &= h_\tau(x) & \text{for } x \in X \\ q_\tau(tu) &= \tfrac{2}{1+\tau} tu & \text{for } 0 \le t \le \tfrac{1+\tau}{2}, \quad u \in \partial D^k \\ q_\tau(tu) &= (h_{2-2t+\tau} \circ f)(u) & \text{for } \tfrac{1+\tau}{2} \le t \le 1, \quad u \in \partial D^k. \end{aligned}$$

Therefore, H has a left homotopy inverse, namely $F \circ G : Y \cup_{hof} D^k \to X \cup_f D^k$, and a similar proof shows that G also has a left homotopy inverse.

We can now complete the proof of the lemma as follows. Since

$$F \circ (G \circ H) \simeq \text{identity},$$

and F is known to have a left homotopy inverse (by Lemma 3.29), it follows that

$$(G \circ H) \circ F \simeq \text{identity}.$$

Similarly,

$$G \circ (H \circ F) \simeq \text{identity}$$

and the fact that G has a left homotopy inverse imply that

$$(H \circ F) \circ G \simeq \text{identity}.$$

Thus, $H \circ (F \circ G) \simeq$ identity and $F \circ G$ is also a right homotopy inverse for H. Therefore, H is a homotopy equivalence.

$\square$

Proof of Theorem 3.28:

Let $c_0 < c_1 < c_2 < \cdots$ be the critical values of $f : M \to \mathbb{R}$. Note that the sequence $\{c_i\}$ has no accumulation point since M^t is compact for all t. The set M^t is vacuous for all $t < c_0$, and M^{t_0} is homotopic to a discrete set of points for all $c_0 < t_0 < c_1$, i.e the set of critical points $\{p \in \mathrm{Cr}(f)| \ f(p) = c_0\}$. Suppose now for the purposes of induction that $c_{i-1} < t_{i-1} < c_i$ for some $i = 1, 2, \ldots$ and there is a homotopy equivalence $h_{i-1} : M^{t_{i-1}} \to X_{i-1}$ where X_{i-1} is some CW-complex.

By Theorem 3.25 and Remark 3.26, there exists an $\varepsilon > 0$ such that $M^{c_i + \varepsilon}$ has the homotopy type of

$$M^{c_i - \varepsilon} \cup_{f_1} D^{\lambda_{p_1}} \cup_{f_2} \cdots \cup_{f_j} D^{\lambda_{p_j}}$$

where $p_1, \ldots, p_j$ are the critical points in $f^{-1}(c_i)$ and $f_1, \ldots, f_j$ are attaching maps. Moreover, by Theorem 3.20 there is a homotopy equivalence $g_{i-1} : M^{c_i - \varepsilon} \to M^{t_{i-1}}$.

For every $k = 1, \ldots, j$, the map $h_{i-1} \circ g_{i-1} \circ f_k : S^{\lambda_{p_k} - 1} \to X_{i-1}$ is homotopic to a map

$$\Psi_k : S^{\lambda_{p_k} - 1} \to X_{i-1}^{(\lambda_{p_k} - 1)}$$

by Theorem 2.20 where $X_{i-1}^{(\lambda_{p_k} - 1)}$ denotes the $(\lambda_{p_k} - 1)$-skeleton of X_{i-1}. Thus,

$$X_i \stackrel{\mathrm{def}}{=} X_{i-1} \cup_{\Psi_1} D^{\lambda_{p_1}} \cup_{\Psi_2} \cdots \cup_{\Psi_j} D^{\lambda_{p_j}}$$

is a CW-complex which has the same homotopy type as $M^{c+\varepsilon}$ by Lemma 3.29 and Lemma 3.30.

By induction and Remarks 3.26 and 3.27, it follows that M^t has the homotopy type of a CW-complex for all $t \in \mathbb{R}$. If M is compact this completes the proof. If M is not compact, but all the critical points lie in M^t for some $t \in \mathbb{R}$, then a proof similar to that of Theorem 3.20 shows that M^t is a deformation retract of M, so the proof is again complete.

If there are infinitely many critical points, then the above construction gives us an infinite sequence of homotopy equivalences

$$
\begin{array}{ccccccccc}
M^{t_0} & \longrightarrow & M^{t_1} & \longrightarrow & M^{t_2} & \longrightarrow & \cdots & & M \\
\downarrow h_0 & & \downarrow h_1 & & \downarrow h_2 & & & & \downarrow h \\
X_0 & \longrightarrow & X_1 & \longrightarrow & X_2 & \longrightarrow & \cdots & & X
\end{array}
$$

where each homotopy equivalence extends the previous one. Let X denote the union $\bigcup_i X_i$ with the direct limit topology, and let $h : M \to X$ be the direct limit map. Then h induces isomorphisms of homotopy groups in all dimensions, and hence, h is a homotopy equivalence (see for instance Theorem 1 of [151]).

$\square$

Remark 3.31 Theorem 1 of [151] can be stated as follows. Let X and Y be connected topological spaces that are dominated by CW-complexes. Then a map $h : X \to Y$ is a homotopy equivalence of and only if $f_n : \pi_n(X) \to \pi_n(Y)$ is an isomorphism for every n such that $1 \leq n \leq N$ where $N = \max(\Delta X, \Delta Y) \leq \infty$.

A space P **dominates** a space X if and only if there are maps $\sigma : X \to P$ and $\sigma' : P \to X$ such that $\sigma' \circ \sigma \simeq 1$. A CW-complex X is dominated by itself, and a finite dimensional manifold M is dominated by the CW-complex consisting of a tubular neighborhood of M embedded in some Euclidean space. If X is dominated by a CW-complex of finite dimension, then ΔX denotes the minimum dimension of all CW-complexes that dominate X. If none of the CW-complexes that dominate X have finite dimension, then $\Delta X = \infty$.

Remark 3.32 There is a version of Theorem 3.28 called the **Handle Presentation Theorem** that keeps track of the differentiable structure. See Chapter VII of [91] for more details.

3.4 The Morse Inequalities

Let M be a compact smooth manifold and F a field. Then $H_k(M; F)$ is a finite dimensional vector space over F, and the kth **Betti number** of M, denoted $b_k(F)$, is defined to be the dimension of $H_k(M; F)$. Similarly, if $F = \mathbb{Z}$ then $H_k(M; \mathbb{Z})$ modulo its torsion subgroup is a finitely generated free $\mathbb{Z}$-module, and $b_k(\mathbb{Z})$ is defined to be the rank of this finitely generated free $\mathbb{Z}$-module. We will write b_k for $b_k(F)$ when $F = \mathbb{Z}$ or F is understood from the context.

Let $f : M \to \mathbb{R}$ be a Morse function on M, and let ν_k be the number of critical points of f of index k for all $k = 0, \ldots, m$. As a consequence of Theorem 2.15 and Theorem 3.28 we have

$$\nu_k \geq b_k(F)$$

for all $k = 0, \ldots, m$ since $\nu_k = \operatorname{rank} \underline{C}_k(M; F)$ and $H_k(M; F)$ is a quotient of this module. These inequalities are known as the **weak Morse inequalities**.

In particular, the weak Morse inequalities imply that the total number of critical points of f is greater than or equal to the sum of the Betti numbers,

$$\sum_{k=0}^{m} \nu_k \geq \sum_{k=0}^{m} b_k(F).$$

It is this observation that motivated The Arnold Conjecture (Conjecture 9.1).

Observe that ν_k depends on the Morse function $f : M \to \mathbb{R}$ (but not on F), and $b_k(F)$ depends on the topology of M and F (but not on the Morse function). The following **strong Morse inequalities** give more relations between ν_k and $b_k(F)$.

Theorem 3.33 (Morse Inequalities) *For any Morse function $f : M \to \mathbb{R}$ on a compact smooth manifold M of dimension m we have the following.*

(a) $\displaystyle\sum_{k=0}^{n}(-1)^{k+n}\nu_k \geq \sum_{k=0}^{n}(-1)^{k+n}b_k(F)$ *for every $n = 0, \ldots, m$.*

(b) $\displaystyle\sum_{k=0}^{m}(-1)^{k}\nu_k = \sum_{k=0}^{m}(-1)^{k}b_k(F).$

Remark 3.34 Part (b) of the preceeding theorem shows that the **Euler characteristic**, $\mathcal{X}(M) = \sum_{k=0}^{m}(-1)^{k}b_k(F)$, is independent of the field F. If F is a field of characteristic zero or $F = \mathbb{Z}$, then the same is true for the Betti numbers. This can be seen as a corollary to the Universal Coefficient Theorem for Homology. (See for instance Section VI.7.19 of [39].)

Definition 3.35 *The **Poincaré polynomial** of M is defined to be*

$$P_t(M) = \sum_{k=0}^{m} b_k(F)t^{k},$$

*and the **Morse polynomial** of f is defined to be*

$$M_t(f) = \sum_{k=0}^{m} \nu_k t^{k}.$$

In Lemma 3.43 we will show that Theorem 3.33 is equivalent to the following theorem. It is this version of the Morse inequalities that generalizes to Morse-Bott functions (see Theorem 3.53).

Theorem 3.36 (Polynomial Morse Inequalities) *For any Morse function $f : M \to \mathbb{R}$ on a smooth manifold M we have*

$$M_t(f) = P_t(M) + (1+t)R(t)$$

where $R(t)$ is a polynomial with non-negative integer coefficients. That is, $R(t) = \sum_{k=0}^{m-1} r_k t^k$ where $r_k \in \mathbb{Z}$ satisfies $r_k \geq 0$ for all $k = 0, \ldots, m-1$.

Remark 3.37 Theorem 3.36 is sometimes misstated in the literature. One mistake is to claim that the polynomial $R(t)$ is always positive. Another mistake is to claim that the Morse polynomial is greater than or equal to the Poincaré polynomial for all $t \in \mathbb{R}$. Although both of these statements are true for $t \geq 0$, they are not always true for $t < 0$. The following example is a counterexample for both of these misstatements.

Example 3.38 (Wiener-Dog Counterexample) Consider the function f defined by projection onto the z axis in the following picture.

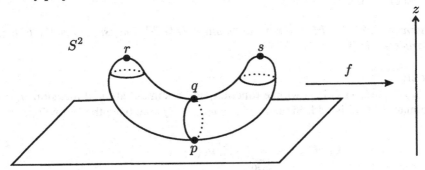

The critical points p, q, r, and s have indices $0, 1, 2$, and 2 respectively. Hence, $M_t(f) = 2t^2 + t + 1$, and since $P_t(S^2) = t^2 + 1$, we see that

$$M_t(f) = 2t^2 + t + 1 = t^2 + 1 + (1+t)t = P_t(M) + (1+t)R(t).$$

Thus, $R(t) = t$, and we see that $R(t) < 0$ for all $t < 0$. Moreover, it is easy to check that $M_t(f) < P_t(f)$ for $-1 < t < 0$.

The following is an immediate consequence of Theorem 3.36.

Theorem 3.39 (Morse's Lacunary Principle) *If $M_t(f)$ has no consecutive powers of t, then*

$$M_t(f) = P_t(M).$$

In fact, $\nu_k = b_k$ and $R(t)$ is identically zero.

Example 3.40 (Bott's perfect Morse functions) Let $f : \mathbb{C}P^n \to \mathbb{R}$ be the Morse function discussed in Example 3.7. We have

$$M_t(f) = 1 + t^2 + \cdots + t^{2n}$$

and the preceeding theorem implies that

$$P_t(\mathbb{C}P^n) = 1 + t^2 + \cdots + t^{2n}.$$

Note that this allows us to find the homology groups of $\mathbb{C}P^n$ by simply counting the number of critical points of each index. We will see that a similar result holds for complex Grassmann manifolds in Chapter 8.

Definition 3.41 *If $f : M \to \mathbb{R}$ is a Morse function such that $M_t(f) = P_t(M)$, then f is called a **perfect Morse function**.*

Note that if a manifold admits a perfect Morse function, then its homology doesn't have any torsion. For more details about perfect Morse functions as well as examples where the Morse inequalities are not sharp see [118].

The next theorem is an important result that follows as an easy consequence of Theorem 3.33.

Theorem 3.42 *Let M be a compact manifold of odd dimension, then the Euler characteristic is zero, i.e. $\mathcal{X}(M) = 0$.*

Proof:
Let $f : M \to \mathbb{R}$ be a Morse function, and assume that the dimension m of the manifold M is odd. Since $\nu_k(f) = \nu_{m-k}(-f)$ we have the following.

$$
\begin{aligned}
\mathcal{X}(M) &= \sum_{k=0}^{m}(-1)^k\nu_k(f) \\
&= \sum_{k=0}^{m}(-1)^k\nu_{m-k}(-f) \\
&= (-1)^m\sum_{k=0}^{m}(-1)^{m-k}\nu_{m-k}(-f) \\
&= (-1)^m\sum_{k=0}^{m}\nu_k(-f) \\
&= (-1)^m\mathcal{X}(M)
\end{aligned}
$$

Hence, $\mathcal{X}(M) = 0$ if m is odd.

$\square$

We will now prove the Morse Inequalities. We will see that Theorem 3.33 and Theorem 3.36 are both easy consequences of Theorem 3.28. However, it should be noted that Morse proved his inequalities in the 1930's before Theorem 3.28 was known [106]. Morse's inequalities provided the first deep relationship between the critical points of a real valued function on a smooth manifold M and the topology of M. For an exposition of Morse's approach to the Morse Inequalities (that does not rely on Theorem 3.28) see Section 5 of [100].

In the following we will first prove Lemma 3.43 which states that the Morse Inequalities are equivalent to the Polynomial Morse Inequalities. The reader should note that the proof of Lemma 3.43 shows explicitly the connection between the inequalities in part (a) of Theorem 3.33 and the coefficients of the polynomial $R(t)$ in Theorem 3.36. Next we will prove both Theorem 3.33 and Theorem 3.36. Of course, Lemma 3.43 shows that it would be sufficient to prove either one of these theorems. However, both proofs are of interest, and so we complete this circle of ideas by including separate proofs for both Theorem 3.33 and Theorem 3.36.

Lemma 3.43 *Theorem 3.33 is equivalent to Theorem 3.36.*

Proof:

By part (b) of Theorem 3.33 we have

$$M_{-1}(f) = \sum_{k=0}^{m}(-1)^k\nu_k = \sum_{k=0}^{m}(-1)^k b_k(F) = P_{-1}(M).$$

Thus $M_t(f) - P_t(M)$ is divisible by $1+t$, and $M_t(f) = P_t(M) + (1+t)R(t)$ for some polynomial $R(t) = \sum_{n=0}^{m-1} r_n t^n$. It's clear that $r_n \in \mathbb{Z}$ for all $n = 0, \ldots, m-1$ since both $M_t(f)$ and $P_t(M)$ have integer coefficients, and it remains to show that $r_n \geq 0$ for all $n = 0, \ldots, m-1$. We will show that this is equivalent to the inequalities in part (a) of Theorem 3.33.

To see this, first note that $M_t(f) = P_t(M) + (1+t)R(t)$ implies that

$$\nu_0 = b_0(F) + r_0.$$

Next, note that $\nu_1 = b_1(F) + r_1 + r_0$, and so $\nu_1 = b_1(F) + r_1 + \nu_0 - b_0(F)$, i.e.

$$\nu_1 - \nu_0 = b_1(F) - b_0(F) + r_1.$$

Continuing in this fashion we see that

$$\nu_n - \nu_{n-1} + \cdots + (-1)^n\nu_0 = b_n(F) - b_{n-1}(F) + \cdots + (-1)^n b_0(F) + r_n$$

for all $n = 0, \ldots, m-1$. Thus, the inequalities in part (a) of Theorem 3.33 imply that $r_n \geq 0$ for all $n = 0, \ldots, m-1$, and Theorem 3.33 implies Theorem 3.36.

Now assume that $M_t(f) = P_t(M) + (1+t)R(t)$ where $R(t) = \sum_{n=0}^{m-1} r_n t^n$ is a polynomial with non-negative integer coefficients. As above, this implies that

$$\nu_n - \nu_{n-1} + \cdots + (-1)^n\nu_0 = b_n(F) - b_{n-1}(F) + \cdots + (-1)^n b_0(F) + r_n$$

for all $n = 0, \ldots, m-1$. Since $r_n \geq 0$ for all $n = 0, \ldots, m-1$ we recover the inequalities in part (a) of Theorem 3.33. To recover part (b) we simply put $t = -1$ in the equation $M_t(f) = P_t(M) + (1+t)R(t)$. Therefore, Theorem 3.36 implies Theorem 3.33.

□

We will now show that Theorem 3.33 is an easy consequence of Theorem 3.28 and The Euler-Poincaré Theorem. Recall that a finitely generated chain complex $(\underline{C}_*, \partial_*)$ is a chain complex such that $\underline{C}_k$ is a finitely generated abelian group for all $k \in \mathbb{Z}_+$, and $\underline{C}_k = 0$ except for a finite set of integers.

Theorem 3.44 (Euler-Poincaré Theorem) *Let $(\underline{C}_*, \partial_*)$ be a finitely generated chain complex, and assume that $\underline{C}_k = 0$ for all $k > m$. Let $c_k = $ rank $\underline{C}_k$ and $b_k = $ rank $H_k(\underline{C}_*)$ for all $k = 0, \ldots, m$. Then,*

$$\sum_{k=0}^{m}(-1)^k c_k = \sum_{k=0}^{m}(-1)^k b_k.$$

Proof:

The exact sequence

$$0 \to \ker \partial_k \to \underline{C}_k(X; F) \overset{\partial_k}{\to} \operatorname{im} \partial_k \to 0$$

shows that $c_k = $ rank ker $\partial_k + $ rank im ∂_k for all $k = 0, \ldots, m$. Similarly,

$$0 \to \operatorname{im} \underline{\partial}_{k+1} \to \ker \partial_k \to H_k(X; F) \to 0$$

shows that rank ker $\partial_k = $ rank im $\underline{\partial}_{k+1} + b_k$, and hence

$$\operatorname{rank} \ker \partial_k = c_k - \operatorname{rank} \operatorname{im} \partial_k = \operatorname{rank} \operatorname{im} \underline{\partial}_{k+1} + b_k$$

for all $k = 0, \ldots, m$. Thus,

$$\sum_{k=0}^{m}(-1)^k (c_k - \operatorname{rank} \operatorname{im} \partial_k) = \sum_{k=0}^{m}(-1)^k (\operatorname{rank} \operatorname{im} \underline{\partial}_{k+1} + b_k)$$

which implies that

$$\sum_{k=0}^{m}(-1)^k c_k = \sum_{k=0}^{m}(-1)^k b_k.$$

□

Now, let $f : M \to \mathbb{R}$ be a Morse function on a compact smooth manifold M, and let X be the associated CW-complex given by Theorem 3.28. Let

$\underline{C}_k(X; F)$ denote the kth chain group in the CW-chain complex of X with coefficients in F where F is either a field or $\mathbb{Z}$, and let $\underline{\partial}_k$ denote the kth boundary operator in the CW-chain complex (as in Theorem 2.15). Note that

$$\begin{array}{rcl} \nu_k & = & \text{rank } \underline{C}_k(X; F) \quad \text{and} \\ b_k & = & \text{rank } H_k(X; F) \end{array}$$

where $b_k = b_k(F)$ is the kth Betti number of M with coefficients in F for all $k = 0, \ldots, m$.

Proof of Theorem 3.33:

Part (b) of Theorem 3.33 is The Euler-Poincaré Theorem (Theorem 3.44) applied to the chain complex $(\underline{C}_*(X; F), \underline{\partial}_*)$. To prove part (a) let $n \leq m$ and consider the truncated chain complex $(\underline{C}_*^{(n)}(X; F), \underline{\partial}_*)$ given by

$$\underline{C}_k^{(n)}(X; F) = \begin{cases} \underline{C}_k(X; F) & \text{if } k \leq n \\ 0 & \text{if } k > n. \end{cases}$$

By The Euler-Poincaré Theorem we have

$$(-1)^n \sum_{k=0}^{n} (-1)^k \text{rank } \underline{C}_k^{(n)}(X; F) = (-1)^n \sum_{k=0}^{n} (-1)^k \text{rank } H_k(\underline{C}_*^{(n)}(X; F)).$$

Since $\nu_k = \text{rank } \underline{C}_k^{(n)}(X; F)$ for all $k = 0, \ldots, n$, $b_k = \text{rank } H_k(\underline{C}_*^{(n)}(X; F))$ for all $k = 0, \ldots, n-1$ and $H_n(\underline{C}_*(X; F))$ is a quotient of $H_n(\underline{C}_*^{(n)}(X; F))$, we have

$$\nu_n - \nu_{n-1} + \cdots + (-1)^n \nu_0 \geq b_n - b_{n-1} + \cdots + (-1)^n b_0.$$

$\square$

Proof of Theorem 3.36:

Let $z_k = \text{rank ker } \underline{\partial}_k$ for all $k = 0, \ldots, m$. As in the proof of Theorem 3.44, the exact sequence

$$0 \to \text{ker } \underline{\partial}_k \to \underline{C}_k(X; F) \xrightarrow{\underline{\partial}_k} \text{im } \underline{\partial}_k \to 0$$

implies that $\nu_k = z_k + \text{rank im } \underline{\partial}_k$ for all $k = 0, \ldots, m$, and

$$0 \to \text{im } \underline{\partial}_{k+1} \to \text{ker } \underline{\partial}_k \to H_k(X; F) \to 0$$

implies that $b_k = z_k - \text{rank im } \underline{\partial}_{k+1}$ for all $k = 0, \dots, m$. Hence,

$$
\begin{aligned}
M_t(f) - P_t(M) &= \sum_{k=0}^{m} \nu_k t^k - \sum_{k=0}^{m} b_k t^k \\
&= \sum_{k=0}^{m} (z_k + \text{rank im } \underline{\partial}_k) t^k - \sum_{k=0}^{m} (z_k - \text{rank im } \underline{\partial}_{k+1}) t^k \\
&= \sum_{k=0}^{m} (\text{rank im } \underline{\partial}_k + \text{rank im } \underline{\partial}_{k+1}) t^k \\
&= \sum_{k=0}^{m} (\nu_k - z_k + \nu_{k+1} - z_{k+1}) t^k \\
&= \sum_{k=0}^{m} (\nu_k - z_k) t^k + \sum_{k=0}^{m} (\nu_{k+1} - z_{k+1}) t^k \\
&= t \sum_{k=1}^{m} (\nu_k - z_k) t^{k-1} + \sum_{k=1}^{m} (\nu_k - z_k) t^{k-1} \quad (\text{since } \nu_0 = z_0) \\
&= (t + 1) \sum_{k=1}^{m} (\nu_k - z_k) t^{k-1}.
\end{aligned}
$$

Therefore, $M_t(f) = P_t(M) + (1 + t)R(t)$ where $R(t) = \sum_{k=0}^{m-1} (\nu_{k+1} - z_{k+1}) t^k$. Note that $\nu_{k+1} - z_{k+1} \geq 0$ for all $k = 0, \dots, m - 1$ because z_{k+1} is the rank of a subgroup of $\underline{C}_{k+1}(X; F)$ and $\nu_{k+1} = \text{rank } \underline{C}_{k+1}(X; F)$.

$\square$

3.5   Morse-Bott functions

In this section we introduce a generalization of the notion of a Morse function due to Bott [21]. The results in this section will not be used elsewhere in the book.

A lot of functions which arise naturally are not Morse functions: their critical sets are not a discrete set of points but rather smooth manifolds (possibly of different dimensions). Consider for instance the torus T^2 as the surface of a doughnut lying on a flat dinner plate.

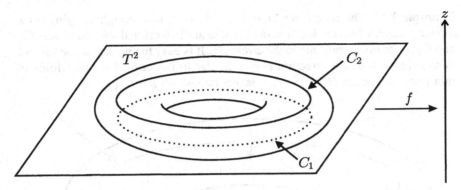

The height function $f : T^2 \to \mathbb{R}$ has 2 "critical circles", C_1, the inverse image of the minimum, and C_2, the inverse image of the maximum. Other common examples come from functions which are invariant under the action of a Lie group, where the whole orbit of a critical point is a critical set.

Let $f : M \to \mathbb{R}$ be a smooth function whose critical set $\mathrm{Cr}(f)$ contains a submanifold C of positive dimension. Pick a Riemannian metric on M and use it to split $T_* M|_C$ as

$$T_* M|_C = T_* C \oplus \nu_* C$$

where $T_* C$ is the tangent space of C and $\nu_* C$ is the normal bundle of C. Let $p \in C$, $V \in T_p C$, $W \in T_p M$, and let $H_p(f)$ be the Hessian of f at p. We have

$$H_p(f)(V, W) = V_p \cdot (\tilde{W} \cdot f) = 0$$

since $V_p \in T_p C$ and any extension of W to a vector field $\tilde{W}$ satisfies $df(\tilde{W})|_C = 0$. Therefore, the Hessian $H_p(f)$ induces a symmetric bilinear form $H_p^\nu(f)$ on $\nu_p C$, whose matrix in local coordinates will be denoted $M_p^\nu(f)$.

Definition 3.45 *A smooth function $f : M \to \mathbb{R}$ on a smooth manifold M is called a **Morse-Bott function** if and only if $\mathrm{Cr}(f)$ is a disjoint union of connected submanifolds, and for each connected submanifold $C \subseteq \mathrm{Cr}(f)$ the bilinear form $H_p^\nu(f)$ is non-degenerate for all $p \in C$.*

Often one says that the Hessian of a Morse-Bott function f is non-degenerate in the direction normal to the critical submanifolds.

Examples of Morse-Bott functions

Example 3.46 Any Morse function $f : M \to \mathbb{R}$ is a Morse-Bott function. The critical submanifolds are just the critical points and $H_p(f) = H_p^\nu(f)$ for all $p \in \mathrm{Cr}(f)$ since the critical submanifolds are all of dimension zero.

Example 3.47 The height function $f : T^2 \to \mathbb{R}$ of a doughnut lying on a dinner plate is a Morse-Bott function. There are two critical submanifolds, C_1 and C_2, each diffeomorphic to the circle S^1. It is easy to check that the second derivative of f in the direction perpendicular to the critical submanifolds is non-zero, and hence $H_p^\nu(f)$ is non-degenerate for all $p \in \mathrm{Cr}(f)$.

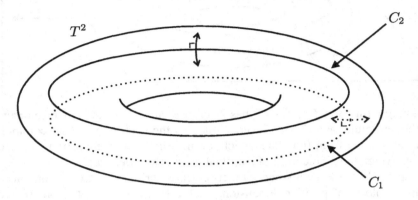

Example 3.48 Any constant function $f \equiv c \in \mathbb{R}$ is a Morse-Bott function since $\mathrm{Cr}(f) = M$ and $\nu_* C = \emptyset$. The trivial remark that $f \equiv 0$ is a Morse-Bott function turns out to be very useful!

Example 3.49 The square of the height function on S^n considered in Example 3.5 is a Morse-Bott function. Notice that even in this simple example the critical submanifolds are of different dimensions, two of the critical submanifolds (N and S) have dimension 0 and the other critical submanifold has dimension $n - 1$. This is not uncommon. There is no reason to expect that the critical submanifolds of a Morse-Bott function will be of the same dimension.

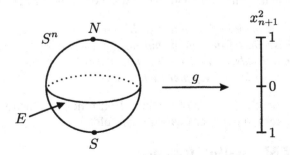

Example 3.50 If $\pi : E \to M$ is a smooth fiber bundle over a smooth manifold M and $f : M \to \mathbb{R}$ is a Morse-Bott function, then the composite $\pi \circ f : E \to \mathbb{R}$ is a Morse-Bott function.

The Morse-Bott Lemma

The following is the analogue of Lemma 3.11 for Morse-Bott functions. As a consequence of this lemma we see that there is a well-defined "Morse-Bott index" $\lambda_C \in \mathbb{Z}_+$ associated to a connected critical submanifold C.

Lemma 3.51 (Morse-Bott Lemma) *Let $f : M \to \mathbb{R}$ be a Morse-Bott function and $C \subseteq Cr(f)$ a connected component. For any $p \in C$ there is a local chart of M around p and a local splitting $\nu_* C = \nu_*^-(C) \oplus \nu_*^+(B)$, identifying a point $x \in M$ in its domain to (u, v, w) where $u \in C$, $v \in \nu_*^- C$, $w \in \nu_*^+ C$ such that within this chart f assumes the form*

$$f(x) = f(u, v, w) = f(C) - |v|^2 + |w|^2.$$

Proof:

Let $C \subseteq M$ be an n-dimensional connected critical submanifold of f. By replacing f with $f - c$, where c is the common value of f on the critical submanifold C, we may assume that $f(p) = 0$ for all $p \in C$. Let $p \in C$, and choose a coordinate chart $\phi : U \xrightarrow{\approx} \mathbb{R}^n \times \mathbb{R}^{m-n}$ defined on an open neighborhood U of p such that $\phi(p) = (0, 0)$ and $\phi(U \cap C) = \mathbb{R}^n \times \{0\}$. By composing this chart with a diffeomorphism of $\mathbb{R}^n \times \mathbb{R}^{m-n}$ that is constant in the first component, we may assume that the matrix of the Hessian in the direction normal to $\mathbb{R}^n \times \{0\}$ at $(0, 0) \in \mathbb{R}^n \times \mathbb{R}^{m-n}$ for the local expression $h(x, y) = (f \circ \phi^{-1})(x, y)$:

$$M_0^\nu(f) = \left(\frac{\partial^2 h}{\partial y_i \partial y_j} \bigg|_{(0,0)} \right)$$

is a diagonal matrix with the first k diagonal entries equal to -1 and the rest equal to $+1$.

The assumption that f is Morse-Bott means that for every $x \in \mathbb{R}^n$ the quadratic form

$$q_x(y) = {}^t y \left(\frac{\partial^2 h}{\partial y_i \partial y_j} \bigg|_{(x,0)} \right) y$$

is non-degenerate. If we fix $x \in \mathbb{R}^n$ and apply Palais' construction in the proof of the Morse Lemma (Lemma 3.11) to the quadratic form $q_x : \mathbb{R}^{m-n} \to \mathbb{R}$, then we get a family ψ_x of diffeomorphisms depending smoothly on x between neighborhoods of $0 \in \mathbb{R}^{m-n}$ such that

$$h(x, \psi_x(y)) = q_x(y).$$

Therefore, $\tilde{\phi}^{-1}(x, y) = \phi^{-1}(x, \psi_x(y)) : \mathbb{R}^n \times \mathbb{R}^{m-n} \to M$ is a chart such that

$$(f \circ \tilde{\phi}^{-1})(x, y) = q_x(y).$$

It's clear that $M_x^\nu(f)$ depends smoothly on x, and hence by Proposition 3.16, for every $x \in \mathbb{R}^n$ sufficiently close to 0 there exists a matrix $Q_x \in GL_{m-n}(\mathbb{R})$ depending smoothly on x such that

$$
{}^t Q_x \left(\frac{\partial^2 h}{\partial y_i \partial y_j} \bigg|_{(x,0)} \right) Q_x = \left(\frac{\partial^2 h}{\partial y_i \partial y_j} \bigg|_{(0,0)} \right).
$$

Therefore, for all $x \in \mathbb{R}^n$ sufficiently close to 0 we have

$$
\begin{aligned}
(f \circ \tilde{\phi}^{-1})(x, Q_x y) &= q_x(Q_x y) \\
&= {}^t y\, {}^t Q_x \left(\frac{\partial^2 h}{\partial y_i \partial y_j} \bigg|_{(x,0)} \right) Q_x y \\
&= {}^t y \left(\frac{\partial^2 h}{\partial y_i \partial y_j} \bigg|_{(0,0)} \right) y \\
&= \sum_{j=1}^{m-n} \delta_j y_j^2
\end{aligned}
$$

where $\delta_j = \frac{\partial^2 h}{\partial y_j^2} \big|_{(0,0)} = -1$ for all $j = 1, \ldots, k$ and $\delta_j = \frac{\partial^2 h}{\partial y_j^2} \big|_{(0,0)} = +1$ for all $j = k+1, \ldots, m$.

$\square$

Definition 3.52 *Let $f : M \to \mathbb{R}$ be a Morse-Bott function on a finite dimensional smooth manifold M, and let C be a critical submanifold of f. For any $p \in C$ let λ_p denote the index of $H_p^\nu(f)$. This integer is the dimension of the normal bundle $\nu_p C$ and is locally constant by the preceeding lemma. If C is connected, then λ_p is constant throughout C and we call $\lambda_p = \lambda_C$ the **Morse-Bott index** of the connected critical submanifold C.*

The Morse-Bott inequalities

We will now show how to generalize the Morse Inequalities from the previous section. Recall that the Morse polynomial $M_t(f)$ of a Morse function $f : M \to \mathbb{R}$ is

$$
M_t(f) = \sum_{k=0}^{m} \nu_k t^k
$$

where ν_k is the number of critical points of index k. The Poincaré polynomial of M is

$$
P_t(M) = \sum_{k=0}^{m} b_k t^k
$$

where b_k is the kth Betti number of M. By Theorem 3.36 we have

$$M_t(f) = P_t(M) + (1 + t)R(t)$$

where $R(t)$ is a polynomial with non-negative integer coefficients.

Let $f : M \rightarrow \mathbb{R}$ be a Morse-Bott function on a finite dimensional compact smooth manifold, and assume that

$$\text{Cr}(f) = \coprod_{j=1}^{l} C_j,$$

where $C_1, \ldots, C_l$ are disjoint connected orientable critical submanifolds. Under these assumptions, we define the Morse-Bott polynomial of f to be

$$MB_t(f) = \sum_{j=1}^{l} P_t(C_j)t^{\lambda_j}$$

where λ_j is the Morse-Bott index of the critical submanifold C_j and $P_t(C_j)$ is the Poincaré polynomial of C_j. Clearly $MB_t(f)$ reduces to $M_t(f)$ when f is a Morse function. For a proof of the following theorem we refer the reader to Section 1 of [9].

Theorem 3.53 (Morse-Bott Inequalities) *Let $f : M \rightarrow \mathbb{R}$ be a Morse-Bott function on a finite dimensional compact smooth manifold, and assume that all the critical submanifolds of f are orientable. Then there exists a polynomial $R(t)$ with non-negative integer coefficients such that*

$$MB_t(f) = P_t(M) + (1 + t)R(t).$$

Remark 3.54 The preceeding theorem can be generalized to the case where the critical submanifolds are not orientable by using the homology of the critical submanifolds with twisted coefficients in an orientation bundle (see Section 1 of [9]). For an interesting approach to proving the preceeding theorem using the Morse Homology Theorem (Theorem 7.4) see [86]. However, as was pointed out to us by a reader of an early version of this manuscript, [86] fails to take into account the necessary assumption that the critical submanifolds are orientable.

Examples of the Morse-Bott inequalities

Example 3.55 If $f \equiv c \in \mathbb{R}$ is a constant function, then $MB_t(f)$ is just $P_t(M)$.

Example 3.56 If $g : S^n \rightarrow \mathbb{R}$ is given by $g(x_1, \ldots, x_{n+1}) = x_{n+1}^2$ (the square of the height function from Examples 3.5 and 3.49), then

$$\text{Cr}(g) = \{(x_1, \ldots, x_{n+1}) \in S^n \mid x_{n+1} = -1, \ x_{n+1} = 0, \ \text{or} \ x_{n+1} = 1\}.$$

Each point $x_{n+1} = \pm 1$ has index n and the equator $x_{n+1} = 0$ has index 0. Hence,

$$
\begin{aligned}
MB_t(g) &= P_t(\{x_{n+1} = 0\})t^0 + P_t(\{x_{n+1} = -1\})t^n + P_t(\{x_{n+1} = +1\})t^n \\
&= P_t(S^{n-1}) + t^n + t^n \\
&= 1 + t^{n-1} + 2t^n.
\end{aligned}
$$

Since $P_t(S^n) = 1 + t^n$ we see that

$$
1 + t^{n-1} + 2t^n = (1 + t^n) + (1 + t)t^{n-1},
$$

and hence, $R(t) = t^{n-1}$.

Remark 3.57 Note that the preceeding example provides a counterexample for Morse-Bott functions to the misstatements discussed in Remark 3.37.

Example 3.58 If $f : T^2 \to \mathbb{R}$ is the height function of a torus lying on plane, then $\mathrm{Cr}(f) = C_1 \cup C_2$ where C_1 is the circle where f takes its minimum and C_2 is the circle where f takes its maximum.

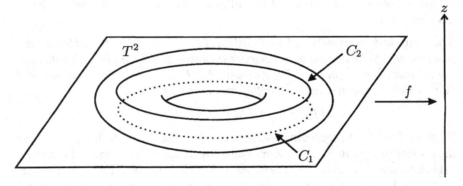

The Morse-Bott index of C_1 is 0 and the Morse-Bott index of C_2 is 1. Since $P_t(C_1) = P_t(C_2) = 1 + t$ we have

$$
MB_t(f) = (1 + t)t^0 + (1 + t)t = 1 + 2t + t^2 = P_t(T^2)
$$

and $R(t) \equiv 0$.

We would like to thank the reader of an early version of this manuscript who provided us with the following example. This example shows that the statement of the Morse-Bott Inequalities in Theorem 3.53 may not hold when the critical submanifolds are not orientable. The correct statement of the Morse-Bott Inequalities in the case where the critical submanifolds are not orientable uses homology with local coefficients (see Section 1 of [9]).

Example 3.59 Consider the Morse-Bott function on T^2 described in the previous example. Pull this back to a function on $T^2 \times [0,1]$. Now glue the ends together by a reflection on T^2 which preserves the Morse-Bott function but reflects each critical circle. We then get a Morse-Bott function on the mapping torus of a reflection on T^2, whose critical submanifolds are Klein bottles, which fails the Morse-Bott Inequalities as stated in Theorem 3.53.

A perturbation technique for Morse-Bott functions

We end this section by giving a useful construction relating Morse-Bott functions and Morse functions. We will see in Theorem 5.27 that given any smooth function f on a smooth manifold there exists a Morse function g arbitrarily close to f. However, the proof of Theorem 5.27 is not constructive since it relies on Sard's Theorem. The following useful construction produces an explicit Morse function arbitrarily close to a given Morse-Bott function [13].

Let T_j be a tubular neighborhood around each connected component $C_j \subseteq \mathrm{Cr}(f)$ for all $j = 1, \ldots, l$. Pick a Morse function f_j on each C_j for all $j = 1, \ldots, l$, and extend f_j to a function on T_j by making f_j constant in the direction normal to C_j. Let ρ_j be a bump function which is equal to 1 near C_j and equal to 0 outside of T_j. Choose $\varepsilon > 0$ and define

$$g = f + \varepsilon \left(\sum_{j=1}^{l} \rho_j f_j \right).$$

For ε small enough, g is a Morse function close to f, and the critical points of g are exactly the critical points of f_j for all $j = 1, \ldots, l$. It is easy to see that if $p \in C_j$ is a critical point of $f_j : C_j \to \mathbb{R}$ of index λ_p^j, then p is a critical point of g of index

$$\lambda_p = \lambda_p^j + \dim C_j$$

for all $j = 1, \ldots, l$.

Problems

1. Show that in local coordinates around a critical point $p \in M$ the matrix for the Hessian $H_p(f)$ is given by the matrix $M_p(f)$ of second partial derivatives.

2. Let $\lambda_1, \ldots, \lambda_n$ be distinct non-zero numbers. The function

$$(x_0, \ldots, x_n) \mapsto \sum_{j=0}^{n} \lambda_j x_j^2$$

on S^n descends to a smooth function $f : \mathbb{R}P^n \to \mathbb{R}$. Show that f is a Morse function and find the indices of its critical points.

3. Let $M_{n\times n}(\mathbb{R})$ denote the set of $n \times n$ matrices with entries in $\mathbb{R}$. Show that when $n = 2$ the determinant function det : $M_{n\times n}(\mathbb{R}) \to \mathbb{R}$ is a Morse function and find the indices of its critical points. What can you say about det when $n > 2$?

4. Find coordinate charts for the height function on the torus T^2 from Example 3.6 that satisfy the conclusion of The Morse Lemma (Lemma 3.11).

5. Prove that a smooth vector field on a smooth manifold M that vanishes outside of a compact set $K \subseteq M$ generates a unique 1-parameter group of diffeomorphisms of M.

6. Give a proof of the Morse Lemma (Lemma 3.11) using Remark 3.13 and Proposition 3.16.

7. Show that any Morse function on a compact surface of genus g has at least $2g + 2$ critical points.

8. Suppose that a Morse function $f : S^n \to \mathbb{R}$ satisfies $f(x) = f(-x)$. Show that f must have at least 2 critical points of index j for all $j = 0, \ldots, n$.

9. Let M be a compact manifold. Prove that there exists a Morse function on M such that no two critical points have the same value.

10. Let M be a manifold with boundary and suppose that $\partial M = M_0 \cup M_1$ where M_0 and M_1 are disjoint and compact. Show that there exists a Morse function $f : M \to [0, 1]$ such that $f^{-1}(0) = M_0$, $f^{-1}(1) = M_1$ and f has no critical points in a neighborhood of ∂M.

11. Let $f : M \to \mathbb{R}$ be a smooth function on a Riemannian manifold (M, g). Let $\gamma(t)$ be a non-constant gradient flow line of f between critical points q and p. Then

$$t \mapsto f(\gamma(t))$$

is a diffeomorphism $h : \mathbb{R} \to (f(p), f(q))$. Show that $f(\gamma(h^{-1}(t)) = t$ for all $t \in (f(p), f(q))$.

12. Let $h : \mathbb{R} \to (f(p), f(q))$ be the diffeomorphism from the previous problem. Show that $\omega = \gamma \circ h^{-1} : (f(p), f(q)) \to M$ satisfies the following differential equation.

$$\frac{d\omega}{ds}(s) = \frac{(\nabla f)(\omega(s))}{\|(\nabla f)(\omega(s))\|^2}$$

13. Let $f : N \to \mathbb{R}$ be a smooth function on a finite dimensional smooth Riemannian manifold (N, g), and let $M \subseteq N$ be an embedded submanifold. Show that for every $x \in M$, $(\nabla f)(x)$ projects orthogonally onto

$(\nabla f|_M)(x)$. That is, the gradient vector of $f|_M$ is the image of the gradient vector of f under the orthogonal projection $T_x N \to T_x M$ determined by the metric g.

14. Let $f : M \to \mathbb{R}$ be a Morse function on a finite dimensional smooth Riemannian manifold (M, g), and let $p \in M$ be a critical point of f. A coordinate chart $\phi : U \to \mathbb{R}^m$ around $p \in U$ is called a **Morse chart** if and only if $f \circ \phi^{-1}$ satisfies the conclusion of the Morse Lemma (Lemma 3.11). Assume that there exists a Morse chart around p that is an **isometry** with respect to the standard Riemannian structure on $\mathbb{R}^m$. Show that the **local stable manifold**

$$W^s(p)|_U = \{x \in U |\lim_{t \to \infty} \varphi_t(x) = p\}.$$

is diffeomorphic to $\mathbb{R}^{m-\lambda_p}$, and the **local unstable manifold** of p

$$W^u(p)|_U = \{x \in U |\lim_{t \to -\infty} \varphi_t(x) = p\}.$$

is diffeomorphic to $\mathbb{R}^{\lambda_p}$ where λ_p is the index of p and φ_t is the gradient flow.

15. Let $f_0, f_1 : M \to \mathbb{R}$ be two Morse functions. Show that there exists a C^∞-homotopy $F : M \times I \to \mathbb{R}$ such that for $F(x, t) = f_t(x)$ we have $F(x, 0) = f_0(x)$, $F(x, 1) = f_1(x)$, and $f_t : M \to \mathbb{R}$ is a Morse function for all $t \in I - \{t_1, \dots, t_r\}$ where $\{t_1, \dots, t_r\}$ is a finite set of points.

16. Show that Corollary 3.22 remains true even if the critical points are degenerate.

17. Show that if a compact closed manifold M of dimension m admits a Morse function with exactly 3 critical points, then the critical points must have index 0, $m/2$, and m. In particular, the manifold must have even dimension.

18. A Morse function $f : M \to \mathbb{R}$ that satisfies the condition $f(p) = \lambda_p$ for all critical points $p \in M$ is called a **self-indexing** Morse function. Show that every finite dimensional compact smooth manifold M possesses a self-indexing Morse function.

19. Let $f : M \to \mathbb{R}$ be a Morse function on a compact closed manifold, and let ν_k denote the number of critical points of f of index k for all $k = 0, \dots, m$. Show that if for some k we have $\nu_{k-1} = \nu_{k+1} = 0$, then $\nu_k = b_k$ where b_k denotes the kth Betti number of M.

20. Let $f : M \to \mathbb{R}$ be a Morse-Bott function on a finite dimensional compact smooth manifold M. Suppose that

$$\mathrm{Cr}(f) = \coprod_{j=1}^{l} C_j$$

where $C_1, \ldots, C_l$ are disjoint compact connected critical submanifolds. Show that

$$\mathcal{X}(M) = \sum_{j=1}^{l} (-1)^{\lambda_j} \mathcal{X}(C_j)$$

where λ_j is the Morse-Bott index of the critical submanifold C_j.

21. Let $\pi : E \to M$ be a smooth fiber bundle with fiber F. Show that $\mathcal{X}(E) = \mathcal{X}(M)\mathcal{X}(F)$.

22. Show that the trace function on $SO(n)$, $U(n)$, and $Sp(n)$ is a Morse-Bott function.

23. Prove the analogue of Theorem 3.25 for Morse-Bott functions [21] [25]: Let $f : M \to \mathbb{R}$ be a Morse-Bott function on a finite dimensional compact smooth Riemannian manifold M and let $C \subseteq M$ be a critical submanifold. Let $[a, b]$ be a closed interval containing $c = f(C)$ and no other critical value. Show that M^b has the homotopy type of $M^a \cup \nu^-(C)$ where $\nu^-(C) \subseteq M$ is the negative normal bundle of C viewed as a subset of a tubular neighborhood of C.

24. Prove the following: If $f : M \to \mathbb{R}$ is a Morse-Bott function on a compact connected manifold M which has no index 1 or index $m - 1$ critical submanifolds (where $m = \dim M$), then it has a unique local minimum and a unique local maximum. Moreover, all of its non-empty level sets are connected.

25. A symplectic manifold is a couple (M, ω) where M is a manifold of dimension $2n$ and ω is a 2-form such that $d\omega = 0$ and

$$\underbrace{\omega \wedge \omega \wedge \cdots \wedge \omega}_{n} \qquad \text{(the } n\text{-fold exterior product)}$$

is everywhere non-zero. A smooth function $f : M \to \mathbb{R}$ on a symplectic manifold (M, ω) defines a vector field X_f, called the **Hamiltonian vector field** of f, defined by the equation

$$\omega(X_f, V) = df(V)$$

for all vector fields V. (See also Section 9.2).

Prove **Frankel's Theorem**: Let (M, ω) be a compact and connected symplectic manifold and $f : M \to \mathbb{R}$ a smooth function such that the closure of the flow of its Hamiltonian vector field in the whole group of diffeomorphisms of M is a torus. Then f is a Morse-Bott function and all the critical submanifolds of f have even indices.

26. Prove the famous **Atiyah-Guillemin-Sternberg Convexity Theorem**: Let $f = (f_1, \ldots, f_n) : M \to \mathbb{R}^n$ be a smooth map from a compact, connected symplectic manifold (M, ω) into $\mathbb{R}^n$. Suppose that the flows of the Hamiltonian vector fields X_{f_i} for $i = 1, \ldots, n$ generate a subgroup in the diffeomorphism group of M whose closure is a torus. Then the image $f(M) \subseteq \mathbb{R}^n$ of f is a convex subset of $\mathbb{R}^n$.

27. Let M be a compact manifold with a family of subsets $\mathcal{F}$ such that for any isotopy $g_t : M \to M$, $F \in \mathcal{F}$ implies that $g_1(F) \in \mathcal{F}$. Prove that for any smooth function $f : M \to \mathbb{R}$

$$\text{minimax}(F, \mathcal{F}) \stackrel{\text{def}}{=} \inf_{F \in \mathcal{F}} \{\sup\{f(x) | \, x \in F\}\}$$

is a critical value of f.

28. The **Ljusternik-Schnirelmann category**, $\text{cat}(M)$, of a topological space M is the smallest number $k \in \mathbb{N}$ such that M can be covered by k closed contractible subsets of M. Prove that any smooth function $f : M \to \mathbb{R}$ on a compact smooth manifold M has at least $\text{cat}(M)$ critical points.

29. The **cup length**, $\text{cl}(M)$, of a topological space M is the largest integer k such that there exists a ring R and cohomology classes $\alpha_1, \ldots, \alpha_{k-1} \in H^*(M; R)$ of positive degree such that $\alpha_1 \cup \cdots \cup \alpha_{k-1} \neq 0$. Show that if M is connected, then $\text{cat}(M) \geq \text{cl}(M)$.

Chapter 4

The Stable/Unstable Manifold Theorem

The main goal of this chapter is to prove the Stable/Unstable Manifold Theorem for a Morse Function (Theorem 4.2). To do this, we first show that a non-degenerate critical point of a smooth function $f : M \to \mathbb{R}$ on a finite dimensional smooth Riemannian manifold (M, g) is a hyperbolic fixed point of the diffeomorphism φ_t coming from the gradient flow (for any fixed $t \neq 0$). This is accomplished by computing local formulas for ∇f, $d\nabla f|_p$, and $d\varphi_t|_p$ with respect to the Riemannian metric g on M (Lemmas 4.3, 4.4, and 4.5).

Next we prove the Local Stable Manifold Theorem in its most general form (Theorem 4.11), i.e. for a hyperbolic fixed point of an automorphism of a Banach space, following the proof of M.C. Irwin [82] [133]. The Global Stable/Unstable Manifold Theorems for a Diffeomorphism (Theorems 4.15 and 4.17) are then proved for a hyperbolic fixed point of a smooth diffeomorphism of a finite dimensional smooth Riemannian manifold using a construction due to Smale [136]. The proof of Theorem 4.2 is completed by Lemma 4.20 which shows that the stable/unstable manifolds of a Morse function on a finite dimensional compact smooth Riemannian manifold are embedded (rather than just injectively immersed) submanifolds.

4.1 The Stable/Unstable Manifold Theorem for a Morse function

Let $f : M \to \mathbb{R}$ be a smooth function on a finite dimensional compact smooth Riemannian manifold (M, g). Recall from Definition 3.17 that the gradient vector field of f with respect to the metric g is the unique smooth vector field ∇f such that

$$g(\nabla f, V) = df(V) = V \cdot f$$

for all smooth vector fields V on M. The gradient vector field determines a smooth flow $\varphi : \mathbb{R} \times M \to M$ by $\varphi_t(x) = \gamma_x(t)$ where $\frac{d}{dt}\gamma_x(t) = -\nabla f|_{\gamma_x(t)}$ and $\gamma_x(0) = x$. Since M is compact, the flow φ_t satisfies the following.

1. $\varphi_t : M \to M$ is a diffeomorphism for all $t \in \mathbb{R}$.

2. $\varphi_{t_1} \circ \varphi_{t_2} = \varphi_{t_1+t_2}$ for any $t_1, t_2 \in \mathbb{R}$.

That is, φ_t is a 1-parameter group of diffeomorphisms defined on $\mathbb{R} \times M$ (see for instance [77] Section 6.2 or [91] Section I.6).

Definition 4.1 *Let $p \in M$ be a non-degenerate critical point of f.*

1. *The **stable manifold** of p is defined to be*

$$W^s(p) = \{x \in M|\ \lim_{t \to \infty} \varphi_t(x) = p\}.$$

2. *The **unstable manifold** of p is defined to be*

$$W^u(p) = \{x \in M|\ \lim_{t \to -\infty} \varphi_t(x) = p\}.$$

The main result of this chapter is the following.

Theorem 4.2 (Stable/Unstable Manifold Theorem for a Morse Function)
Let $f : M \to \mathbb{R}$ be a Morse function on a compact smooth Riemannian manifold (M, g) of dimension $m < \infty$. If $p \in M$ is a critical point of f, then the tangent space at p splits as

$$T_p M = T_p^s M \oplus T_p^u M$$

where the Hessian is positive definite on $T_p^s M$ and negative definite on $T_p^u M$. Moreover, the stable and unstable manifolds are surjective images of smooth embeddings

$$\begin{aligned} E^s : T_p^s M &\to W^s(p) \subseteq M \\ E^u : T_p^u M &\to W^u(p) \subseteq M. \end{aligned}$$

Hence, $W^s(p)$ is a smoothly embedded open disk of dimension $m - \lambda_p$, and $W^u(p)$ is a smoothly embedded open disk of dimension λ_p, where λ_p is the index of the critical point p.

The proof of this theorem will take most of the rest of this chapter.

Local formulas for ∇f, $d\nabla f|_p$, and $d\varphi_t|_p$

Let $x_1, \ldots, x_m$ be local coordinates on $U \subseteq M$. Then $\frac{\partial}{\partial x_1}, \ldots, \frac{\partial}{\partial x_m}$ span $T_x U$ for all $x \in U$, and the components of the metric g are defined to be the smooth functions $g_{ij} : U \to \mathbb{R}$ given by $g_{ij} = g(\frac{\partial}{\partial x_i}, \frac{\partial}{\partial x_j})$ for all $1 \leq i, j \leq m$. So if $V = \sum_{i=1}^m V_i \frac{\partial}{\partial x_i}$ and $W = \sum_{i=1}^m W_j \frac{\partial}{\partial x_j}$ are smooth vector fields on U, then

$$g(V, W) = \sum_{i,j} g_{ij} V_i W_j : U \to \mathbb{R}.$$

Since g is non-degenerate, at each point $x \in U$ the matrix $(g_{ij}(x))$ is invertible. We denote this inverse by (g^{ij}). Note that since g is symmetric,

$$g_{ij} = g_{ji} \quad \text{and} \quad g^{ij} = g^{ji}$$

for all $1 \leq i, j \leq m$. The cotangent bundle T^*U has $dx^1, \ldots, dx^m$ as a basis where $dx^i(\frac{\partial}{\partial x_j}) = \delta_i^j$, and with respect to this basis we have

$$df = \sum_{i=1}^m \frac{\partial f}{\partial x_i} dx^i.$$

Lemma 4.3 *In the local coordinates $x_1, \ldots, x_m$ on $U \subseteq M$ we have*

$$\nabla f = \sum_{i,j} g^{ij} \frac{\partial f}{\partial x_i} \frac{\partial}{\partial x_j}.$$

Proof:
   Suppose that $\nabla f = \sum_{j=1}^m V_j \frac{\partial}{\partial x_j}$ in the local coordinates $x_1, \ldots, x_m$ on U. Then for any $j = 1, \ldots, m$ we have

$$\frac{\partial f}{\partial x_j} = g(\nabla f, \frac{\partial}{\partial x_j}) = \sum_{i=1}^m g_{ij} V_i.$$

Hence,

$$(V_1, \ldots, V_m)(g_{ij}) = (\frac{\partial f}{\partial x_1}, \ldots, \frac{\partial f}{\partial x_m})$$

which implies that

$$(V_1, \ldots, V_m) = (\frac{\partial f}{\partial x_1}, \ldots, \frac{\partial f}{\partial x_m})(g^{ij})$$

since (g_{ij}) is invertible with inverse (g^{ij}), and

$$V_j = \sum_{i=1}^{m} \frac{\partial f}{\partial x_i} g^{ij}.$$

□

Lemma 4.4 *If $x_1, \ldots, x_m$ is a local coordinate system around a critical point $p \in U \subseteq M$ such that $\frac{\partial}{\partial x_1}, \ldots, \frac{\partial}{\partial x_m}$ is an orthonormal basis for $T_p U$ with respect to the metric g, then the matrix for the differential of $\nabla f : U \to \mathbb{R}^m$ is equal to the matrix of the Hessian at p, i.e.*

$$\frac{\partial}{\partial \vec{x}} \nabla f \bigg|_p = M_p(f).$$

Proof:
    In the local coordinates $x_1, \ldots, x_m$ on U we have

$$\nabla f = \left(\sum_k g^{k1} \frac{\partial f}{\partial x_k}, \sum_k g^{k2} \frac{\partial f}{\partial x_k}, \ldots, \sum_k g^{km} \frac{\partial f}{\partial x_k} \right)$$

by Lemma 4.3, and we can compute the matrix for the differential of $\nabla f : U \to \mathbb{R}^m$ as follows.

$$\frac{\partial}{\partial \vec{x}} \nabla f = \left(\frac{\partial}{\partial x_j} \sum_k g^{ki} \frac{\partial f}{\partial x_k} \right) = \left(\sum_k \left[\frac{\partial g^{ki}}{\partial x_j} \frac{\partial f}{\partial x_k} + g^{ki} \frac{\partial^2 f}{\partial x_j \partial x_k} \right] \right)$$

Hence, at a critical point $p \in U$ we have,

$$\frac{\partial}{\partial \vec{x}} \nabla f \bigg|_p = \left(\sum_k g^{ki} \frac{\partial^2 f}{\partial x_j \partial x_k} \right),$$

and if $(\frac{\partial}{\partial x_1}, \ldots, \frac{\partial}{\partial x_m})$ is an orthonormal basis at p, then

$$\frac{\partial}{\partial \vec{x}} \nabla f \bigg|_p = \left(\frac{\partial^2 f}{\partial x_j \partial x_i} \right) = M_p(f).$$

□

Lemma 4.5 *If $x_1, \ldots, x_m$ is a local coordinate system around a critical point $p \in M$ such that $\frac{\partial}{\partial x_1}, \ldots, \frac{\partial}{\partial x_m}$ is an orthonormal basis for the tangent space at p, then for any $t \in \mathbb{R}$ the matrix for the differential of φ_t at p is the exponential of minus the matrix for the Hessian at p, i.e.*

$$\frac{\partial}{\partial \vec{x}} \varphi_t \bigg|_p = e^{-M_p(f)t}.$$

Proof:

In the coordinate chart around p, we know by the existence and uniqueness theorem for O.D.E.s (see for instance [94] or [117]) that

$$\varphi : (-\varepsilon, \varepsilon) \times U \to U$$

is smooth in both $t \in (-\varepsilon, \varepsilon)$ and $\vec{x} \in U$ for some neighborhood U of p and some small $\varepsilon > 0$. Since

$$\frac{d}{dt}\varphi(t, \vec{x}) = -(\nabla f)(\varphi(t, \vec{x}))$$

for any $\vec{x} \in U$, we can interchange the order of differentiation and we have

$$\frac{d}{dt}\frac{\partial}{\partial \vec{x}}\varphi(t, \vec{x}) = -\frac{\partial}{\partial \vec{x}}(\nabla f)(\varphi(t, \vec{x})) = -(\frac{\partial}{\partial \vec{x}}\nabla f)\frac{\partial}{\partial \vec{x}}\varphi(t, \vec{x}).$$

Thus, $\Phi(t, \vec{x}) = \frac{\partial}{\partial \vec{x}}\varphi(t, \vec{x})$ is a solution to the following linear system of O.D.E.s.

$$\frac{d}{dt}\Phi(t, \vec{x}) = -(\frac{\partial}{\partial \vec{x}}\nabla f)\Phi(t, \vec{x})$$
$$\Phi(0, \vec{x}) = I_{m \times m}$$

Since $e^{-(\frac{\partial}{\partial \vec{x}}\nabla f)t}$ is also a solution to the above system, we have

$$\Phi(t, \vec{x}) = e^{-(\frac{\partial}{\partial \vec{x}}\nabla f)t}$$

because the solution is unique. Hence, at the critical point p we have

$$\frac{\partial}{\partial \vec{x}}\varphi(t, \vec{x})\Big|_p = e^{-M_p(f)t}$$

by Lemma 4.4.

$\square$

If p is a non-degenerate critical point, then there is a basis for T_pU such that $M_p(f)$ is a diagonal matrix with λ_p of the diagonal entries negative, $\alpha_1, \ldots, \alpha_{\lambda_p}$, and $m - \lambda_p$ of the diagonal entries positive, $\beta_{\lambda_p+1}, \ldots, \beta_m$. With respect to this basis we have

$$e^{-M_p(f)t} = \begin{pmatrix} e^{-\alpha_1 t} & & & & & & \\ & \ddots & & & & 0 & \\ & & e^{-\alpha_{\lambda_p} t} & & & & \\ & & & e^{-\beta_{\lambda_p+1}t} & & & \\ & 0 & & & \ddots & & \\ & & & & & e^{-\beta_m t} \end{pmatrix}$$

where λ_p of the diagonal entries have length greater than 1, and $m - \lambda_p$ of the diagonal entries have length less than 1 when $t > 0$. Thus, Lemma 4.5 implies that for $t \neq 0$, $d\varphi_t|_p : T_pM \to T_pM$ has no eigenvalues of length 1, i.e. p is a **hyperbolic** fixed point of $\varphi_t : M \to M$. Moreover, the tangent space at p splits as $T_pM \approx T_p^sM \oplus T_p^uM$ where

$$d\varphi_t|_p : T_p^sM \quad \to \quad T_p^sM \quad \text{is contracting}$$
$$d\varphi_t|_p : T_p^uM \quad \to \quad T_p^uM \quad \text{is expanding.}$$

If $t > 0$, then the dimension of T_p^uM is λ_p, and if $t < 0$, then the dimension of T_p^sM is λ_p. In the next section we will show that the existence of such a splitting implies the existence of local stable and unstable manifolds.

4.2   The Local Stable Manifold Theorem

In this section we prove the Local Stable Manifold Theorem following the proof of M.C. Irwin [82] found in Appendix II to Chapter 5 of [133]. Other proofs of this theorem can be found in [1], [83], [96], [114], [143], and [149].

This section contains the most difficult real analysis in the entire book. Readers who have an aversion to real analysis can skip this section without hindering their understanding of the rest of the book.

Let $E = E_s \times E_u$ be a Banach space and $T : E_s \times E_u \to E_s \times E_u$ a linear automorphism that preserves the splitting. Assume that there exists a $\lambda \in \mathbb{R}$ with $0 < \lambda < 1$ such that

$$\|T_s\| < \lambda \quad \text{and} \quad \|T_u^{-1}\| < \lambda$$

where $T_s = T|_{E_s}$ and $T_u = T|_{E_u}$. The map T_s is said to be **contracting** and T_u is said to be **expanding**. Recall that

$$\|T_s\| = \sup\{|T_s(x)| \mid x \in E_s \text{ with } |x| \leq 1\}$$

and similarly,

$$\|T_u\| = \sup\{|T_u(x)| \mid x \in E_u \text{ with } |x| \leq 1\}.$$

Let $p_s : E \to E_s$ and $p_u : E \to E_u$ denote projections, and for any $x \in E$ define

$$\|x\| = \sup\{\|p_s(x)\|, \|p_u(x)\|\} \quad \text{(the box norm).}$$

For any map $\varphi : E \to E$, we will denote the composite $p_s \circ \varphi : E \to E_s$ by φ_s and the composite $p_u \circ \varphi : E \to E_u$ by φ_u. For any $r \geq 0$, $E_s(r)$ denotes the closed ball of radius r about 0 in E_s, and $E_u(r)$ denotes the closed ball of radius r about 0 in $E_u(r)$.

Definition 4.6 *A map $G : E \to E'$ between Banach spaces E and E' is called* **Lipschitz** *if and only if there exists a number $k \geq 0$ such that*

$$\|G(x) - G(y)\| \leq k\|x - y\|$$

for all $x, y \in E$. The **Lipschitz constant**, $Lip(G)$, *is defined to be the least such k, i.e.*

$$Lip(G) = \inf\{k \in \mathbb{R}_+ | \|G(x) - G(y)\| \leq k\|x - y\| \text{ for all } x, y \in E\}.$$

Clearly every Lipschitz map is **uniformly continuous**. *If $\varphi : E \to E'$ is a map such that $\varphi - G$ is Lipschitz, then we will call φ a* **Lipschitz perturbation** *of G.*

Note that $T : E \to E$ is Lipschitz with $Lip(T) = \|T\|$, $T_s : E_s \to E_s$ is Lipschitz with $Lip(T_s) \leq \lambda$, and $T_u^{-1} : E_u \to E_u$ is Lipschitz with $Lip(T_u^{-1}) \leq \lambda$.

Lemma 4.7 *If $\varphi : E_s(r) \times E_u(r) \to E$ is a Lipschitz perturbation of T with $Lip(\varphi - T) \leq \varepsilon < 1 - \lambda$, then φ preserves the family of cones parallel to $E_u(r)$. That is, if $x = (x_s, x_u) \in E_s \times E_u$ and $y = (y_s, y_u) \in E_s \times E_u$ satisfy*

$$\|x_s - y_s\| < \|x_u - y_u\|$$

then

$$\begin{aligned}
\|\varphi_s(x) - \varphi_s(y)\| &\leq (\lambda + \varepsilon)\|x_u - y_u\| \\
&< \|x_u - y_u\| \\
&< (\lambda^{-1} - \varepsilon)\|x_u - y_u\| \\
&\leq \|\varphi_u(x) - \varphi_u(y)\|.
\end{aligned}$$

Proof:

We have,

$$\begin{aligned}
\|\varphi_s(x) - \varphi_s(y)\| &\leq \|p_s(\varphi - T)(x) - p_s(\varphi - T)(y)\| + \|T_s x_s - T_s y_s\| \\
&\leq \varepsilon\|x - y\| + \lambda\|x_s - y_s\| \\
&\leq (\lambda + \varepsilon)\|x_u - y_u\| \\
&< \|x_u - y_u\|
\end{aligned}$$

where the first inequality comes from the triangle inequality and the fact that $p_s T x = p_s(T_s x_s, T_u x_u) = T_s x_s$, and the third inequality comes from the box norm and the assumption that $\|x_s - y_s\| < \|x_u - y_u\|$.

Since $\|T_u^{-1} x_u - T_u^{-1} y_u\| < \lambda\|x_u - y_u\|$ implies that $\|x_u - y_u\| < \lambda\|T_u x_u - T_u y_u\|$, we also have,

$$\begin{aligned}
\|\varphi_u(x) - \varphi_u(y)\| &\geq \|T_u x_u - T_u y_u\| - \|p_u(\varphi - T)x - p_u(\varphi - T)y\| \\
&\geq \lambda^{-1}\|x_u - y_u\| - \varepsilon\|x - y\| \\
&= (\lambda^{-1} - \varepsilon)\|x_u - y_u\|.
\end{aligned}$$

This proves the lemma since $\varepsilon < 1 - \lambda$ implies $\lambda^{-1} > 1/(1 - \varepsilon)$, and hence $\lambda^{-1} - \varepsilon > 1/(1 - \varepsilon) - \varepsilon > 1$ for $\varepsilon < 1$.

□

Definition 4.8 *For any map* $\varphi : E_s(r) \times E_u(r) \to E$ *we define*

$$W_r^s(\varphi) = \{x \in E_s(r) \times E_u(r)| \, \forall \, n \geq 0, \, \varphi^n(x) \text{ is defined and } \|\varphi^n(x)\| \leq r\}$$

*to be the **stable set** of* φ. *The **unstable set** of* φ *is defined to be* $W_r^u(\varphi) = W_r^s(\varphi^{-1})$.

Proposition 4.9 (Lipschitz Local Stable Manifold Theorem) *For any map* $\varphi : E_s(r) \times E_u(r) \to E$ *with* $Lip(\varphi - T) \leq \varepsilon < 1 - \lambda$, $W_r^s(\varphi)$ *is the graph of a function* $g : A \to E_u(r)$ *where* $A = p_s(W_r^s(\varphi))$. *Moreover, g is Lipschitz with* $Lip(g) \leq 1$, *and* φ *restricted to* $W_r^s(\varphi)$ *contracts distances. Therefore,* φ *has at most one fixed point which, if it exists, attracts all other points of* $W_r^s(\varphi)$.

Proof:

Let $x = (x_s, x_u) \in W_r^s(\varphi)$ and $y = (y_s, y_u) \in W_r^s(\varphi)$. Suppose that $\|x_u - y_u\| > \|x_s - y_s\|$. Then by Lemma 4.7,

$$\|\varphi_u(x) - \varphi_u(y)\| \geq (\lambda^{-1} - \varepsilon)\|x_u - y_u\| > \|x_u - y_u\|,$$

and so for all $n \geq 1$

$$\|\varphi_u^n(x) - \varphi_u^n(y)\| \geq (\lambda^{-1} - \varepsilon)^n \|x_u - y_u\|.$$

But $\|\varphi_u^n(x) - \varphi_u^n(y)\| < \|\varphi_u^n(x)\| + \|\varphi_u^n(y)\| \leq 2r$ and $(\lambda^{-1} - \varepsilon)^n \to \infty$ as $n \to \infty$ imply that $\|x_u - y_u\| = 0$. This contradicts the assumption that $\|x_u - y_u\| > \|x_s - y_s\|$, and therefore,

$$\|x_u - y_u\| \leq \|x_s - y_s\|.$$

The preceeding inequality shows that if $x_s = y_s$, then $x_u = y_u$. Thus, there is a function $g : A \to E_u(r)$ such that $g(x_s) = x_u$, and g satisfies

$$\|g(x_s) - g(x_s)\| \leq \|x_s - y_s\|$$

i.e. $Lip(g) \leq 1$. By the box norm and the above inequality we also have $\|x - y\| = \|x_s - y_s\|$ for all $x, y \in W_r^s(\varphi)$, and since $\varphi(W_r^s(\varphi)) \subseteq W_r^s(\varphi)$ we have $\|\varphi(x) - \varphi(y)\| = \|\varphi_s(x) - \varphi_s(y)\|$. Hence,

$$\begin{aligned}
\|\varphi(x) - \varphi(y)\| &= \|\varphi_s(x) - \varphi_s(y)\| \\
&\leq \|p_s(\varphi - T)(x) - p_s(\varphi - T)(y)\| + \|T_s x_s - T_s y_s\| \\
&\leq \varepsilon\|x - y\| + \lambda\|x_s - y_s\| \\
&\leq (\lambda + \varepsilon)\|x - y\|,
\end{aligned}$$

and since $\lambda + \varepsilon < 1$ we see that φ restricted to $W_r^s(\varphi)$ is a contraction.

$\square$

The preceeding theorem shows that the stable set is the graph of a Lipschitz function $g : p_s(W_r^s(\varphi)) \rightarrow E_u(r)$ and hence a Lipschitz manifold. In order to show that $W_r^s(\varphi)$ has the structure of a smooth manifold whenever φ is a smooth diffeomorphism we need a more explicit description of g. To do this we will use the following.

Theorem 4.10 (Lipschitz Inverse Function Theorem) *Let the map $h : U \rightarrow V$ be a homeomorphism where $U \subseteq E$ and $V \subseteq E'$ are open subsets of the Banach spaces E and E', and assume that h^{-1} is Lipschitz. If $k : U \rightarrow E'$ is a map such that $k - h$ is Lipschitz with*

$$Lip(k - h) < [Lip(h^{-1})]^{-1},$$

then k is injective, k^{-1} is Lipschitz with

$$Lip(k^{-1}) \leq ([Lip(h^{-1})]^{-1} - Lip(k - h))^{-1},$$

and k is a homeomorphism onto its image.

Proof:

$$
\begin{aligned}
\|k(x) - k(y)\| &= \|h(x) - h(y) - [h(x) - k(x) - (h(y) - k(y))]\| \\
&\geq \|h(x) - h(y)\| - \|(h - k)(x) - (h - k)(y)\| \\
&= \|h(x) - h(y)\| - \|(k - h)(x) - (k - h)(y)\| \\
&\geq ([Lip(h^{-1})]^{-1} - Lip(k - h))\|x - y\|
\end{aligned}
$$

where the last inequality follows from $\|x - y\| \leq \mathrm{Lip}(h^{-1})\|h(x) - h(y)\|$. Since $[\mathrm{Lip}(h^{-1})]^{-1} - \mathrm{Lip}(k - h) > 0$ we see that $k(x) \neq k(y)$ if $x \neq y$. Substituting $k^{-1}(x)$ for x and $k^{-1}(y)$ for y we have

$$\|k^{-1}(x) - k^{-1}(y)\| \leq ([\mathrm{Lip}(h^{-1})]^{-1} - \mathrm{Lip}(k - h))^{-1}\|x - y\|.$$

Since k^{-1} is Lipschitz, it is continuous, and k is continuous because $k = h + (k - h)$ where $k - h$ is Lipschitz and thus continuous. Therefore, k is a homeomorphism onto its image.

$\square$

Theorem 4.11 (Local Stable Manifold Theorem) *There exists an $\varepsilon > 0$, depending only on λ, and for every $r > 0$ a $\delta > 0$ such that if $\varphi : E_s(r) \times$*

$E_u(r) \to E$ satisfies $Lip(\varphi - T) < \varepsilon$, and $\|\varphi(0)\| < \delta$, then $W_r^s(\varphi)$ is the graph of a Lipschitz function $g : E_s(r) \to E_u(r)$ with $Lip(g) \leq 1$. Moreover, we have the following.

1. If φ is C^k, then so is g.

2. If φ is C^1, $\varphi(0) = 0$, and $d\varphi_0 = T$, then $g(0) = 0$ and $dg_0 = 0$. Hence, the tangent space to $W_r^s(\varphi)$ at 0 is E_s.

Main Idea: A point $x \in E(r)$ determines a sequence

$$\gamma_x = \{\gamma_x(n)| \, \gamma_x(n) = \varphi^n(x) \text{ for all } n \geq 1\},$$

that satisfies the following three properties whenever $x \in W_r^s(\varphi)$.

1. $\|\gamma_x(n)\| \leq r$ for all $n \geq 1$.

2. $\varphi(\gamma_x(n)) - \gamma_x(n+1) = 0$ for all $n \geq 1$.

3. $\varphi(x) - \gamma_x(1) = 0$.

Conversely, any sequence $\gamma = \{\gamma(n)\}_{n\geq 1}$ of points in $E(r)$ that satisfies these three conditions must be of the form $\gamma = \gamma_x$ for some $x \in W_r^s(\varphi)$. This observation suggests that we identify $W_r^s(\varphi)$ as the inverse image of a smooth map into the Banach space of all bounded sequences in E.

Proof of Theorem 4.11:

Let

$$B = \{\gamma = \{\gamma(n)\}_{n\geq 1}| \, \gamma(n) \in E, \, \sup_{n\geq 1} \|\gamma(n)\| < \infty\}$$

be the Banach space consisting of all bounded sequences in E with norm $\|\gamma\| = \sup_{n\geq 1} \|\gamma(n)\|$. Let $B(r)$ be the closed ball of radius r in B, and endow $E_s \times B$, $E_u \times B$, and $E_s \times E_u \times B$ with the box norm. Define

$$F : E_s(r) \times E_u(r) \times B(r) \to E_s \times B$$

by $F(x_s, x_u, \gamma) = (x_s, F_{x_s}(x_u, \gamma))$ where $F_{x_s}(x_u, \gamma) = \upsilon(x_s, x_u, \gamma) \in B$ is defined by

$$\upsilon(x_s, x_u, \gamma)(n) = \begin{cases} \varphi(x_s, x_u) - \gamma(1) & \text{if } n = 1 \\ \varphi(\gamma(n-1)) - \gamma(n) & \text{if } n \geq 2 \end{cases}$$

We will show that F is a homeomorphism onto its image and its image contains $E_s(r) \times 0$. We can then define

$$g = \pi_u F^{-1}|_{E_s(r) \times 0}$$

where $\pi_u : E_s(r) \times E_u(r) \times B(r) \to E_u(r)$ is the projection. It's clear that the graph of g is $W_r^s(\varphi)$.

Step 1: F is a homeomorphism onto its image.

Define $G : E_s(r) \times E_u(r) \times B(r) \to E_s \times B$ analogous to F but with the linear map T in place of φ. That is, $G(x_s, x_u, \gamma) = (x_s, G_{x_s}(x_u, \gamma))$ where $G_{x_s}(x_u, \gamma) = \nu(x_s, x_u, \gamma) \in B$ is defined by

$$\nu(x_s, x_u, \gamma)(n) = \begin{cases} T(x_s, x_u) - \gamma(1) & \text{if } n = 1 \\ T(\gamma(n-1)) - \gamma(n) & \text{if } n \geq 2 \end{cases}$$

It's clear that G is a continuous linear operator with $\|G\| \leq 1 + \|T\|$. We will show that G is invertible and $\operatorname{Lip}(F - G) \leq \|G^{-1}\|^{-1}$. Thus, F is homeomorphism onto its image by Theorem 4.10.

We can solve for x_u and γ in terms of x_s and ν as follows. Let $\gamma(n) = (\gamma_s(n), \gamma_u(n)) \in E_s \times E_u$ and $\nu(n) = (\nu_s(n), \nu_u(n)) \in E_s \times E_u$. With this notation we have the following.

$$\begin{align} \nu_s(1) &= T_s(x_s) - \gamma_s(1) & (4.1) \\ \nu_u(1) &= T_u(x_u) - \gamma_u(1) & (4.2) \\ \nu_s(n) &= T_s(\gamma_s(n-1)) - \gamma_s(n) & n \geq 2 & (4.3) \\ \nu_u(n) &= T_u(\gamma_u(n-1)) - \gamma_u(n) & n \geq 2 & (4.4) \end{align}$$

Equations (4.1) and (4.3) give

$$\begin{align} \gamma_s(1) &= T_s(x_s) - \nu_s(1) \\ \gamma_s(n) &= T_s(\gamma_s(n-1)) - \nu_s(n) & n \geq 2 \end{align}$$

which imply

$$\gamma_s(n) = T_s^n(x_s) - \sum_{j=1}^n T_s^{n-j}(\nu_s(j)).$$

Equation (4.4) gives

$$\begin{align} \gamma_u(n-1) &= T_u^{-1}[\nu_u(n) + \gamma_u(n)] \\ &= T_u^{-1}[\nu_u(n) + T_u^{-1}[\nu_u(n+1) + \gamma_u(n+1)]] \\ &= \cdots \end{align}$$

and hence,

$$\gamma_u(n) = \sum_{j=1}^{\infty} T_u^{-j}(\nu_u(n+j)).$$

By equation (4.2), $x_u = T_u^{-1}\nu_u(1) + T_u^{-1}\gamma(1)$, and thus the preceeding equation implies,

$$x_u = \sum_{j=1}^{\infty} T_u^{-j}(\nu_u(j)).$$

These series converge since $\|T_u^{-1}\| < \lambda < 1$ implies that $\|T_u^{-j}(\nu_u(j))\| < \lambda^j \|\nu_u(j)\|$ for all $j \geq 1$ and $\sup_{n \geq 1} \|\gamma(n)\| < \infty$ implies that

$$\sup_{j \geq 1} \|\nu_u(j)\| < \infty.$$

Thus $G^{-1} : E_s \times B \to E_s \times E_u \times B$ exists and is given by $G^{-1}(x_s, \nu) = (x_s, x_u, \gamma)$ where

$$x_u = \sum_{j=1}^{\infty} T_u^{-j}(\nu_u(j))$$

$$\gamma_s(n) = T_s^n(x_s) - \sum_{j=1}^{\infty} T_s^{n-j}(\nu_s(j))$$

$$\gamma_u(n) = \sum_{j=1}^{\infty} T_u^{-j}(\nu_u(n+j)).$$

Using the box norm and comparing with a geometric series we see that we have $\|G^{-1}\| \leq (1-\lambda)^{-1}$. From the definitions one can check that

$$\mathrm{Lip}(F - G) \leq \mathrm{Lip}(\varphi - T),$$

and thus if we choose $\varepsilon > 0$ such that $\mathrm{Lip}(\varphi - T) < \varepsilon < 1 - \lambda$ we have $\mathrm{Lip}(F - G) < \|G^{-1}\|^{-1}$. Therefore, F is a homeomorphism onto its image by Theorem 4.10.

Step 2: The image of F contains $E_s(r) \times 0$ when $\varepsilon > 0$ and $\varphi(0)$ are sufficiently small.

First note that $(x_s, 0) \in \mathrm{Image}(F)$ if and only if $0 \in \mathrm{Image}(F_{x_s})$. Moreover, $\mathrm{Lip}(F_{x_s} - G_{x_s}) \leq \mathrm{Lip}(\varphi - T)$, and since G_{x_s} differs from G_0 by translation by the sequence $\{T(x_s, x_u), 0, 0, \ldots\}$ we have $\mathrm{Lip}(F_{x_s} - G_0) \leq \mathrm{Lip}(\varphi - T)$. The function $G_0 : E_u \times B \to B$ is the linear map defined by

$$G_0(x_u, \gamma)(n) = \begin{cases} T(0, x_u) - \gamma(1) & \text{if } n = 1 \\ T(\gamma(n-1)) - \gamma(n) & \text{if } n \geq 2 \end{cases}$$

and, as in Step 1, G_0 is invertible with $G_0^{-1} : B \to E_u \times B$ given by $G_0^{-1}(\nu) = (x_u, \gamma)$ where

$$x_u = \sum_{j=1}^{\infty} T_u^{-j}(\nu_u(j))$$

$$\gamma_s(n) = -\sum_{j=1}^{\infty} T_s^{n-j}(\nu_s(j))$$

$$\gamma_u(n) = \sum_{j=1}^{\infty} T_u^{-j}(\nu_u(n+j)).$$

Moreover, $\|G_0^{-1}\| \leq (1-\lambda)^{-1}$, and if $\mathrm{Lip}(\varphi - T) < \varepsilon < 1 - \lambda$, then

$$\mathrm{Lip}(G_0^{-1}F_{x_s} - \mathrm{id}) = \mathrm{Lip}(G_0^{-1})\mathrm{Lip}(F_{x_s} - G_0) \leq \frac{\varepsilon}{1-\lambda} < 1.$$

By Theorem 4.10 we conclude that $k = G_0^{-1}F_{x_s} : E_u \times B \to E_u \times B$ is a homeomorphism onto its image, and k^{-1} is Lipschitz with

$$\mathrm{Lip}(k^{-1}) \leq \left(1 - \frac{\varepsilon}{1-\lambda}\right)^{-1}.$$

Since

$$
\begin{aligned}
\|k^{-1}(k(0) + (x_u,\gamma))\| &= \|k^{-1}(k(0)) - k^{-1}(k(0) + (x_u,\gamma))\| \\
&\leq \mathrm{Lip}(k^{-1})\|k(0) - (k(0) + (x_u,\gamma))\| \\
&= \mathrm{Lip}(k^{-1})\|(x_u,\gamma))\|
\end{aligned}
$$

we have $k^{-1}(k(0)+E_u(\tilde{r})\times B(\tilde{r})) \subset E_u(r)\times B(r)$ where $\tilde{r} = r\left(1 - \frac{\varepsilon}{1-\lambda}\right)$.

That is,

$$G_0^{-1}F_{x_s}[E_u(r) \times B(r)] \supset G_0^{-1}F_{x_s}(0) + E_u(\tilde{r}) \times B(\tilde{r}).$$

We will now estimate $\|G_0^{-1}F_{x_s}(0)\|$.

$$v(x_s,0,0)(n) = F_{x_s}(0)(n) = \begin{cases} \varphi(x_s,0) & \text{if } n = 1 \\ \varphi(0,0) & \text{if } n \geq 2 \end{cases}$$

So, $G_0^{-1}F_{x_s}(0) = (x_u,\gamma)$ where

$$x_u = T_u^{-1}[\varphi_u(x_s,0)] + \sum_{j=2}^{\infty} T_u^{-j}(\varphi_u(0,0))$$

$$\gamma_s(n) = T_s^{n-1}(\varphi_s(x_s,0)) - \sum_{j=2}^{\infty} T_s^{n-j}(\varphi_s(0,0))$$

$$\gamma_u(n) = \sum_{j=1}^{\infty} T_u^{-j}(\varphi_u(0,0)).$$

We can estimate

$$
\begin{aligned}
\|\varphi_s(x_s,0)\| &\leq \|p_s(\varphi - T)(x_s,0)\| + \|T_s(x_s)\| \\
&\leq \|(\varphi - T)(x_s,0)\| + \lambda\|x_s\|
\end{aligned}
$$

and

$$\begin{aligned}
\|\varphi_u(x_s, 0)\| &\leq \|p_u(\varphi - T)(x_s, 0)\| + \|T_u(0)\| \\
&\leq \|(\varphi - T)(x_s, 0)\|
\end{aligned}$$

and

$$\begin{aligned}
\|(\varphi - T)(x_s, 0)\| &\leq \|(\varphi - T)(0)\| + \|(\varphi - T)(x_s, 0) - (\varphi - T)(0)\| \\
&\leq \|\varphi(0)\| + \varepsilon\|x_s\|.
\end{aligned}$$

Thus we have the following estimates.

$$\begin{aligned}
\|x_u\| &\leq \lambda(\|\varphi(0)\| + \varepsilon\|x_s\|) + \sum_{j=2}^{\infty} \lambda^j\|\varphi(0)\| \\
&\leq \varepsilon\lambda\|x_s\| + \frac{\lambda}{1-\lambda}\|\varphi(0)\|
\end{aligned}$$

$$\begin{aligned}
\|\gamma_s(n)\| &\leq \lambda^n\|x_s\| + \varepsilon\lambda^{n-1}\|x_s\| + \lambda^{n-1}\|\varphi(0)\| + \sum_{j=2}^{n} \lambda^{n-j}\|\varphi(0)\| \\
&\leq \|x_s\|(\lambda + \varepsilon) + \sum_{j=0}^{\infty} \lambda^j\|\varphi(0)\| \\
&= \|x_s\|(\lambda + \varepsilon) + \frac{1}{1-\lambda}\|\varphi(0)\|
\end{aligned}$$

$$\begin{aligned}
\|\gamma_u(n)\| &\leq \sum_{j=1}^{\infty} \lambda^j\|\varphi(0)\| \\
&= \frac{\lambda}{1-\lambda}\|\varphi(0)\|
\end{aligned}$$

Hence,

$$\|G_0^{-1}F_{x_s}(0)\| \leq (\lambda + \varepsilon)r + \frac{1}{1-\lambda}\|\varphi(0)\|$$

if $x_s \in E_s(r)$. Since the image of $G_0^{-1}F_{x_s}$ contains the ball about $G_0^{-1}F_{x_s}(0)$ of radius $\tilde{r} = r\left(1 - \dfrac{\varepsilon}{1-\lambda}\right)$, it contains 0 if

$$(\lambda + \varepsilon)r + \frac{1}{1-\lambda}\|\varphi(0)\| < r\left(1 - \frac{\varepsilon}{1-\lambda}\right).$$

That is,

$$
\begin{aligned}
\|\varphi(0)\| \;&<\; (1-\lambda)r(1 - \frac{\varepsilon}{1-\lambda} - (\lambda+\varepsilon)) \\
&=\; r[(1-\lambda)^2 - \varepsilon - \varepsilon(1-\lambda)] \\
&=\; r[(1-\lambda)^2 - \varepsilon(2-\lambda)].
\end{aligned}
$$

Now if we choose $\varepsilon > 0$ such that

$$
\varepsilon < \frac{(1-\lambda)^2}{2-\lambda} < 1-\lambda,
$$

then we can choose a $\delta > 0$ such that

$$
\delta < r[(1-\lambda)^2 - \varepsilon(2-\lambda)].
$$

If $\mathrm{Lip}(\varphi - T) < \varepsilon$ and $\|\varphi(0)\| < \delta$, then $G_0^{-1}F_{x_s}$ contains 0, and since G_0 is linear, the image of F_{x_s} will also contain 0. Therefore, the image of F contains $E_s(r) \times 0$, and $W_r^s(\varphi)$ is the graph of

$$
g = \pi_u F^{-1}|_{E_s(r)\times 0}
$$

where $\pi_u : E_s(r) \times E_u(r) \times B(r) \to E_u(r)$ is the projection. Moreover, $\mathrm{Lip}(g) \le 1$ by Theorem 4.9.

Step 3: g is C^k whenever φ is C^k.

The proof of the following lemma is based on Appendix A of [146].

Lemma 4.12 *Let U be an open subset of a Banach space E, and let $f : U \to E'$ be a Lipschitz map into some Banach space E'. If f is differentiable at $x \in U$, then*

$$
\|Df_x\| \le \mathrm{Lip}(f).
$$

Proof:

Recall that f is differentiable at $x \in U$ if and only if there exists a linear map $Df_x : E \to E'$ such that

$$
\lim_{h\to 0} \frac{f(x+h) - f(x) - Df_x(h)}{\|h\|} = 0.
$$

So, for any $\alpha > 0$ there is a $\beta > 0$ such that $\|h\| < \beta$ implies

$$
\|f(x+h) - f(x) - Df_x(h)\| < \alpha\|h|.
$$

Hence,

$$
\begin{aligned}
\|Df_x(h)\| \;&\le\; \|f(x) - f(x+h)\| + \|f(x+h) - f(x) - Df_x(f)\| \\
&<\; \mathrm{Lip}(f)\|h\| + \alpha\|h\| \\
&=\; (\mathrm{Lip}(f) + \alpha)\|h\|
\end{aligned}
$$

This shows that $\|Df_x\| \leq \mathrm{Lip}(f)+\alpha$ for any $\alpha > 0$. Hence, $\|Df_x\| \leq \mathrm{Lip}(f)$.

□

Note: One can also show that $\mathrm{Lip}(f) \leq \sup_x \|Df_x\|$ where the sup is taken over all points in the domain of f. This follows from the Mean Value Theorem (see for instance Corollary 5.4.3 of [94]). In particular, every C^1 map is **locally Lipschitz**, i.e. around every point x_0 in the domain of f there is an open ball B_{x_0} and a constant k_0 such that

$$|f(x) - f(y)| \leq k_0|x - y|$$

for all $x, y \in B_{x_0}$.

For any $(x, \gamma) \in E(r) \times B(r)$, the lemma implies that

$$\|DF_{(x,\gamma)} - G\| < \|G^{-1}\|^{-1}$$

since $\mathrm{Lip}(F - G) < \|G^{-1}\|^{-1}$. Therefore, $DF_{(x,\gamma)}$ is a continuous linear isomorphism by Theorem 4.10, and we can apply the Inverse Function Theorem for differentiable maps (see Theorem 6.1.2 of [94]) to conclude that if F is C^k, then so is F^{-1}. Thus, to show that g is C^k it suffices to show that F is C^k.

For any $(x, \gamma) = (x_s, x_u, \gamma) \in E_s(r) \times E_u(r) \times B(r)$ define a linear map $L : E_s \times E_u \times B \to E_s \times B$ by $L(y_s, y_u, \nu) = (y_s, \zeta)$ where

$$\zeta(x_s, x_u, \gamma)(n) = \begin{cases} D\varphi_x(y_s, y_u) - \nu(1) & \text{if } n = 1 \\ D\varphi_{\gamma(n-1)}(\nu(n - 1)) - \nu(n) & \text{if } n \geq 2. \end{cases}$$

Note that $F(x + y, \gamma + \nu) - F(x, \gamma) - L(y, \nu) = (0, \xi) \in E_s \times B$ where

$$\begin{aligned} \xi(1) &= \varphi(x + y) - \varphi(y) - D\varphi_x(y) \\ &= \int_0^1 (D\varphi_{x+ty} - D\varphi_x)(y) \, dt \\ \xi(n) &= \varphi(\gamma(n - 1) + \nu(n - 1)) - \varphi(\gamma(n - 1)) - D\varphi_{\gamma(n-1)}(\nu(n - 1)) \\ &= \int_0^1 (D\varphi_{\gamma(n-1)+t\nu(n-1)} - D\varphi_{\gamma(n-1)})(\nu(n - 1)) \, dt \end{aligned}$$

and so

$$\frac{\|F(x + y, \gamma + \nu) - F(x, \gamma) - L(y, \nu)\|}{\|(y, \nu)\|}$$

is bounded above by

$$\max\left\{ \int_0^1 \|D\varphi_{x+ty} - D\varphi_x\| dt, \sup_{n \geq 1} \int_0^1 \|D\varphi_{\gamma(n)+t\nu(n)} - D\varphi_{\gamma(n)}\| \right\}.$$

Unfortunately, this doesn't necessarily go to zero as $\|(y, \nu)\|$ goes to zero since B is not locally compact and $D\varphi$ is not necessarily uniformly continuous. However, if $\gamma(n)$ converges, then the above will go to zero as $\|(y, \nu)\|$ goes to zero and L is the derivative of F at (x, γ).

Consider $C = \{\gamma \in B \mid \lim_{n \to \infty} \gamma(n) \text{ exists}\}$, i.e. the subspace of all Cauchy sequences. C is a closed subspace of B and thus a Banach space. Clearly,

$$F(E_s(r) \times E_u(r) \times C(r)) \subseteq E_s(r) \times C$$

since

$$\|\varphi(\gamma(n-1)) - \gamma(n) - [\varphi(\gamma(n)) - \gamma(n+1)]\| \leq$$
$$\|\varphi(\gamma(n-1)) - \varphi(\gamma(n))\| + \|\gamma(n+1) - \gamma(n)\|$$

and so we can restrict F to a map

$$\tilde{F} : E_s(r) \times E_u(r) \times C(r) \to E_s(r) \times C.$$

Similarly, G can be restricted to a map

$$\tilde{G} : E_s(r) \times E_u(r) \times C(r) \to E_s(r) \times C.$$

By the above argument, $\tilde{F}$ is differentiable. Moreover, g can be defined as

$$g = \pi_u \tilde{F}^{-1}|_{E_s(r) \times 0}$$

as long as $\tilde{G}^{-1}(E_s \times C) \subseteq E_s \times E_u \times C$, i.e. Steps 1) and 2) hold for $\tilde{F}$ and $\tilde{G}$. We will now check this.

Let $(x, \nu) \in E_s \times C$ and recall $G^{-1}(x, \nu) = (x_s, x_u, \gamma)$ where

$$x_u = \sum_{j=1}^{\infty} T_u^{-j}(\nu_u(j))$$

$$\gamma_s(n) = T_s^n(x_s) - \sum_{j=1}^{\infty} T_s^{n-j}(\nu_s(j))$$

$$\gamma_u(n) = \sum_{j=1}^{\infty} T_u^{-j}(\nu_u(n+j)).$$

We need to show that γ_s and γ_u are Cauchy sequences. First note that since

$$\|\gamma_u(n) - \gamma_u(m)\| \leq \left(\sum_{j=1}^{\infty} \lambda^j \right) \sup_{j \geq 1} \|\nu_u(n+j) - \nu_u(m+j)\|$$

ν_u is a Cauchy sequence implies that γ_u is a Cauchy sequence.

For γ_s we have

$$\|\gamma_s(n)-\gamma_s(m)\| \leq \|\sum_{j=1}^{n} T^{n-j}(\nu_s(j))-\sum_{j=1}^{m} T^{m-j}(\nu_s(j))\|+\|T_s^n(x_s)-T_s^m(x_s)\|.$$

The second term goes to zero as $n,m \to \infty$ because $\|T_s\| < \lambda < 1$. For the first term we have

$$\|\sum_{j=1}^{n} T_s^{n-j}(\nu_s(j)) - \sum_{j=1}^{m} T_s^{n-j}(\nu_s(j))\|$$

$$= \|\sum_{k=0}^{n-1} T^k(\nu_s(n-k)) - \sum_{k=0}^{m-1} T^k(\nu_s(m-k))\|$$

$$\leq \left(\sum_{k=0}^{\infty} \lambda^k\right) \sup_{\substack{h \geq n-N \\ j > m-N}} \|\nu_s(h) - \nu_s(l)\| + 2\left(\sum_{k=N+1}^{\infty} \lambda^k\right) \|\nu_s\|$$

For any $\mu > 0$ we can pick N_0 large enough so that

$$\sup_{\substack{h \geq n-N \\ j > m-N}} \|\nu_s(h) - \nu_s(l)\| \leq \frac{1-\lambda}{2}\mu$$

and

$$\sum_{k=N_0+1}^{\infty} \lambda^k < \frac{\mu}{4\|\nu_s\|}.$$

Then if $m,n \geq 2N_0$ we have

$$\|\sum_{j=1}^{n} T_s^{n-j}(\nu_s(j)) - \sum_{j=1}^{m} T_s^{m-j}(\nu_s(j))\| \leq (\sum_{k=0}^{\infty} \lambda^k)\frac{1-\lambda}{2}\mu + \frac{2\mu\|\nu_s\|}{4\|\nu_s\|} = \mu.$$

This shows that γ_s is also Cauchy, and thus $\tilde{G}^{-1}(E_s \times C) \subseteq E_s \times E_u \times C$, and g can be defined as

$$g = \pi_u \tilde{F}^{-1}|_{E_s(r) \times 0}.$$

It's clear that L depends continuously on (x,γ) whenever φ is of class C^1, and therefore g is of class C^1 whenever φ is of class C^1. The proof that g is of class C^k whenever φ is of class C^k is similar.

Now assume that φ is of class C^1, $\varphi(0) = 0$, and $D\varphi_0 = T$. Then $g(0) = 0$, and $Dg_0 = \pi_u D\tilde{F}_0^{-1}|_{E_s \times 0} = \pi_u (D\tilde{F}_0)^{-1}|_{E_s \times 0}$ by the Inverse Function Theorem. The above formula for $L = D\tilde{F}$ shows that $D\tilde{F}_0 = \tilde{G}_0$ since $D\varphi_0 = T$, and hence for $\nu_s \in E_s$ we have

$$Dg_0(\nu_s) = \pi_u \tilde{G}^{-1}(\nu_s, 0) = \sum_{j=1}^{\infty} T_u^{-j}(0) = 0.$$

This proves that $E_s = T_0 W_r^s(\varphi)$.

$\square$

4.3 The Global Stable/Unstable Manifold Theorem

Let M be a finite dimensional smooth manifold.

Definition 4.13 *A fixed point $p \in M$ of a smooth diffeomorphism $\varphi : M \to M$ is called **hyperbolic** if and only if none of the complex eigenvalues of $d\varphi_p : T_pM \to T_pM$ have length 1.*

Remark 4.14 If $p \in M$ if a hyperbolic fixed point of $\varphi : M \to M$, then there is a splitting of T_pM that is preserved by $d\varphi_p$

$$d\varphi_p : T_p^s M \oplus T_p^u M \to T_p^s M \oplus T_p^u M$$

and norms on $T_p^s M$ and $T_p^u M$ (that induce the standard topology) such that $d\varphi_p : T_p^s M \to T_p^s M$ is contracting and $d\varphi_p : T_p^u M \to T_p^u M$ is expanding. This is obvious when $d\varphi_p$ is diagonalizable, e.g. when p is a non-degenerate critical point of a gradient vector field and φ is induced from the gradient flow.

In general, the existence of the splitting follows from the Spectral Decomposition Theorem, and the existence of the norms follows from arguments involving the **spectral radius** (see for instance the Appendix to Chapter 4 of [83] or Chapter VII of [44]). In finite dimensions, the existence of the splitting can also be shown by considering the **generalized eigenvectors** of $d\varphi_p$ (see for instance Section 1.9 of [117]).

Theorem 4.15 (Global Stable Manifold Theorem for a Diffeomorphism)
If $\varphi : M \to M$ is a smooth diffeomorphism of a finite dimensional smooth manifold M and p is a hyperbolic fixed point of φ, then

$$W_p^s(\varphi) = \{x \in M \mid \lim_{n \to \infty} \varphi^n(x) = p\}$$

is an immersed submanifold of M with $T_pW_p^s(\varphi) = T_p^s M$. Moreover, $W_p^s(\varphi)$ is the surjective image of a smooth injective immersion

$$E^s : T_p^s M \to W_p^s(\varphi) \subseteq M.$$

Hence, $W_p^s(\varphi)$ is a smooth injectively immersed open disk in M.

Proof:
   In light of the preceeding remark and the fact that every C^1 map is locally Lipschitz, we can apply Theorem 4.11 to conclude that there is an open ball $B \subset M$ centered at p, and a coordinate chart $\tilde{\phi} : B \to T_pM$, such that

$\tilde{\phi}(W_p^s(\varphi|_B))$ is the graph of a smooth function $g : T_p^s M \to T_p^u M$. Restricting $\tilde{\phi} : B \to T_p M$ to $W_p^s(\varphi|_B)$ and composing with the projection $p_s : T_p^s M \times T_p^u M \to T_p^s M$ gives a coordinate chart $\phi : U \to T_p^s M$ on an open neighborhood $U \subseteq W_p^s(\varphi)$ of p. Theorem 4.11 implies that with respect to the smooth structure determined by ϕ we have $T_p W_p^s(\varphi) = T_p^s M$.

To define a smooth structure on $W_p^s(\varphi)$ note that

$$W_p^s(\varphi) = \bigcup_{n=0}^{\infty} \varphi^{-n}(U).$$

Define $U_n = \varphi^{-n}(U)$ for all $n \geq 0$, and let $\phi_n : U_n \to T_p^s M$ be given by $\phi_n(x) = \phi(\varphi^n(x))$ for all $x \in U_n$. It's clear that (U_n, ϕ_n) is a smooth atlas on $W_p^s(\varphi)$ such that the inclusion $W_p^s(\varphi) \hookrightarrow M$ is an immersion, i.e. $W_p^s(\varphi)$ is an immersed submanifold of M.

Let $U' = \phi(U) \subseteq T_p^s M$, and define a smooth map $h : U' \to U'$ by $h = \phi \circ \varphi \circ \phi^{-1}$. The map h is a diffeomorphism onto its image, and since

$$dh_0 = d\phi_p \circ d\varphi_p \circ d\phi_0^{-1}$$

dh_0 and $d\varphi_p|_{T_p^s M}$ have the same eigenvalues. In particular, the eigenvalues of dh_0 all have norm less than 1, and we can introduce an inner product on $T_p M$ such that $\|dh_0\| < 1$. Choose $\alpha \in \mathbb{R}$ such that $\|dh_0\| < \alpha < 1$, and let $B_0 \subseteq U'$ be an open ball centered at 0 such that $\|dh_x\| \leq \alpha < 1$ for all $x \in B_0$. Extend $h|_{B_0}$ to a C^1 diffeomorphism $\tilde{h} : T_p^s M \to T_p^s M$ such that $\|d\tilde{h}_x\| \leq \alpha$ for all $x \in T_p^s M$. Since $\alpha < 1$ the Mean Value Theorem implies that $\tilde{h} : T_p^s M \to T_p^s M$ is a contraction. (See for instance Corollary 5.4.3 of [94].)

Define a map $E^s : T_p^s M \to W_p^s(\varphi)$ by

$$E^s(x) = (\varphi^{-n} \circ \phi^{-1} \circ \tilde{h}^n)(x)$$

where n is any positive integer such that $\tilde{h}^n(x) \in B_0$. Note that E^s is independent of the choice of n because if $\tilde{h}^n(x) \in B_0$, then

$$\begin{aligned}
(\varphi^{-(n+1)} \circ \phi^{-1} \circ \tilde{h}^{n+1})(x) &= (\varphi^{(-n+1)} \circ \phi^{-1} \circ \tilde{h} \circ \tilde{h}^n)(x)) \\
&= (\varphi^{(-n+1)} \circ \phi^{-1} \circ \phi \circ \varphi \circ \phi^{-1} \circ \tilde{h}^n)(x) \\
&= (\varphi^{-n} \circ \phi^{-1} \circ \tilde{h}^n)(x).
\end{aligned}$$

It's clear that E^s is smooth and injective. To see that E^s is surjective, note that for any $y \in W_p^s(\varphi)$ there exists a positive integer n such that $\varphi^n(x) \in \phi^{-1}(B_0)$, and the point $x = \tilde{h}^{-n}(\phi(\varphi^n(y))) \in T_p^s M$ satisfies $E^s(x) = y$. Since $d\tilde{h}_x$ and $d\phi_x^{-1}$ are injective for all $x \in T_p^s M$ and $d\varphi_y$ is injective for all

$y \in M$, it's clear that dE_x^s is injective for all $x \in T_p^s M$. Hence, E^s is a smooth injective immersion of $T_p^s M$ onto $W_p^s(\varphi) \subseteq M$.

$\square$

Remark 4.16 The construction of the smooth immersion E^s in the preceeding theorem is due to Smale [136].

If we replace φ with φ^{-1} in the preceeding theorem, then we have the following.

Theorem 4.17 (Global Unstable Manifold Theorem for a Diffeomorphism)
If $\varphi : M \to M$ is a smooth diffeomorphism of a finite dimensional smooth manifold M and p is a hyperbolic fixed point of φ, then

$$W_p^u(\varphi) = \{x \in M \mid \lim_{n \to -\infty} \varphi^n(x) = p\}$$

is an immersed submanifold of M with $T_p W_p^u(\varphi) = T_p^u M$. Moreover, $W_p^u(\varphi)$ is the surjective image of a smooth injective immersion

$$E^u : T_p^u M \to W_p^s(\varphi) \subseteq M.$$

Hence, $W_p^u(\varphi)$ is a smooth injectively immersed open disk in M.

Stable and unstable manifolds for vector fields

Definition 4.18 *Let ξ be a smooth vector field on a finite dimensional smooth manifold M. A critical point $p \in M$ of ξ, i.e. a point p where $\xi_p = 0$, is called* **hyperbolic** *if and only if $d\xi_p : T_p M \to T_p M$ does not have any complex eigenvalues whose real part is zero.*

Note that Lemma 4.4 implies that a non-degenerate critical point $p \in M$ of a smooth function $f : M \to \mathbb{R}$ on a finite dimensional smooth Riemannian manifold (M, g) is a hyperbolic critical point of the gradient vector field ∇f (since a symmetric matrix has real eigenvalues and the determinant of $M_p(f)$ is non-zero). In Section 4.1 we used Lemma 4.5 to conclude that a non-degenerate critical point p is also a hyperbolic critical point for the diffeomorphism φ_t associated to the gradient flow (for any $t \neq 0$). By extending the proof of Lemma 4.5 to general vector fields we get the following.

Lemma 4.19 *Let ξ be a smooth vector field on a finite dimensional smooth manifold M, and let $\varphi_t : M \to M$ be the associated 1-parameter group of diffeomorphisms. A point $p \in M$ is a hyperbolic critical point of ξ if and only if p is a hyperbolic critical point of φ_t for some (and hence every) $t \neq 0$.*

Proof:

In a coordinate chart around p, we know by the existence and uniqueness theorem for O.D.E.s that $\varphi : (-\varepsilon, \varepsilon) \times U \to U$ is smooth in both $t \in (-\varepsilon, \varepsilon)$ and $\vec{x} \in U$ for some neighborhood U of p and some small $\varepsilon > 0$. Since

$$\frac{d}{dt}\varphi(t, \vec{x}) = \xi(\varphi(\vec{x}, t))$$

for any $\vec{x} \in U$, we can interchange the order of differentiation and we have

$$\frac{d}{dt}\frac{\partial}{\partial \vec{x}}\varphi(t, \vec{x}) = \frac{\partial}{\partial \vec{x}}\xi(\varphi(\vec{x}, t)) = (\frac{\partial}{\partial \vec{x}}\xi)\frac{\partial}{\partial \vec{x}}\varphi(t, \vec{x}).$$

Thus, $\Phi(t, \vec{x}) = \frac{\partial}{\partial \vec{x}}\varphi(t, \vec{x})$ is a solution to the following linear system of O.D.E.s.

$$\begin{aligned} \frac{d}{dt}\Phi(t, \vec{x}) &= (\frac{\partial}{\partial \vec{x}}\xi)\Phi(t, \vec{x}) \\ \Phi(0, \vec{x}) &= I_{m \times m} \end{aligned}$$

Since $e^{(\frac{\partial}{\partial \vec{x}}\xi)t}$ is also a solution to the above system, we have

$$\Phi(t, \vec{x}) = e^{(\frac{\partial}{\partial \vec{x}}\xi)t} = \frac{\partial}{\partial \vec{x}}\varphi(t, \vec{x})$$

because the solution is unique. Hence, at the critical point p we have

$$\frac{\partial}{\partial \vec{x}}\varphi(t, \vec{x})\Big|_{p} = e^{(\frac{\partial}{\partial \vec{x}}\xi|_{p})t}$$

and it is clear that for any $t \neq 0$, $\frac{\partial}{\partial \vec{x}}\varphi(t, \vec{x})|_{p}$ has no eigenvalues of length 1 if and only if $\frac{\partial}{\partial \vec{x}}\xi|_{p}$ does not have any complex eigenvalues whose real part is zero.

$\square$

Since $\varphi_t^n = \varphi_{tn}$ for all $n \in \mathbb{N}$, it's clear that for any critical point $p \in M$ and any fixed $t > 0$ we have

$$\begin{aligned} W_p^s(\varphi_t) &= \{x \in M \,|\, \lim_{n \to \infty} \varphi_t^n(x) = p\} \\ &= \{x \in M \,|\, \lim_{\lambda \to \infty} \varphi_\lambda(x) = p\} \\ &= W_p^s(\xi) \end{aligned}$$

and

$$\begin{aligned} W_p^u(\varphi_t) &= \{x \in M \,|\, \lim_{n \to -\infty} \varphi_t^n(x) = p\} \\ &= \{x \in M \,|\, \lim_{\lambda \to -\infty} \varphi_\lambda(x) = p\} \\ &= W_p^u(\xi). \end{aligned}$$

Thus, Theorem 4.15 and Theorem 4.17 hold for the stable and unstable manifolds of a smooth vector field ξ on M. Taking $\xi = \nabla f$, Theorem 4.2 follows from Theorem 4.15, Theorem 4.17, the results in Section 4.1, and the following lemma.

Lemma 4.20 *If* $f : M \rightarrow \mathbb{R}$ *is a Morse function on a finite dimensional compact smooth Riemannian manifold* (M, g), *and* p *is a critical point of* f, *then*

$$E^s : T_p^s M \quad \rightarrow \quad W^s(p) \subseteq M \quad and$$
$$E^u : T_p^u M \quad \rightarrow \quad W^u(p) \subseteq M$$

are homeomorphisms onto their images.

Proof:

It suffices to prove the lemma for $W^s(p)$ because the unstable manifold of f is the stable manifold of $-f$. E^s is continuous because it is smooth. Hence, it suffices to prove that $(E^s)^{-1} : W^s(p) \rightarrow T_p^s M$ is continuous where $W^s(p) \subseteq M$ is given the subspace topology.

By Theorem 3.2 we can choose an open set $U \subset M$ containing p such that p is the only critical point in U. Since E^s is a local diffeomorphism, we can choose an open set $V \in T_p^s M$ around $0 \in T_p^s M$ such that $E^s(V) \subset U$ and $E^s|_V$ is a homeomorphism onto its image. Let $x_j \in W^s(p)$ be a sequence converging to some point $x \in W^s(p)$, and let $v_j = (E^s)^{-1}(x_j)$ and $v = (E^s)^{-1}(x)$. Suppose that v_j does not converge to v as $j \rightarrow \infty$.

For any $t \in \mathbb{R}$, $x_j^t = \varphi_t(x_j)$ is a sequence that converges to $x^t = \varphi_t(x)$, and if we choose t sufficiently large, then x^t and x_j^t are in U for all j sufficiently large. Let $v_j^t = (E^s)^{-1}(x_j^t)$ and $v^t = (E^s)^{-1}(x^t)$. The assumption that v_j does not converge to v implies that v_j^t does not converge to v^t. Hence, we must have $\|v_j^t\| \rightarrow \infty$ as $j \rightarrow \infty$. Otherwise, there would be some value of $t > 0$ such that $v_j^t \in V$ for all j sufficiently large, and since $E^s|_V$ is a homeomorphism onto its image we would have $v_j^t \rightarrow v^t$ as $j \rightarrow \infty$.

But if $\|v_j^t\| \rightarrow \infty$ as $j \rightarrow \infty$, then there is a subsequence of x_j^t that converges to some critical point q by Proposition 3.19 and Corollary 3.3. Since $x_j^t \in U$ for all j sufficiently large and p is the only critical point in U we must have $q = p$. This would imply that there is a gradient flow line from p to itself, which contradicts Proposition 3.18. Therefore, $v_j^t \rightarrow v^t$ as $j \rightarrow \infty$, and hence, $v_j \rightarrow v$ as $j \rightarrow \infty$.

$\square$

The proof of Theorem 4.2 is now complete.

Remark 4.21 The reader should compare the following proposition to Theorem 3.28. Note that this proposition **does not say** that the stable and unstable manifolds of a Morse function are necessarily cells in some CW-structure. Theorem 4.2 says that the stable and unstable manifolds of a Morse function are smoothly embedded **open** disks, but a CW-structure requires **closed** disks and attaching maps. See the problems at the end of Chapter 6 for an example of a Morse function whose unstable manifolds do not form a CW-structure.

In Chapter 6 (Corollary 6.27) we will describe the closure of the stable (unstable) manifolds of a "Morse-Smale" function (see Definition 6.1) in terms of the stable (unstable) manifolds of lower dimension. Thus, Corollary 6.27 describes what the **image** of the attaching maps for a CW-structure should be, but of course, knowing what the image of the attaching maps should be is not enough to define a CW-structure. The relationship between the stable (unstable) manifolds of a Morse-Smale function and the cells in the CW-complex described by Theorem 3.28 is explored further in Section 6.4 using an additional assumption on the Riemannian metric.

Proposition 4.22 *If $f : M \to \mathbb{R}$ is a Morse function on a compact smooth closed Riemannian manifold (M, g), then M is a disjoint union of the unstable manifolds of f, i.e.*

$$M = \coprod_{p \in Cr(f)} W^u(p).$$

Similarly,

$$M = \coprod_{q \in Cr(f)} W^s(q).$$

Proof:

By the existence and uniqueness theorem for O.D.E.s, every point $x \in M$ lies on a unique gradient flow line γ_x. By Proposition 3.19, $\lim_{t \to -\infty} \gamma_x(t) = p$ exists where p is a critical point of f. Similarly, $\lim_{t \to \infty} \gamma_x(t) = q$ exists where q is a critical point.

$\square$

4.4   Examples of stable/unstable manifolds

In this section we give some examples of stable and unstable manifolds for functions on Riemannian manifolds, vector fields, and diffeomorphisms.

Example 4.23 (The height function on the sphere) Let

$$S^n = \{(x_1, \ldots, x_{n+1}) \in \mathbb{R}^{n+1} \mid x_1^2 + \cdots + x_{n+1}^2 = 1\}$$

be the n-sphere, and define $f : S^n \rightarrow \mathbb{R}$ by $f(x_1, \ldots, x_{n+1}) = x_{n+1}$. The function f is a smooth Morse function on S^n with two critical points, the north pole $N = (0, \ldots, 0, +1)$ (the maximum) and the south pole $S = (0, \ldots, 0, -1)$ (the minimum).

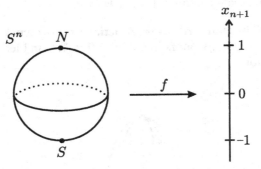

The critical points N and S are both non-degenerate and are of index $\lambda_N = n$ and $\lambda_S = 0$ respectively. With respect to the standard metric on $\mathbb{R}^{n+1}$ we have the following.

$$
\begin{aligned}
W^u(N) &= S^n - \{S\} \\
W^s(N) &= \{N\} \\
W^u(S) &= \{S\} \\
W^s(S) &= S^n - \{N\}
\end{aligned}
$$

Note that $W^u(N)$ is diffeomorphic to an open disk of dimension λ_N, and $W^u(S)$ is diffeomorphic to an open disk of dimension λ_S.

Example 4.24 (A function with degenerate critical points)   The function $g : S^n \rightarrow [0, 1]$ given by $g(x_1, \ldots, x_{n+1}) = x_{n+1}^2$ is not a Morse function because it has infinitely many critical points.

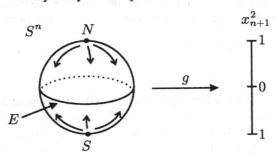

The north pole N, the south pole S, and any point on the equator

$$E = \{(x_1, \ldots, x_{n+1}) \in S^n \mid x_{n+1} = 0\}$$

is a critical point. However, Theorem 4.2 can still be applied to the non-degenerate critical points N and S. With respect to the standard metric on $\mathbb{R}^{n+1}$, $W^u(N)$ and $W^u(S)$ are diffeomorphic to open n-disks, and $W^s(N)$ and $W^s(S)$ are 0-disks. The points on the equator are all degenerate critical points and are not included in $W^u(N)$ or $W^u(S)$.

Example 4.25 (The standard height function on the torus)   Consider the torus T^2 resting vertically on the plane $z = 0$ in $\mathbb{R}^3$, and let $f : T^2 \to \mathbb{R}$ be the height function.

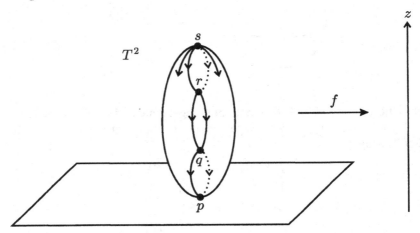

The function f is a Morse function with 4 critical points: the minimum p, the saddle points q and r, and the maximum s. These are all non-degenerate and have indices 0, 1, 1, and 2 respectively. The unstable manifold $W^u(p)$ is just the critical point p, the unstable manifold $W^u(q)$ is the circle through q and p minus the critical point p, and the unstable manifold $W^u(r)$ is a circle around the inside of the hole minus the critical point q. Every other point lies in $W^u(s)$, and hence T^2 can be written as the following disjoint union.

$$T^2 = W^u(p) \amalg W^u(q) \amalg W^u(r) \amalg W^u(s)$$

Example 4.26 (Bott's perfect Morse functions) In Example 3.7 we showed that Bott's perfect Morse function $f : \mathbb{C}P^n \to \mathbb{R}$ is a Morse function with $n+1$ critical points, namely $c_j = [e_j]$ for $j = 0, \dots, n$. Moreover, we showed that c_j has index $2j$ for all $j = 0, \dots, n$. In order to discuss the stable and unstable manifolds of f, we need to introduce a Riemannian metric on $\mathbb{C}P^n$. One way to do this is to identify $\mathbb{C}P^n$ as an adjoint orbit in the Lie algebra $\mathfrak{u}(n+1)$ of skew Hermitian matrices.

In Chapter 8 we will consider the more general case of a complex Grassmann manifold. We will show how to embed a complex Grassmann manifold as an adjoint orbit in the Lie algebra of skew-Hermitian matrices, and we will define a Morse function on the adjoint orbit analogous to the Morse functions constructed in Theorem 3.8 using the trace form. We will also prove the rather remarkable result that the unstable manifolds of the Morse function are the Schubert cells of the complex Grassmann manifold.

Example 4.27 (A vector field with closed orbits) Consider the vector field $\xi(x, y) = (2y, 4x - 4x^3)$ on $\mathbb{R}^2$. The flow of ξ is determined by the solution to the following system of non-linear differential equations.

$$\frac{dx}{dt} = 2y$$

$$\frac{dy}{dt} = 4x - 4x^3$$

This is a **Hamiltonian system** of differential equations, i.e.

$$\frac{dx}{dt} = \frac{\partial H}{\partial y}$$

$$\frac{dy}{dt} = -\frac{\partial H}{\partial x}$$

where $H(x, y) = y^2 - 2x^2 + x^4 + C$ for some constant $C \in \mathbb{R}$. Every Hamiltonian system is **conservative** in the sense that the Hamiltonian $H(x, y)$ stays constant along the flow lines of the system, i.e.

$$\begin{aligned}
\frac{dH}{dt} &= \frac{\partial H}{\partial x}\frac{dx}{dt} + \frac{\partial H}{\partial y}\frac{dy}{dt} \\
&= \frac{\partial H}{\partial x}\frac{\partial H}{\partial y} - \frac{\partial H}{\partial y}\frac{\partial H}{\partial x} \\
&= 0.
\end{aligned}$$

(See Section 1.9 of [6] or Section 2.12 of [117] for more details on Hamiltonian systems.)

The vector field ξ has 3 critical points, $(-1, 0)$, $(1, 0)$, and $(0, 0)$. Looking at the proof of Lemma 4.19 we see that at a critical point p the differential of the flow φ_t is given by e^{Bt} where

$$B = \begin{pmatrix} 0 & 2 \\ 4 - 12x^2 & 0 \end{pmatrix}.$$

At $(0, 0)$ the eigenvalues of B are $\pm 2\sqrt{2}$, and at $(\pm 1, 0)$ the eigenvalues of B are $\pm 4i$. Thus, $(0, 0)$ is a hyperbolic critical point, and $(\pm 1, 0)$ are not hyperbolic critical points. The integral curves $y^2 - 2x^2 + x^4 = c$ of ξ are as follows.

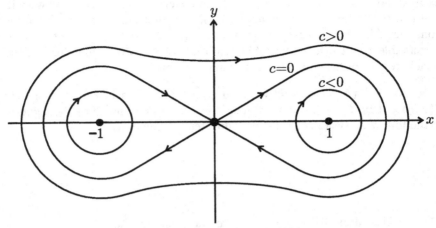

(Figure adapted from [143].)

Note that $W_p^s(\xi) = W_p^u(\xi)$ for the hyperbolic critical point $p = (0,0)$, i.e. the integral curve $c = 0$. This is not possible for a gradient vector field because the function is strictly decreasing along flow lines by Proposition 3.18. Also note that $W_p^s(\xi)$ and $W_p^u(\xi)$ are injectively immersed open 1-disks that are **not embedded**, i.e. the topology of the open 1-disk $(-1, 1)$ does not coincide with the topology $W_p^s(\xi)$ and $W_p^u(\xi)$ inherit as subspaces of the manifold $\mathbb{R}^2$.

Example 4.28 (An algebraic automorphism of the torus) Let $\varphi : \mathbb{R}^2 \to \mathbb{R}^2$ be a linear mapping given by the matrix

$$\varphi = \begin{pmatrix} a & b \\ c & d \end{pmatrix}$$

where $a, b, c, d \in \mathbb{Z}$. Then φ induces a mapping of the torus $T^2 = \mathbb{R}^2/\Gamma$ where Γ is the group of all translations of $\mathbb{R}^2$ with integer coordinates.

$$\Gamma = \{\gamma : \mathbb{R}^2 \to \mathbb{R}^2 | \gamma(x, y) = (x + n, y + m) \text{ for some } n, m \in \mathbb{Z}\}$$

If $ad - cb = \pm 1$, then φ^{-1} also induces a mapping of the torus, and we call φ an **algebraic automorphism** of T^2.

   If φ is an algebraic automorphism of T^2 with no eigenvalue of length 1, then a straightforward computation shows that φ must have two real eigenvalues λ_s and λ_u with $|\lambda_s| < 1$ and $|\lambda_u| > 1$. Let $\vec{x}_s$ and $\vec{x}_u$ be the corresponding eigenvectors. Clearly, $p = (0,0)$ is a hyperbolic fixed point of φ, and in $\mathbb{R}^2$ the stable and unstable manifolds at p are the eigenspaces.

$$\begin{aligned} W_p^s(\varphi) &= \{t\vec{x}_s|\, t \in \mathbb{R}\} \\ W_p^u(\varphi) &= \{t\vec{x}_u|\, t \in \mathbb{R}\} \end{aligned}$$

The stable and unstable manifolds for $\varphi : T^2 \to T^2$ are obtained by projecting these lines onto T^2. One can show that both $W_p^s(\varphi)$ and $W_p^u(\varphi)$ are dense in T^2 (see for instance p. 22 of [143]). As in the previous example, this cannot happen for diffeomorphisms induced from gradient vector fields because of Proposition 3.18.

Problems

1. Prove that the graph of a smooth function $f : M \to N$ between smooth manifolds M and N is a smooth submanifold of $M \times N$.

2. Suppose that $f : U \to \mathbb{R}^m$ is C^1 where U is an open subset of $\mathbb{R}^m$. Prove that f is locally Lipschitz.

3. Let $\Phi : \mathbb{R} \times \mathbb{R}^m \to \mathbb{R}^m$ be the solution to the linear system of differential equations $\frac{d}{dt}\vec{x} = A\vec{x}$ where A is an $m \times m$ matrix of real numbers. Show that $\Phi_t(\vec{x}) = \Phi(t, \vec{x})$ satisfies the following properties for all $\vec{x} \in \mathbb{R}^m$:

   1. $\Phi_0(\vec{x}) = \vec{x}$
   2. $(\Phi_s \circ \Phi_t)(\vec{x}) = \Phi_{s+t}(\vec{x})$ for all $s, t \in \mathbb{R}$.

4. Show that $0 \in \mathbb{R}^m$ is a hyperbolic fixed point for the linear diffeomorphisms determined by the following matrices, and find the corresponding stable and unstable manifolds.

   (1) $A = \begin{pmatrix} 1 & 3 \\ 4 & 2 \end{pmatrix}$

   (2) $B = \begin{pmatrix} 4 & 1 \\ 1 & 4 \end{pmatrix}$

   (3) $C = \begin{pmatrix} -2 & 12 & 28 \\ -3 & -29 & -59 \\ 2 & 14 & 28 \end{pmatrix}$

5. Let A be an $m \times m$ matrix with entries in $\mathbb{R}$, and let $\vec{u}_j + i\vec{v}_j$ be a generalized eigenvector of A corresponding to an eigenvalue $a_j + ib_j$. Note that if $b_j = 0$, then $\vec{u}_j = 0$. Suppose that

   $$\{\vec{u}_1, \ldots, \vec{u}_k, \vec{u}_{k+1}, \vec{v}_{k+1}, \ldots, \vec{u}_n, \vec{v}_n\}$$

is a basis for $\mathbb{R}^m$ (where $m = 2n - k$), and define

$$
\begin{aligned}
E^s &= \text{span}\{\vec{u}_j, \vec{v}_j | \, a_j < 0\} \\
E^c &= \text{span}\{\vec{u}_j, \vec{v}_j | \, a_j = 0\} \\
E^u &= \text{span}\{\vec{u}_j, \vec{v}_j | \, a_j > 0\}.
\end{aligned}
$$

Show that $\mathbb{R}^m = E^s \oplus E^c \oplus E^u$ is a decomposition of $\mathbb{R}^m$ into subspaces that are invariant under the flow determined by the linear system of differential equations $\frac{d}{dt}\vec{x} = A\vec{x}$.

6. Let $\vec{x} : \mathbb{R} \to \mathbb{R}^m$ and $\vec{b} : \mathbb{R} \to \mathbb{R}^m$ be vector valued functions where $\vec{x}(t)$ is differentiable and $\vec{b}(t)$ is continuous. Let A be an $m \times m$ matrix with entries in $\mathbb{R}$, and let $\Phi(t) = e^{At}$. Show that

$$
\vec{x}(t) = \Phi(t)\Phi^{-1}(0)\vec{x}_0 + \int_0^t \Phi(t)\Phi^{-1}(\tau)\vec{b}(\tau) \, d\tau
$$

is a solution to the nonhomogeneous linear system of differential equations

$$
\frac{d}{dt}\vec{x} = A\vec{x} + \vec{b}
$$

with initial condition $\vec{x}(0) = \vec{x}_0$.

7. Let A be an $m \times m$ matrix with entries in $\mathbb{R}$, and let $\vec{y} \in \mathbb{R}^m$. Show that if $\vec{x}(t, \vec{y})$ is a solution to the linear system of differential equations

$$
\begin{aligned}
\frac{d}{dt}\vec{x} &= A\vec{x} \\
\vec{x}(0) &= \vec{y}
\end{aligned}
$$

then

$$
\Phi(t, \vec{y}) = \frac{\partial \vec{x}}{\partial \vec{y}}(t, \vec{y})
$$

is a solution to

$$
\begin{aligned}
\frac{d}{dt}\Phi &= A\Phi \\
\Phi(0) &= I_{m \times m}.
\end{aligned}
$$

8. Prove the **Contraction Mapping Theorem**: Let (X, d) be a complete metric space, and suppose that $T : X \to X$ is a contraction, i.e. assume that there exists a number K with $0 < K < 1$ such that

$$
d(Tx, Ty) \le K \, d(x, y)
$$

for all $x, y \in X$. Then T has a unique fixed point x_0, and for any $x \in X$ we have $x_0 = \lim_{n \to \infty} T^n x$.

9. Let $T : \mathbb{R}^m \to \mathbb{R}^m$ be a linear transformation, and recall that

$$\|T\| = \sup\{|T(\vec{x})| \mid \vec{x} \in \mathbb{R}^m \text{ and } |\vec{x}| \leq 1\}.$$

Show that

$$\|T\| = \sup_{|\vec{x}|=1} |T(\vec{x})| = \sup_{\vec{x} \neq 0} \frac{|T(\vec{x})|}{|\vec{x}|},$$

and if T is invertible, then

$$\|T^{-1}\| \geq \frac{1}{\|T\|}.$$

10. Give an example of an invertible linear transformation $T : \mathbb{R}^2 \to \mathbb{R}^2$ whose complex eigenvalues do not have length 1 that is neither expanding nor contracting. Why does this not contradict Remark 4.14?

11. The **spectral radius** $r(T)$ of a linear transformation $T : \mathbb{R}^m \to \mathbb{R}^m$ is defined to be

$$r(T) = \sup\{|\lambda| \mid \lambda \text{ is a complex eigenvalue of } T\}.$$

Prove that the spectral radius of T is strictly less than 1 if and only if there is a norm $|\ \ |'$ on $\mathbb{R}^m$, equivalent to the standard norm, such that T is contracting with respect to $|\ \ |'$. Similarly, show that if $T : \mathbb{R}^m \to \mathbb{R}^m$ is invertible, then the complex eigenvalues of T all have length strictly greater than 1 if and only if there is a norm $|\ \ |'$ on $\mathbb{R}^m$, equivalent to the standard norm, such that T is expanding with respect to $|\ \ |'$.

12. Give an example of a smooth injective immersion that is not an embedding.

13. Show that a smooth injective immersion of a compact manifold is an embedding.

14. Describe a CW-structure on S^n such that the open cells in the CW-structure agree with the unstable manifolds described in Example 4.23.

15. Describe a CW-structure on T^2 such that the open cells in the CW-structure agree with the unstable manifolds described in Example 4.25.

16. For any linear transformation $T^2 : \mathbb{R}^2 \to \mathbb{R}^2$ it is possible to choose a basis of $\mathbb{R}^2$ such that the matrix A of T with respect to this basis has one of the following three Jordan forms:

$$(1) \begin{pmatrix} a & 0 \\ 0 & b \end{pmatrix}, \qquad (2) \begin{pmatrix} a & 1 \\ 0 & a \end{pmatrix}, \qquad (3) \begin{pmatrix} a & -b \\ b & a \end{pmatrix}.$$

If T is an invertible linear transformation, then what conditions are needed in each of these three cases for the origin to be a hyperbolic fixed point of the diffeomorphism $T : \mathbb{R}^2 \to \mathbb{R}^2$? The matrix A also determines a vector field ξ on $\mathbb{R}^2$ via the system of linear differential equations $\frac{d}{dt}\vec{x} = A\vec{x}$. What conditions are needed in each of the these three cases for the origin to be a hyperbolic critical point of the vector field ξ?

17. Show that the following map is a diffeomorphism of $\mathbb{R}^2$ and determine whether or not its fixed points are hyperbolic

$$\varphi(x, y) = (y, x - y + y^2).$$

18. Show that each of the following functions determines a local diffeomorphism around $0 \in \mathbb{R}^m$, and determine whether or not 0 is a hyperbolic fixed point for the local diffeomorphism.

   1. $\varphi(x, y) = (y + y^2, 2x + x^2)$
   2. $\varphi(x, y) = (2\sin x + xy, x - \frac{1}{2}y\cos y)$
   3. $\varphi(x, y) = (ye^y, e^{2x} - 1)$
   4. $\varphi(x, y, z) = (-x, y + x^3, z + y^2)$
   5. $\varphi(x, y, z) = (y^2, x - z^2, z^2 - x^2)$

19. Find the critical points of the following vector fields and determine whether or not the critical points are hyperbolic.

   1. $\xi(x, y) = (y + y^2, 2x + x^2)$
   2. $\xi(x, y) = (x + y^2, 1 - xy)$
   3. $\xi(x, y) = (y, -y + \sin x)$
   4. $\xi(x, y, z) = (-x, y + x^3, z + y^2)$
   5. $\xi(x, y, z) = (y^2, x - z^2, z^2 - x^2)$

20. Let $\varphi : M \to M$ be a diffeomorphism of a smooth manifold M. Show that $p \in M$ is a hyperbolic fixed point of φ if and only if it is a hyperbolic fixed point of φ^{-1}.

21. Show that if $\varphi_t : M \to M$ is flow on a differentiable manifold M, then $(\varphi_t)^{-1} = \varphi_{-t}$ for all $t \in \mathbb{R}$.

22. Let $\varphi_t : M \to M$ be a flow on a differentiable manifold M. Show that if two orbits intersect, then they coincide.

23. Show that both $W_p^s(\varphi)$ and $W_p^u(\varphi)$ in Example 4.28 are dense in the torus T^2.

24. Two diffeomorphisms $\varphi, \psi : M \to M$ of a differentiable manifold M are called **topologically conjugate** if and only if there exists a homeomorphism $h : M \to M$ such that $h \circ \varphi = \psi \circ h$. If $\varphi_t, \psi_t : M \to M$ are two flows on M, then φ_t and ψ_t are called **topologically conjugate** if and only if there exists a homeomorphism $h : M \to M$ such that $h \circ \varphi_t = \psi_t \circ h$ for all $t \in \mathbb{R}$. Suppose that $f : \mathbb{R} \to \mathbb{R}$ is a diffeomorphism with $df(x) > 0$ for some $x \in \mathbb{R}$. Given that the differential equation $\frac{d}{dt}x = f(x) - x$ defines a flow $\varphi_t : \mathbb{R} \to \mathbb{R}$, show that f is topologically conjugate to φ_1.

25. Two diffeomorphisms $\varphi, \psi : M \to M$ of a smooth manifold M are called C^r **conjugate** if and only if the map h in the above definition of topological conjugacy can be chosen to be of class C^r, i.e. $h \circ \varphi = \psi \circ h$ where $h : M \to M$ is of class C^r. Show that if φ and ψ are C^1 conjugate and p is a hyperbolic fixed point for φ, then $h(p)$ is a hyperbolic fixed point for ψ.

26. An **ordinary point** of a flow φ_t is any point that is not a fixed point. Prove the **Tubular Flow Theorem**: If $\vec{x}_0 \in \mathbb{R}^m$ is an ordinary point of a flow $\varphi_t : \mathbb{R}^m \to \mathbb{R}^m$, then in every sufficiently small neighborhood of $\vec{x}_0$, the flow φ_t is C^1 conjugate to the flow $\psi_t(\vec{x}) = \vec{x} + te_1$ where e_1 is the first element in the standard basis for $\mathbb{R}^m$.

27. Prove the Hartman-Grobman Theorem: If p is a hyperbolic fixed point of a C^1 diffeomorphism $\varphi : \mathbb{R}^m \to \mathbb{R}^m$, then there is a neighborhood $U \subseteq \mathbb{R}^m$ of p and a neighborhood $U' \subseteq \mathbb{R}^m$ containing the origin such that $\varphi|_U$ is topologically conjugate to the fixed point $0 \in \mathbb{R}^m$ of $d\varphi_p|_{U'}$.

28. Let U be an open set in $\mathbb{R}^m$, and let $f \in C^2(U)$. A system of differential equations of the form

$$\frac{d}{dt}\vec{x} = -\text{grad } f(\vec{x})$$

where

$$\text{grad } f = \left(\frac{\partial f}{\partial x_1}, \ldots, \frac{\partial f}{\partial x_m} \right)$$

is called a **gradient system** on U. Show that at any regular point of grad f the flow lines of the above gradient system are orthogonal to the level sets $f(\vec{x}) = \text{constant}$.

29. Let U be an open subset of $\mathbb{R}^{2m}$, and let $H \in C^2(U)$ be given by $H(\vec{x}, \vec{y})$ where $\vec{x}, \vec{y} \in \mathbb{R}^m$. Find a gradient system of differential equations whose flow lines are orthogonal to the flow lines of the following Hamiltonian system of differential equations

$$\frac{d\vec{x}}{dt} = \frac{\partial H}{\partial \vec{y}}$$

$$\frac{d\vec{y}}{dt} = -\frac{\partial H}{\partial \vec{x}}$$

where

$$\frac{\partial H}{\partial \vec{x}} = \left(\frac{\partial H}{\partial x_1}, \ldots, \frac{\partial H}{\partial x_m} \right) \quad \text{and}$$

$$\frac{\partial H}{\partial \vec{y}} = \left(\frac{\partial H}{\partial y_1}, \ldots, \frac{\partial H}{\partial y_m} \right).$$

Chapter 5

Basic Differential Topology

In this chapter we prove some results on transversality, general position, orientations, and intersection numbers that will be used in later chapters, including the Inverse Image Theorem (Theorem 5.11) and the Homotopy Transversality Theorems (Theorem 5.17 and Theorem 5.19). As an application of these results we show that the class of Morse functions on a finite dimensional smooth manifold M is locally stable (Corollary 5.24) and dense as a subspace of the space of all smooth functions on M with the uniform topology (Theorem 5.27). We also show that the set of Morse functions on M is an open and dense subspace of the space of all smooth functions on M with the smooth topology (Theorem 5.31). In the last two sections of this chapter we define orientations and intersection numbers, and we prove the Lefschetz Fixed Point Theorem (Theorem 5.50).

5.1 Immersions and submersions

Definition 5.1 *Suppose that M and N are smooth manifolds of dimension m and n respectively. A map $f : M \to N$ is said to be an **immersion at a point** $x \in M$ if and only if $df_x : T_x M \to T_y N$ is injective where $y = f(x) \in N$. If f is an immersion at x for all $x \in M$, then f is called an **immersion**.*

Example: Consider the map $i : \mathbb{R}^m \to \mathbb{R}^n$ for $n \geq m$ given by

$$i(x_1, \ldots, x_m) = (x_1, \ldots, x_m, 0, \ldots, 0).$$

The inclusion i is called the **canonical immersion**. The next theorem says that up to diffeomorphism this is the only immersion.

Theorem 5.2 (Local Immersion Theorem) *Suppose that* $f : M \to N$ *is an immersion at* $x \in M$ *and* $y = f(x) \in N$, *then there exists local coordinates around* x *and* y *such that*

$$f(x_1, \ldots, x_m) = (x_1, \ldots, x_m, 0, \ldots, 0),$$

i.e. f *is locally equivalent to an inclusion.*

Proof:

Choose local coordinates

$$
\begin{array}{ccc}
M & \xrightarrow{\ f\ } & N \\
{\scriptstyle \phi^{-1}}\big\uparrow & & \big\uparrow{\scriptstyle \psi^{-1}} \\
U & \xrightarrow[\ g\ ]{} & V
\end{array}
$$

such that $\phi(x) = 0$, $\psi(y) = 0$, and the above diagram commutes. Since $dg_0 : \mathbb{R}^m \to \mathbb{R}^m$ is injective, we can change the coordinate system so that the matrix of dg_0 is the following $n \times m$ matrix.

$$
\begin{pmatrix}
1 & & 0 \\
 & \ddots & \\
0 & & 1 \\
0 & \cdots & 0 \\
\vdots & & \vdots \\
0 & \cdots & 0
\end{pmatrix}
$$

Define $G : U \times \mathbb{R}^{n-m} \to \mathbb{R}^n$ by

$$G(x, z) = g(x) + (0, z).$$

Clearly, $dG_0 = \mathrm{id}$, and thus G is a local diffeomorphism of $\mathbb{R}^n$ at 0 by the Inverse Function Theorem (see for instance Theorem 6.1.2 of [94]). Note that $g = G \circ i$ where $i : \mathbb{R}^m \to \mathbb{R}^n$ is the canonical immersion. If we shrink U and V to open sets $\tilde{U}$ and $\tilde{V}$ so that we can use $\psi^{-1} \circ G$ as the chart around y, then

$$
\begin{array}{ccc}
M & \xrightarrow{\ f\ } & N \\
{\scriptstyle \phi^{-1}}\big\uparrow & & \big\uparrow{\scriptstyle \psi^{-1}\circ G} \\
\tilde{U} & \xrightarrow[\ i\ ]{} & \tilde{V}
\end{array}
$$

commutes and we have local coordinates in which f is equal to the canonical immersion.

$\square$

Corollary 5.3 *If* $f : M \to N$ *is an immersion at* $x \in M$, *then* f *is an immersion on some open neighborhood of* x.

Definition 5.4 *Suppose that* M *and* N *are smooth manifolds of dimension* m *and* n *respectively. A smooth map* $f : M \to N$ *is called a* **submersion at a point** $x \in M$ *if any only if* $df_x : T_x M \to T_y M$ *is surjective where* $y = f(x) \in N$. *If* f *is a submersion at every point* $x \in M$, *then* f *is called a* **submersion**.

Theorem 5.5 (Local Submersion Theorem) *Suppose that* $f : M \to N$ *is a submersion at* $x \in M$ *and* $y = f(x) \in N$, *then there exists local coordinates around* x *and* y *such that*

$$f(x_1, \ldots, x_m) = (x_1, \ldots, x_n),$$

i.e. f *is locally equivalent to a projection.*

Proof:

Choose local coordinates

$$
\begin{array}{ccc}
M & \xrightarrow{\ f\ } & N \\
{\scriptstyle \phi^{-1}}\big\uparrow & & \big\uparrow{\scriptstyle \psi^{-1}} \\
U & \xrightarrow{\ g\ } & V
\end{array}
$$

such that $\phi(x) = 0$, $\psi(y) = 0$, and the above diagram commutes. Since $dg_0 : \mathbb{R}^m \to \mathbb{R}^m$ is surjective, we can modify the chart ϕ so that dg_0 is the following $n \times m$ matrix.

$$
\begin{pmatrix}
1 & & 0 & 0 & \cdots & 0 \\
 & \ddots & & \vdots & & \vdots \\
0 & & 1 & 0 & \cdots & 0
\end{pmatrix}
$$

Define a map $G : U \to \mathbb{R}^m$ by $G(z) = (g(z), z_{n+1}, \ldots, z_m)$ where $z = (z_1, \ldots, z_m)$. The matrix of dG_0 is the identity matrix, and so G is a local diffeomorphism at 0 by the Inverse Function Theorem. Thus, $G^{-1} : \tilde{U} \to U$ is defined for some open neighborhood $\tilde{U}$ of $0 \in \mathbb{R}^n$, and $g = p \circ G$ where $p : \mathbb{R}^m \to \mathbb{R}^n$ is the projection $p(x_1, \ldots, x_n, x_{n+1}, \ldots, x_m) = (x_1, \ldots, x_n)$. Thus,

$$
\begin{array}{ccc}
M & \xrightarrow{\ f\ } & N \\
{\scriptstyle \phi^{-1} \circ G^{-1}}\big\uparrow & & \big\uparrow{\scriptstyle \psi^{-1}} \\
\tilde{U} & \xrightarrow{\ p\ } & V
\end{array}
$$

commutes where p is the projection onto the first n coordinates.

$\square$

Corollary 5.6 *If $f : M \to N$ is a submersion at $x \in M$, then f is a submersion on some open neighborhood of x.*

Definition 5.7 *A point $x \in M$ is called a **critical point** of a smooth map $f : M \to N$ if and only if f is not a submersion at x, i.e. if $df_x : T_xM \to T_{f(x)}N$ fails to be surjective. A point $x \in M$ that is not a critical point is called a **regular point**. A point $y \in N$ is called a **critical value** of a smooth map $f : M \to N$ if and only if $f^{-1}(y)$ contains at least one critical point. A point $y \in N$ is called a **regular value** of a smooth map $f : M \to N$ if it is not a critical value, i.e. if $df_x : T_xM \to T_yN$ is surjective for all $x \in M$ such that $f(x) = y$.*

Note: Every point $y \in N$ that is not in the image of $f : M \to N$ is a regular value because the above condition is true vacuously.

Remark 5.8 If dim M = dim N, then $y \in f(M) \subseteq N$ is a regular value of $f : M \to N$ if and only if df_x is non-singular for all $x \in f^{-1}(y)$. So when dim M = dim N, the Inverse Function Theorem (see for instance Theorem 6.1.2 of [94]) implies that if $y \in N$ is a regular value of $f : M \to N$, then f is injective in a neighborhood of each point of $f^{-1}(y)$. Hence, $f^{-1}(y)$ is discrete, and if M is compact, then $f^{-1}(y)$ is a finite set (possibly empty) for any regular value $y \in N$ and $\#f^{-1}(y)$ is locally constant.

Corollary 5.9 (The Preimage Theorem) *If $y \in N$ is a regular value in the image of $f : M \to N$, then $f^{-1}(y)$ is a submanifold of M of dimension $m-n$.*

Proof:
   Let $x \in f^{-1}(y)$. Since $df_x : T_xM \to T_yM$ is surjective we can choose local coordinates around x such that

$$f(x_1, \ldots, x_n, x_{n+1}, \ldots, x_m) = (x_1, \ldots, x_n)$$

and y corresponds to $(0, \ldots, 0) \in \mathbb{R}^n$. Thus in some neighborhood U of $x \in M$, $f^{-1}(y) \cap U$ is the set of points of the form

$$(0, \ldots, 0, x_{n+1}, \ldots, x_m),$$

and $f^{-1}(y) \cap U$ is an open neighborhood of x in $f^{-1}(y)$ with coordinates $(x_{n+1}, \ldots, x_m)$.

$\square$

5.2 Transversality

Definition 5.10 *Let* $f : M \to N$ *and* $g : Z \to N$ *be smooth maps where* M, N, *and* Z *are smooth manifolds. We say that* f *is* **transverse** *to* g, $f \pitchfork g$, *if and only if whenever* $f(x) = g(z) = y$ *we have*

$$df_x(T_x M) + dg_z(T_z Z) = T_y N.$$

If $Z \subseteq N$ *and* $g : Z \to N$ *is the inclusion map, then we will denote* $f \pitchfork g$ *by* $f \pitchfork Z$.

Note: If $f(M) \cap g(Z) = \emptyset$, then $f \pitchfork g$ because the above condition is true vacuously. See Section 1.5 of [71] for a nice collection of pictures showing intersections of submanifolds that are transverse and intersections that are not transverse.

Theorem 5.11 (Inverse Image Theorem) *Let* $Z \subseteq N$ *be an immersed submanifold and* $f : M \to N$ *a smooth map. If* $f \pitchfork Z$, *then* $f^{-1}(Z)$ *is a submanifold of* M *whose codimension in* M *is the same as the codimension of* Z *in* N, *i.e.*

$$dim\, M - dim\, f^{-1}(Z) = dim\, N - dim\, Z.$$

Moreover, the normal bundle of Z *in* N *pulls back to the normal bundle of* $f^{-1}(Z)$ *in* M, *i.e.* $\nu f^{-1}(Z) = f^*(\nu Z)$.

Proof:

It suffices to prove the first part of the theorem locally. By the Local Immersion Theorem (Theorem 5.2) we can choose a chart on N such that the pair (N, Z) is represented by $(U \times V, U \times 0)$ where $(U \times V) \subseteq \mathbb{R}^z \times \mathbb{R}^{n-z}$ is an open neighborhood of $(0, 0)$. The map $f : f^{-1}(U \times V) \to U \times V$ is transverse to $U \times 0$ if and only if $g : f^{-1}(U \times V) \xrightarrow{f} U \times V \xrightarrow{\pi} V$ has 0 as a regular value. Since $f^{-1}(U \times 0) = g^{-1}(0)$, $f^{-1}(Z)$ is a submanifold of dimension $(n - z) - m$ by Corollary 5.9.

Now let d be the dimension of the kernel of the composite map

$$TM|_W \xrightarrow{df} TN|_Z \xrightarrow{\pi} \nu Z = TN|_Z / TZ$$

where $W = f^{-1}(Z)$. Since $f \pitchfork Z$, this map is surjective, and hence $m - d \geq$ codim Z. That is, $d \leq m -$ codim $Z = $ dim W. But $TW \subseteq \ker(\pi \circ df)$ since $f_W : W \to Z$ implies that $df_W : TW \to TZ$. Hence, $d \geq$ dim W which implies $\ker(\pi \circ df) = TW$. This implies that $f : W \to Z$ induces a bundle map

$$TM|_W / TW = \nu W \to \nu Z = TN_Z / TZ$$

which is an isomorphism on each fiber. Thus, the pullback of νZ is isomorphic to νW (see for instance Lemma 3.1 of [104].)

$\square$

Corollary 5.12 *If M and Z are immersed submanifolds of N of dimension m, z, and n respectively and $M \pitchfork Z$, then $M \cap Z$ is an immersed submanifold of N of dimension $m + z - n$*

Proof:
    Apply the previous theorem to the inclusion $i : M \to N$ and note that

$$m - \dim (M \cap Z) = n - z.$$

$\square$

5.3   Stability

Definition 5.13 *Let $f_0, f_1 : M \to N$ be smooth maps between smooth manifolds M and N. The maps f_0 and f_1 are said to be **smoothly homotopic** if and only if there exists a smooth map $H : M \times [0, 1] \to N$ such that*

$$\begin{aligned} H(x, 0) &= f_0(x) \\ H(x, 1) &= f_1(x) \end{aligned}$$

*for all $x \in X$. The map H is called a **smooth homotopy** from f_0 to f_1.*

Remark 5.14 *If $f_0 : M \to N$ and $f_1 : M \to N$ are smooth maps that are homotopic (as continuous maps), then they are smoothly homotopic (see for instance Corollary III.2.6 of [91]). Also, if $f_0 : M \to N$ and $f_1 : M \to N$ are continuous maps that are sufficiently close pointwise, then they are homotopic (see for instance Theorem III.2.5 of [91]).*

Definition 5.15 *We will call a property of a class of smooth maps $f : M \to N$ **locally stable** provided that for every $x \in M$ there is a neighborhood $U \subseteq M$ of x such that whenever $f|_U : U \to N$ possesses the property and $H : U \times [0, 1] \to N$ is a smooth homotopy of $f|_U$, then for some $\varepsilon > 0$ each $f_t = H(\cdot, t) : U \to N$ with $t < \varepsilon$ also possesses the property. We will call the property **globally stable** if the above condition holds for $U = M$.*

Theorem 5.16 (Stability Theorem) *If M and N are smooth manifolds, then following classes of smooth maps $f : M \to N$ are locally stable:*

1. *immersions*

2. *submersions*

3. *local diffeomorphisms*

4. *maps transverse to a specified closed submanifold* $Z \subseteq N$.

If M is compact, then the preceeding classes are globally stable.

Proof:
  Assume $f : M \to N$ is an immersion. By choosing local coordinates we may assume that $f : U \to V$ where U and V are open sets with $x \in U \subseteq \mathbb{R}^m$ and $V \subseteq \mathbb{R}^n$. Let $H : U \times [0, 1] \to V$ be a smooth homotopy of $f|_U$, and let $f_t(x) = H(x, t)$. Since $f_0 : U \to V$ is an immersion

$$(df_0)_x : T_x U \to T_{f_0(x)} V$$

is injective for all $x \in U$. We want to show that there is an $\varepsilon > 0$ such that

$$(df_t)_x : T_x U \to T_{f_t(x)} V$$

is injective for all $x \in U$ and for all $t < \varepsilon$.
  Since df_x is injective, the matrix

$$\left(\frac{\partial (f_0)_i}{\partial x_j}(x) \right)$$

contains an $m \times m$ minor

$$\left(\frac{\partial (f_0)_i}{\partial x_j}(x) \right)_I$$

with non-zero determinant. Since each partial derivative $\frac{\partial (f_t)_i}{\partial x_j}(x)$ is continuous in both x and t and the determinant function is continuous, there exists an $\varepsilon > 0$ and an open set $\tilde{U} \subseteq U$ such that the determinant of the $m \times m$ minor

$$\left(\frac{\partial (f_t)_i}{\partial x_j}(x) \right)_I$$

is non-zero for all $t < \varepsilon$ and for all $x \in \tilde{U}$. That is, $(df_t)_x$ is injective for all $x \in \tilde{U}$ and for all $t < \varepsilon$. Thus, the class of immersions is locally stable.
  Now suppose that $H : M \times [0, 1] \to N$ is a smooth homotopy of an immersion $f : M \to N$. The above argument shows that for every point $x \in M$ there is an open neighborhood $U \subseteq M$ of x and an $\varepsilon > 0$ such that $H_t|_U$ is an immersion for all $t < \varepsilon$. If M is compact, then every open cover has a finite subcover, and we can take the smallest $\varepsilon > 0$ for the finite subcover. Thus, the class of immersions is globally stable when M is compact.
  For the case of a submersion repeat the above argument considering an $n \times n$ minor. Since a local diffeomorphism is just an immersion where $m = n$, 3)

follows from 1). For 4) recall that locally $f : f^{-1}(U \times V) \to U \times V$ is transverse to $U \times 0$ if and only if

$$g : f^{-1}(U \times V) \xrightarrow{f} U \times V \xrightarrow{\pi} V$$

restricted to $g^{-1}(0)$ is a submersion.

□

5.4   General position

In this section we prove two theorems on general position following [30] and [91].

Theorem 5.17 (Homotopy Transversality Theorem for Smooth Maps)
Let $f : M \to N$ and $g : Z \to N$ be smooth maps where M, N, and Z are smooth manifolds. Then there is an arbitrarily small smooth homotopy g_t of g such that $g_0 = g$ and $g_1 \pitchfork f$.

Definition 5.18 *A smooth injective immersion $f : M \to N$ between smooth manifolds M and N is called an **embedding** if and only if f is a homeomorphism onto its image.*

Theorem 5.19 (Homotopy Transversality Theorem for Embeddings)
Let M, N, and Z be smooth manifolds, and assume that Z is compact. Let $f : M \to N$ be smooth and let $g : Z \to N$ be a smooth embedding. Then there is an arbitrarily small smooth homotopy g_t of g to a smooth embedding $g_1 : Z \to N$ such that $g_0 = g$ and $g_1 \pitchfork f$. Moreover, the smooth homotopy can be chosen such that $g_t : Z \to N$ an embedding for all t, i.e. g_0 is isotopic to g_1.

To prove these theorems we start with the following lemma.

Lemma 5.20 *If ξ is a smooth vector bundle over a smooth manifold N, then there exists a vector bundle a over N such that $\xi \oplus a$ is trivial.*

Proof:
Identify N with the zero section of the total space $E(\xi)$ and embed some neighborhood $U \subseteq E(\xi)$ of N in $\mathbb{R}^k$ for some $k < \infty$ (see for instance Theorem II.10.8 of [30]). Then

$$\tau_N \oplus \nu_{N,\mathbb{R}^k}$$

is a trivial bundle over N where $\nu_{N,\mathbb{R}^k}$ is the normal bundle of N in $\mathbb{R}^k$ and τ_N is the tangent bundle of N. Since

$$\nu_{N,U} \oplus \nu_{U,\mathbb{R}^k}\big|_N \approx \nu_{N,\mathbb{R}^k}$$

and ξ is isomorphic to $\nu_{N,U}$ we see that $\xi \oplus a$ is trivial where $a = \nu_{U,\mathbb{R}^k}\big|_N \oplus \tau_N$.

$\square$

Proposition 5.21 *Suppose ξ is a smooth vector bundle over N and $f : M \to N$ and $\tilde{g} : Z \to E(\xi)$ are smooth maps where M, N, and Z are smooth manifolds. Then we have the following commutative diagram*

$$
\begin{array}{ccc}
E(f^*(\xi)) & \xrightarrow{\tilde{f}} & E(\xi) \xleftarrow{\tilde{g}} Z \\
\downarrow{\scriptstyle \pi^*} & & \downarrow{\scriptstyle \pi} \searrow{\scriptstyle g} \\
M & \xrightarrow{f} & N
\end{array}
$$

and $\tilde{f} \pitchfork \tilde{g}$ implies that $f \pitchfork g$.

Proof:

Let $\tilde{M} = E(f^*(\xi))$, and suppose that $f(x) = g(z)$. Then there exists an $\tilde{x} \in E(f^*(\xi))$ such that $\pi^*(\tilde{x}) = x$ and $\tilde{f}(\tilde{x}) = \tilde{g}(z)$, i.e. $\tilde{x} = (x, \tilde{g}(z))$. Since $\tilde{f} \pitchfork \tilde{g}$ we have

$$d\tilde{f}(T_{\tilde{x}}\tilde{M}) + d\tilde{g}(T_z Z) = T_{\tilde{f}(\tilde{x})}E(\xi),$$

and since both $d\pi$ and $d\pi^*$ are surjective,

$$
\begin{aligned}
T_{f(x)}N &= d\pi(T_{\tilde{f}(\tilde{x})}E(\xi)) \\
&= d\pi(d\tilde{f}(T_{\tilde{x}}\tilde{M})) + d\pi(d\tilde{g}(T_z Z)) \\
&= df(T_x M) + dg(T_z Z).
\end{aligned}
$$

Hence, $f \pitchfork g$.

$\square$

Theorem 5.22 *Let ξ be a smooth vector bundle over a smooth manifold N. Let M be a smooth manifold and $f : M \to E(\xi)$ a smooth map. Then there exists a smooth cross section $s : N \to \xi$ as close to the zero section as desired such that $f \pitchfork s$.*

Proof:

By Lemma 5.20 there exists a smooth vector bundle over N such that $\xi \oplus a$ is trivial. Consider the commutative diagram

$$
\begin{array}{ccc}
E(f^*(\xi \oplus a)) & \xrightarrow{\tilde{f}} & E(\xi \oplus a) \xrightarrow[\approx]{\phi} N \times \mathbb{R}^k \xrightarrow{p} \mathbb{R}^k \\
\downarrow{\scriptstyle \pi *} & & \downarrow{\scriptstyle \pi} \\
M & \xrightarrow{f} & E(\xi)
\end{array}
$$

where $\phi : E(\xi \oplus a) \to N \times \mathbb{R}^k$ is a diffeomorphism and $p : N \times \mathbb{R}^k \to \mathbb{R}^k$ is projection onto the second component.

By Sard's Theorem (see for instance Theorem II.6.2 of [30]) we can find a regular value $q \in \mathbb{R}^k$ of $p \circ \phi \circ \tilde{f}$ as close to zero as desired. If $\tilde{x} \in E(f^*(\xi \oplus a))$ satisfies $p(\phi(\tilde{f}(\tilde{x}))) = q$, then $d(p \circ \phi \circ \tilde{f})$ maps the tangent space at $\tilde{x}$ onto $T_q \mathbb{R}^k$. Thus, $d\tilde{f}$ must span the complement of the tangent space to $\phi^{-1}(N \times \{q\})$ at $(\tilde{f}(\tilde{x}), q)$, i.e. $\tilde{f}$ is transverse to the section $\tilde{s} : N \to E(\xi \oplus a)$ given in terms of the trivialization by $\tilde{s}(y) = (y, q)$. If we define $s : N \to E(\xi)$ by $s(y) = \pi(\tilde{s}(y))$, then the following diagram commutes,

$$
\begin{array}{ccccc}
E(f^*(\xi \oplus a)) & \xrightarrow{\tilde{f}} & E(\xi \oplus a) & \xleftarrow{\tilde{s}} & N \\
\downarrow{\pi*} & & \downarrow{\pi} & \nearrow{s} & \\
M & \xrightarrow{f} & E(\xi) & &
\end{array}
$$

and $f \pitchfork s$ by Proposition 5.21.

□

Proof of Theorem 5.19:

Let ν be the normal bundle of $g(Z) \subseteq N$, and let $E(\nu) \hookrightarrow N$ be a tubular neighborhood (see for instance Section II.11 of [30] or Section III.2 of [91]). $M' = f^{-1}(E(\nu))$ is an open submanifold of M, and we can apply Theorem 5.22 to the following diagram.

$$
\begin{array}{ccc}
M' & \xrightarrow{f|_{M'}} & E(\nu) \\
& & \downarrow \\
& & g(Z)
\end{array}
$$

That is, there exists a section $s : g(Z) \to E(\nu)$ as close to the zero section as desired with $f|_{M'} \pitchfork s$. Clearly s is smoothly homotopic to the zero section via the smooth homotopy $H_t(x) = ts(x)$, and by Theorem 5.16, $g_t : Z \xrightarrow{g} g(Z) \xrightarrow{H_t} E(\nu) \subseteq N$ will be an embedding for all $0 \le t \le 1$ as long as s is chosen sufficiently close to the zero section.

□

Proof of Theorem 5.17:

Consider the commutative diagram

$$\begin{array}{ccccc} M \times Z & \xrightarrow{\tilde{f}} & N \times Z & \xleftarrow{\tilde{g}} & Z \\ \downarrow & & \downarrow{\scriptstyle \pi} & \swarrow{\scriptstyle g} & \\ M & \xrightarrow{f} & N & & \end{array}$$

where $\tilde{g}(z) = (g(z), z)$, $\tilde{f}(x, z) = (f(x), z)$, and the vertical maps are projections. $\tilde{g}(Z)$ is the graph of g and hence a closed embedded submanifold of $N \times Z$. Thus, $\tilde{g}(Z)$ has a tubular neighborhood $E(\nu)$ (see for instance Corollary III.2.3 of [91]), and by Theorem 5.22 there exists a smooth section $s_1 : \tilde{g}(Z) \to E(\nu) \subseteq N \times Z$ of the tubular neighborhood $E(\nu)$ as close to the zero section as desired with $\tilde{f} \pitchfork s_1$. If H_t is a smooth homotopy from s_0 to s_1, then $H_t \circ \tilde{g}$ is a smooth homotopy from $\tilde{g}$ to $H_1 \circ \tilde{g}$ with $(H_1 \circ \tilde{g}) \pitchfork \tilde{f}$. Hence by Proposition 5.21, $g_t = \pi \circ H_t \circ \tilde{g}$ is a homotopy from g to a map $g_1 = \pi \circ H_1 \circ \tilde{g}$ which satisfies $g_1 \pitchfork f$.

$\square$

5.5 Stability and density for Morse functions

In this section we show that if M is a finite dimensional smooth manifold, then the space of Morse functions on M is a dense locally stable subspace of the space of all smooth functions on M. Our proof follows [91]. We also show that the space of Morse functions on M is open and dense in the C^r topology for any $r \geq 2$ following [102].

Lemma 5.23 *Let* $f : M \to \mathbb{R}$ *be a smooth function on a finite dimensional smooth manifold* M. *Then* $p \in M$ *is a non-degenerate critical point of* f *if and only if* df *is transverse to the zero section of* T^*M *at* p.

Proof:

Let $p \in M$ be a critical point of f, and let $\phi : U \to \mathbb{R}^m$ be a coordinate neighborhood around p such that $\phi(p) = 0 \in \mathbb{R}^m$. Recall from Definition 3.1 that p is a non-degenerate critical point of f if and only if the $m \times m$ matrix

$$M_p(f) = \left(\frac{\partial^2 (f \circ \phi^{-1})}{\partial x_i \partial x_j} \phi(p) \right)$$

has rank m.

The coordinate chart ϕ gives a trivialization of $T^*M|_U \approx U \times \mathbb{R}^m$, and hence a projection $\pi : T^*M|_U \to \mathbb{R}^m$. The section df is transverse to the zero section of T^*M at p if and only if the differential of $\pi \circ df$ is surjective at p. In local coordinates the matrix for the differential of $\pi \circ df$ at p is exactly $M_p(f)$, and hence the differential of $\pi \circ df$ is surjective at p if and only if $M_p(f)$ has rank m, i.e. p is non-degenerate.

$\square$

Corollary 5.24 *The class of Morse functions on a finite dimensional smooth manifold M is locally stable. If M is compact, then the class of Morse functions on M is globally stable.*

Proof:

Lemma 5.23 shows that a smooth function $f : M \to \mathbb{R}$ is a Morse function if and only if df is transverse to the zero section of T^*M, and Theorem 5.16 implies that the class of maps transverse to the zero section in T^*M is locally stable in general and globally stable when M is compact.

$\square$

The following corollary is weaker than Lemma 3.2, but the proof is a nice application of Theorem 5.11 and Lemma 5.23.

Corollary 5.25 *The critical points of a Morse function on a finite dimensional compact smooth manifold are isolated.*

Proof:

Since $f : M \to \mathbb{R}$ is a Morse function, $df : M \to T^*M$ is transverse to the zero section $Z \subseteq T^*M$. Hence, $df^{-1}(Z)$ is a submanifold of M with

$$\dim M - \dim df^{-1}(Z) = \dim T^*M - \dim Z$$

by Theorem 5.11, and the dimension of $df^{-1}(Z)$ is zero. Thus, the critical points of f are isolated since M is compact.

$\square$

The reader should compare the following with Lemma 3.9.

Proposition 5.26 *Let M be a submanifold of $\mathbb{R}^k$ for some k, and let $f : M \to \mathbb{R}$ be a smooth function on M. Then there is a dense set of linear functions $L : \mathbb{R}^k \to \mathbb{R}$ such that $f - L : M \to \mathbb{R}$ is a Morse function.*

Proof:

Consider the following commutative diagram.

$$
\begin{array}{ccccc}
E & \xrightarrow{\;g\;} & T^*\mathbb{R}^k\big|_M & \xleftarrow{dL|_M} & M \\
\downarrow & & \downarrow{\scriptstyle \pi} & \swarrow{\scriptstyle d(L|_M)} & \\
M & \xrightarrow{\;df\;} & T^*M & &
\end{array}
$$

where $E = df^*(T^*\mathbb{R}^k|_M)$, g is defined by commutativity, and $L : \mathbb{R}^k \to \mathbb{R}$ is a linear map chosen as follows. The bundle $T^*\mathbb{R}^k|_M \approx M \times \mathbb{R}^k$ is trivial, and by Sard's Theorem (see for instance Theorem II.6.2 of [30]) the composite map

$$E \xrightarrow{\ g\ } T^*\mathbb{R}^k|_M \xrightarrow[\approx]{\ \phi\ } M \times \mathbb{R}^k \xrightarrow{\ p\ } \mathbb{R}^k$$

has a dense set of regular values $q \in \mathbb{R}^k$. If $x \in E$ satisfies $p(\phi(g(x))) = q$, then the image of $d(\phi \circ g)$ must span the complement of the tangent space to $M \times \{q\}$ at $(\phi(g(x)), q)$. That is, $d(\phi \circ g)$ is transverse to the constant section $M \times \{q\} \subseteq M \times \mathbb{R}^k$ over M. Every constant $q \in \mathbb{R}^k$ is the differential of some linear map $L : \mathbb{R}^k \to \mathbb{R}$, e.g. $L(x) = x \cdot q$, and thus there is a dense set of linear functions $L : \mathbb{R}^k \to \mathbb{R}$ such that $dL|_M \pitchfork g$. Since $\pi \circ dL|_M = d(L|_M)$, we can apply Proposition 5.21 to conclude that $df \pitchfork d(L|_M)$, i.e. $d(f - L|_M)$ is transverse to the zero section.

$\square$

The following theorem is due to M. Morse and first appeared in the 1932 edition of [106].

Theorem 5.27 *Let M be a finite dimensional smooth manifold. Given any smooth function $f : M \to \mathbb{R}$ and any $\varepsilon > 0$, there is a Morse function $g : M \to \mathbb{R}$ such that $\sup\{|f(x) - g(x)| \mid x \in M\} < \varepsilon$.*

Proof:

Embed M as a submanifold of the unit ball in $\mathbb{R}^k$, and choose a linear function $L : \mathbb{R}^k \to \mathbb{R}$ satisfying the conclusion of Proposition 5.26 with $\|L\| < \varepsilon$. Then $g = f - L$.

$\square$

The preceeding theorem shows that the space of Morse functions on a finite dimensional smooth manifold M is a dense subset of the space of all smooth functions on M in the uniform topology. If M is compact, then the uniform topology is the same as the topology of compact convergence, and since $\mathbb{R}$ is a metric space, the topology of compact convergence is the same as the compact open topology. For more details on the topology of mapping spaces see Chapter 7 of [108].

We will now prove a theorem concerning the abundance of Morse functions on finite dimensional compact smooth manifolds that is stronger than Corollary 5.24 and Theorem 5.27. Before stating the theorem we first define the C^r topology. The basic idea behind the C^r topology is this: two functions f and g should be considered "close" if and only if all their partial derivatives up to and including order r are "close" pointwise.

Definition 5.28 *For $0 \leq r < \infty$, let $C^r(M, N)$ denote the space of C^r maps between two C^r manifolds M and N. Let $f \in C^r(M, N)$, and let (ϕ, U) and (ψ, V) be charts on M and N respectively. Let $K \subseteq U$ be a compact set such that $f(K) \subseteq V$, and let $0 < \varepsilon \leq \infty$. Define the subbasis neighborhood*

$$\mathcal{N}^r(f; (\phi, U), (\psi, V), \varepsilon)$$

to be the set of C^r maps $g : M \to N$ such that $g(K) \subseteq V$ and

$$\|D^k(\psi f \phi^{-1})(x) - D^k(\psi g \phi^{-1})(x)\| < \varepsilon$$

for all $x \in \varphi(K)$ and $k = 0, \ldots, r$, i.e. the local representations of f and g, together with their first r derivatives, are within ε of each other at each point of K. The C^r topology on $C^r(M, N)$ is defined to be the topology generated by the subbasis elements $\mathcal{N}^r(f; (\phi, U), (\psi, V), \varepsilon)$. The C^∞ topology on $C^\infty(M, N)$ is defined to be the union of the topologies induced by the inclusion maps $C^\infty(M, N) \subset C^r(M, N)$ for all $0 \leq r < \infty$.

Remark 5.29 For more details concerning C^r topologies see Chapter 2 of [77]. In particular, it is shown in Theorem 2.2.6 that when M is compact and $1 \leq s \leq \infty$, $C^s(M, N)$ is dense in $C^r(M, N)$ for any $0 \leq r < s$. Also, in Theorem 2.2.10 it is shown that any for any $1 \leq r < \infty$, every C^r manifold is C^r diffeomorphic to a C^∞ manifold, and in Theorem 2.4.4 it is shown that for every $0 \leq r \leq \infty$, the space $C^r(M, N)$ is a complete metric space.

Definition 5.30 *We will call a C^r function $f : M \to \mathbb{R}$ a C^r **Morse function** if and only if all the critical points of f are non-degenerate.*

Theorem 5.31 *Let M be a finite dimensional compact smooth manifold. The space of all C^r Morse functions on M is an open dense subspace of $C^r(M, \mathbb{R})$ for any $2 \leq r \leq \infty$ where $C^r(M, \mathbb{R})$ denotes the space of all C^r functions on M with the C^r topology.*

Our proof of this theorem will follow [102]. We will need the following easy result.

Lemma 5.32 *Let K be a compact subset of an open set $U \subseteq \mathbb{R}^m$, and let $f : U \to \mathbb{R}$ be a C^2 function with no degenerate critical points in K. There exists a positive number $\delta > 0$ such that if $g : U \to \mathbb{R}$ is a C^2 function that satisfies*

$$\left| \frac{\partial f}{\partial x_i} - \frac{\partial g}{\partial x_i} \right| < \delta \quad and \tag{5.1}$$

$$\left| \frac{\partial^2 f}{\partial x_i \partial x_j} - \frac{\partial^2 g}{\partial x_i \partial x_j} \right| < \delta \tag{5.2}$$

on K for all $i, j = 1, \ldots, m$, then g also has no degenerate critical points in K.

Proof:

A function $h : U \to \mathbb{R}$ has no non-degenerate critical points in K if and only if

$$\rho(h) = |dh| + \left| \det \left(\frac{\partial^2 h}{\partial x_i \partial x_j} \right) \right| > 0$$

on K where

$$|dh|^2 = \sum_{j=1}^{m} \left| \frac{\partial h}{\partial x_j} \right|^2.$$

By assumption, $\rho(f) > 0$ on K. Let $\mu > 0$ satisfy $\rho_{\min} > \mu > 0$ where $\rho_{\min}$ denotes the minimum of $\rho(f)$ on K. Choose $\delta > 0$ small enough so that (5.1) implies that $| |df| - |dg| | < \mu/2$ and (5.2) implies that

$$\left| \left| \det \left(\frac{\partial^2 f}{\partial x_i \partial x_j} \right) \right| - \left| \det \left(\frac{\partial^2 g}{\partial x_i \partial x_j} \right) \right| \right| < \mu/2,$$

i.e. $|dg| > |df| - \mu/2$ and

$$\left| \det \left(\frac{\partial^2 g}{\partial x_i \partial x_j} \right) \right| > \left| \det \left(\frac{\partial^2 f}{\partial x_i \partial x_j} \right) \right| - \mu/2.$$

Adding these inequalities we get

$$\rho(g) > \rho(f) - \mu > 0.$$

$\square$

Proof of Theorem 5.31:

Let $(U_j, \phi_j)_{j=1,\ldots,k}$ be a finite atlas on M and $\{C_j\}_{j=1,\ldots,k}$ a family of compact sets $C_j \subseteq U_j$ that cover M. We will say that a function f on M is "good" on a set $S \subset M$ if and only if it has no degenerate critical points in S.

Let $r \geq 2$, and let $f : M \to \mathbb{R}$ be a C^r Morse function. By Lemma 5.32 there is a neighborhood $\mathcal{N}_j$ of f in $C^r(M, \mathbb{R})$ such that every function in $\mathcal{N}_j$ is good on C_j for every $j = 1, \ldots, k$. Hence, in the neighborhood $\mathcal{N} = \mathcal{N}_1 \cap \cdots \cap \mathcal{N}_k$ of f every function $g \in \mathcal{N}$ is good on $C_1 \cup \cdots \cup C_k = M$, i.e. g is a Morse function. This shows that the space of Morse functions is open in the space $C^r(M, \mathbb{R})$ endowed with the C^r topology for any $2 \leq r \leq \infty$.

Now let $2 \leq r < \infty$. To show that the set of Morse functions is dense in $C^r(M, N)$ we start with a neighborhood $\mathcal{N}$ of $f \in C^r(M, \mathbb{R})$ and improve f in k steps to arrive at a Morse function g that lies in a neighborhood $\mathcal{N}_k \subseteq \mathcal{N}$ of f.

Let $\lambda : M \to [0,1]$ be a smooth bump function equal to 1 on C_1 with support $K \subset U_1$, i.e. K is a compact set and λ vanishes on $M - K$. For almost every linear map $L : \mathbb{R}^m \to \mathbb{R}$, the function

$$f_1(x) = f(x) + \lambda(x)L(\phi_1(x))$$

is good on C_1 by Proposition 5.26. Note that

$$f_1(\phi_1^{-1}(y)) - f(\phi_1^{-1}(y)) = \lambda(\phi_1^{-1}(y))\left(\sum_{j=1}^m a_j y_j\right)$$

on the compact set $\phi_1(K)$, and $f_1(x) = f(x)$ on $M - K$.

Since $r < \infty$, there are only finite many partial derivatives of the function $\lambda(\phi_1^{-1}(y))$, and since $\phi_1(K)$ is compact, there exists some constant $A_r < \infty$ which is a uniform bound on the set $\phi_1(K)$ for the magnitude of all the partial derivatives of $\lambda(\phi_1^{-1}(y))$ up to and including order r. The partial derivatives of

$$\sum_{j=1}^m a_j \lambda(\phi_1^{-1}(y))y_j$$

are polynomial functions of the partial derivatives of $\lambda(\phi_1^{-1}(y))$, and hence, given any $\varepsilon > 0$ there is a $\delta > 0$ such that if $(a_1, \ldots, a_m) \in \mathbb{R}^m$ satisfies $\|(a_1, \ldots, a_m)\| < \delta$, then all the partial derivatives up to and including order r of $f_1(\phi_1^{-1}(y)) - f(\phi_1^{-1}(y))$ have magnitude less than ε on the compact set $\phi_1(K)$.

Therefore, f_1 belongs to $\mathcal{N}$ if $\varepsilon > 0$ is small enough, and we can find $f \in \mathcal{N}$ which is good on C_1. By Lemma 5.32 there is a neighborhood $\mathcal{N}_1$ of f_1 with $\mathcal{N}_1 \subseteq \mathcal{N}$ such that any function in $\mathcal{N}_1$ is still good on C_1. Repeating the above argument for the neighborhood $\mathcal{N}_1$, we get a function f_2 in $\mathcal{N}_1$ which is good on C_2 and a neighborhood $\mathcal{N}_2 \subseteq \mathcal{N}_1$ of f_2 such that any function in $\mathcal{N}_2$ is good on C_2. The function f_2 is automatically good on C_1 since $f_2 \in \mathcal{N}_1$. Continuing, we finally get

$$f_k \in \mathcal{N}_k \subseteq \mathcal{N}_{k-1} \subseteq \cdots \subseteq \mathcal{N}_1 \subseteq \mathcal{N}$$

which is good on $C_1 \cup \cdots \cup C_k = M$.

This shows that the space of C^r Morse functions is dense in $C^r(M, \mathbb{R})$ for any $2 \leq r < \infty$. Since the topology of $C^\infty(M, \mathbb{R})$ is generated by open sets in $C^r(M, \mathbb{R})$ restricted to $C^\infty(M, \mathbb{R})$, this also shows that the space of C^∞ Morse functions is dense in $C^\infty(M, \mathbb{R})$ in the C^∞ topology.

$\square$

Remark 5.33 Theorem 5.31 is still true even if M is not compact, but the proof is more difficult. See for instance Theorem 6.1.2 of [77].

5.6   Orientations and intersection numbers

Orientations

Let V be a real vector space of finite dimension $m > 0$. On the set of ordered bases of V define a relation $\mathcal{R}$ by

$$v\mathcal{R}w$$

if and only if the change of basis matrix $C = (c_{ij})$ from $v = (v_1, \ldots, v_m)$ to $w = (w_1, \ldots, w_m)$, i.e. $w_i = \sum_{j=1}^{m} c_{ij}v_j$, has positive determinant. The relation $\mathcal{R}$ is an equivalence relation, and there are exactly two equivalence classes.

Definition 5.34 *An **orientation** of a real vector space V is a choice of one of the equivalence classes θ of the relation $\mathcal{R}$, which we call the **positive orientation**. The couple (V, θ) is called an **oriented vector space**. If $\dim V = 0$, then an orientation is an assignment of $+1$ or -1 to the point $V = \{0\}$. If (V, θ) and (V', θ') are two oriented vector spaces of the same positive dimension and $L : V \to V'$ a linear isomorphism, then L is said to be **orientation preserving** if and only if for all $v = (v_1, \ldots, v_m) \in \theta$ we have $(L(v_1), \ldots, L(v_m)) \in \theta'$. A linear isomorphism that is not orientation preserving is said to be **orientation reversing**.*

Note that the matrix of a linear isomorphism $L : V \to V'$ with respect to the bases v and v' in θ and θ' respectively has positive determinant if and only if L is orientation preserving.

Example 5.35 The standard orientation on the vector space $\mathbb{R}^m$ is the orientation θ with $v = (e_1, \ldots, e_m) \in \theta$ where $e_1, \ldots, e_m$ is the standard basis.

Now let M be a differentiable manifold with boundary of dimension m. Around every point $x \in \text{int } M$ there exists a coordinate chart $\phi : U \to \mathbb{R}^m$, and around every point $x \in \partial M$ there exists a coordinate chart $\phi : U \to \mathbb{R}^m_+ = \{(x_1, \ldots, x_m) | x_m \geq 0\}$.

Definition 5.36 *An **orientation** of a differentiable manifold with boundary M of dimension m is a choice of orientation θ_x for each tangent space $T_x M$ that satisfies the following compatability requirement: Around every point in M there is a coordinate chart $\phi : U \to \mathbb{R}^m$ (or $\phi : U \to \mathbb{R}^m_+$) which is **orientation preserving**, i.e. for every point $x \in U$ the linear isomorphism*

$$d\phi_x : T_x M \to \mathbb{R}^m$$

*is orientation preserving where $\mathbb{R}^m$ is given its standard orientation. A differentiable manifold M that possesses an orientation is called **orientable**.*

Note: If M is a differentiable manifold with boundary of dimension m, then $T_x M \approx \mathbb{R}^m$ for all $x \in M$, including $x \in \partial M$.

Example 5.37 The standard orientation on $\mathbb{R}^2$ descends to the torus $T^2 = \mathbb{R}^2/\mathbb{Z}^2$, and hence, T^2 is orientable. However, the Klein bottle K^2, which can be viewed as the space obtained by identifying opposite sides of the unit square in $\mathbb{R}^2$, but with a "flip" along one of the sides, is not orientable.

For a proof of the following theorem see Proposition 4.2 of [148].

Theorem 5.38 *Let M be a differentiable manifold with boundary of dimension m. Then the following are equivalent.*

1. *M is orientable.*

2. *There is a collection $\Phi = \{(U, \phi)\}$ of coordinate systems on M such that*

$$M = \bigcup_{(U,\phi) \in \Phi} U \text{ and } \det\left(d_y(\phi_j \circ \phi_i^{-1})\right) > 0 \text{ on } \phi_i(U_i) \cap \phi_j(U_j)$$

whenever (U_i, ϕ_i) and (U_j, ϕ_j) belong to Φ.

3. *There is a no-where vanishing m-form on M.*

Remark 5.39 If $V = W \oplus W'$, then orientations θ_W and θ'_W on W and W' respectively determine a unique orientation θ on V as follows: If $(w_1, \ldots, w_l) \in \theta_W$ and $w' = (w'_1, \ldots, w'_{m-l}) \in \theta'_W$, then

$$v = (w_1, \ldots, w_l, w'_1, \ldots, w'_{m-l}) \in \theta.$$

Hence, if M is an oriented differentiable manifold with boundary and N is an oriented differentiable manifold, then there is an induced orientation on the manifold with boundary $M \times N$.

Similarly, if (V, θ) is an oriented vector space and (W, θ_W) an oriented subspace, then any complementary subspace W', i.e. a subspace of V such that $V = W \oplus W'$, can be given an orientation θ'_W such that if $w = (w_1, \ldots, w_l) \in \theta_W$ and $w' = (w'_1, \ldots, w'_{m-l}) \in \theta'_W$, then

$$v = (w_1, \ldots, w_l, w'_1, \ldots, w'_{m-l}) \in \theta.$$

The orientation θ'_W is uniquely determined by the orientations θ and θ_W. Thus, if $M \times N$ and M are oriented manifolds with boundary, then there is an induced orientation on N.

If M is a finite dimensional smooth manifold with boundary, then an embedding

$$f : [0, 1) \times \partial M \to M$$

onto a neighborhood of ∂M in M such that $f(x) = x$ for all $x \in \partial M$ is called a **collar** on ∂M, and ∂M is said to be **collared** in M. The **Collaring Theorem** says that the boundary of any smooth manifold with boundary M is collared (see for instance Theorem 6.1 of [77] or Theorem I.7.3 of [91]). For a similar result that applies to topological manifolds and general metric spaces see [32].

Definition 5.40 *Let M be a finite dimensional smooth manifold with non-empty boundary ∂M. The **induced orientation** on ∂M is defined to be the orientation on ∂M that makes a collar*

$$f : [0, 1) \times \partial M \to M$$

orientation preserving, where $[0, 1)$ is oriented in the positive direction and $[0, 1) \times \partial M$ is oriented as in Remark 5.39.

Remark 5.41 If dim $M = 1$, then ∂M is a collection of points, and these points are assigned $+1$ or -1 depending on whether the maps $[0, 1) \to M$ induced locally by the collar are orientation preserving or orientation reversing. It is easy to show that the induced orientation in the preceeding definition does not depend on the choice of the collar $f : [0, 1) \times \partial M \to M$. One often says that the normal bundle to ∂M in M is oriented using "inward" pointing tangent vectors. For other definitions of the induced orientation on ∂M that do not use a collar see Section 3.2 of [71], Section I.1 of [91], or Section 5 of [103].

Intersection numbers

Consider two immersed submanifolds M and Z of a smooth manifold N which meet transversally, i.e. Z is transverse to the inclusion $i : M \to N$. We denote this situation by $M \pitchfork Z$. Since $i^{-1}(Z) = M \cap Z$, this intersection is an immersed submanifold of N of dimension

$$\dim(M \cap Z) = \dim(M) + \dim(Z) - \dim(N)$$

by Corollary 5.12. For the intersection $M \cap Z$ to be nonempty we must have $\dim(M \cap Z) \geq 0$, i.e.

$$\dim(M) + \dim(Z) \geq \dim(N).$$

If $\dim(M) + \dim(Z) < \dim(N)$ and $M \cap Z \neq \emptyset$, then M and Z do not meet transversally. By the Homotopy Transversality Theorem (Theorem 5.17), there exists an arbitrarily small smooth homotopy $g_t : M \to N$ from the inclusion $i : M \to N$, i.e. $g_0 = i$, to a map $g_1 : M \to N$ which is transverse to Z. Therefore, $g_1^{-1}(Z)$ is a submanifold of M of dimension

$$\dim(g_1^{-1}(Z)) = \dim(M) + \dim(Z) - \dim(N) < 0.$$

Hence, $g_1^{-1}(Z)$ is empty, which implies that $g_1(M) \cap Z = \emptyset$. Thus, we see that by perturbing M we can make it disjoint from Z. Therefore, there will be no "intersection theory" for submanifolds M and Z of N that satisfy $\dim(M) + \dim(Z) < \dim(N)$.

When $\dim(M) + \dim(Z) = \dim(N)$ we have $\dim(M \cap Z) = 0$, which means that $M \cap Z$ is a collection of points. If M and Z are closed submanifolds of N and at least one of them is compact, then $M \cap Z$ is a finite collection of points. Likewise, if $f : M \to N$ is a smooth map, Z is a submanifold of N, $\dim(M) + \dim(Z) = \dim(N)$, and $f \pitchfork Z$, then $f^{-1}(Z)$ is a submanifold of M of dimension zero. That is, $f^{-1}(Z)$ is a collection of points, which is finite provided that Z is closed and M is compact.

We will assume from now on that M, N, and Z are oriented, that Z is a closed submanifold of N, and that M is compact. We assume furthermore that $\dim(M) + \dim(Z) = \dim(N)$ and $f : M \to N$ is a smooth map transverse to Z. Then for each $x \in f^{-1}(Z)$ we have $df_x(T_x M) \oplus T_y Z = T_y N$ where $y = f(x)$. Let v_x, v_y, w_y be positive oriented bases of the vector spaces $T_x M$, $T_y N$, and $T_y Z$ respectively. Then $v_y' = (df_x(v_x), w_y)$ is a new basis of $T_y N$. If the bases v_y and v_y' have the same orientation, i.e. if the transition matrix from one to another has positive determinant, then we assign $+1$ to the point $x \in f^{-1}(Z)$ and we write $\text{sign}(x) = +1$. Otherwise, we assign -1 to the point $x \in f^{-1}(Z)$ and we write $\text{sign}(x) = -1$.

Definition 5.42 *The oriented **intersection number**, $I(f, Z) \in \mathbb{Z}$, is defined to be*

$$I(f, Z) = \sum_{x \in f^{-1}(Z)} \text{sign}(x).$$

Remark 5.43 This sum is finite since $f^{-1}(Z)$ is a closed subset of a compact space and hence compact (see for instance Theorem 3.5.2 of [108]). Observe that if Z is a point, then $f \pitchfork Z$ means that the point is a regular value of f. If the point Z has orientation $+1$, then $v_y' = df_x(v_x)$, and the sign of x is $+1$ or -1 depending on whether df_x is orientation preserving or not. Also, note that if $f^{-1}(Z) = \emptyset$, then $I(f, Z) = 0$.

The following fact is crucial. The proof is fairly straightforward if we apply generalizations of Theorem 5.11 and Theorem 5.17 that apply to manifolds with boundary.

Theorem 5.44 *Let $f, g : M \to N$ be two smooth maps which are both transverse to $Z \subset N$. If f is homotopic to g, then, then $I(f, Z) = I(g, Z)$.*

Proof:

Since f and g are smooth maps that are homotopic as continuous maps, they are also smoothly homotopic (see for instance Corollary III.2.6 of [91]). Let $F : [0, 1] \times M \to N$ be a smooth homotopy from f to g. Since both $f \pitchfork Z$ and $g \pitchfork Z$, the smooth homotopy F can be chosen so that $F \pitchfork Z$. (This follows from a slightly stronger version of Theorem 5.17. See for instance Section 2.3 of [71].)

We have $F \pitchfork Z$ and $\partial F \pitchfork Z$, and hence $F^{-1}(Z)$ is an immersed submanifold of $[0, 1] \times M$ whose dimension satisfies

$$1 + \dim(M) - \dim F^{-1}(Z) = \dim(N) - \dim(Z).$$

(This follows from a version of Theorem 5.11 that applies to manifolds with boundary. The proof is essentially the same as the proof of Theorem 5.11. See for instance Section 2.1 of [71] or Proposition IV.1.4 of [91].) Hence, $\dim F^{-1}(Z) = 1$ since $\dim(M) + \dim(Z) = \dim(N)$.

Now, any compact 1-dimensional manifold is orientable, and the sum of the orientation numbers at the boundary points is zero (see for instance Section 3.2 of [71] or the Appendix of [101]). Therefore,

$$\sum_{x \in \partial F^{-1}(Z)} \text{sign}(x) = 0.$$

Since $\partial F^{-1}(Z) = f^{-1}(Z) \cup -g^{-1}(Z)$, where the orientation on $\partial F^{-1}(Z)$ is chosen appropriately and the negative sign means that the orientation on $g^{-1}(Z)$ is reversed, we have

$$\sum_{x \in f^{-1}(Z)} \text{sign}(x) = \sum_{x \in g^{-1}(Z)} \text{sign}(x).$$

$\square$

Remark 5.45 Remark 5.39 implies that if Z is an oriented immersed submanifold of an oriented smooth manifold N, then there is an induced orientation on the normal bundle of Z in N. Thus, if $f : M \to N$ is transverse to Z, then there is an induced orientation on the normal bundle of $f^{-1}(Z)$ by Theorem 5.11. Moreover, if M is oriented, then Remark 5.39 implies that there is an induced orientation on $f^{-1}(Z)$. Similar results hold when M is an oriented manifold with boundary, and for these induced orientations we have $\partial F^{-1}(Z) = f^{-1}(Z) \cup -g^{-1}(Z)$ in the proof of the preceeding theorem.

We are now in a position to define an intersection number $I(f, Z)$ for any smooth map $f : M \to N$ and any closed submanifold Z, which is no longer assumed to be transverse to f. (However, all the other assumptions are still required.)

Definition 5.46 *Assume that M, N, and Z are oriented smooth manifolds, Z is a closed submanifold of N, M is compact, and $dim(M) + dim(Z) = dim(N)$. For any smooth map $f : M \to N$, the Homotopy Transversality Theorem (Theorem 5.17) implies that there is a smooth map f_1 transverse to Z and homotopic to f, and we define the oriented* **intersection number** $I(f, Z)$ *to be $I(f_1, Z)$. This number is well define because $I(f_1, Z)$ is independent of the choice of f_1 by Theorem 5.44.*

Remark 5.47 Every continuous map $f : M \to N$ between smooth manifolds M and N is homotopic to a smooth map (see for instance Theorem III.2.5 of [91]). Hence, the preceeding definition of the oriented intersection number $I(f, Z)$ also applies to a continuous map $f : M \to N$.

5.7   The Lefschetz Fixed Point Theorem

Let $g : M \to M$ be a smooth map of an m-dimensional compact connected smooth oriented manifold to itself. The **graph** $\Gamma(g) \subset M \times M$ of g is the m-dimensional submanifold

$$\Gamma(g) = \{(x, g(x)) \in M \times M | \ x \in M\}$$

of the $2m$-dimensional oriented manifold $M \times M$. The projection onto the first component $\pi_1 : \Gamma(g) \to M$ is a diffeomorphism, and thus there is an orientation on the manifold $\Gamma(g)$ determined by requiring π_1 to be orientation preserving.

Given two smooth maps $f, g : M \to M$ of a compact connected oriented smooth manifold M, we have the oriented closed submanifold $\Gamma(g) \subset M \times M$ and the smooth map $\Gamma_f : M \to M \times M$ determined by

$$\Gamma_f(x) = (x, f(x)).$$

Hence, the oriented intersection number $I(\Gamma_f, \Gamma(g))$ from Definition 5.46 is defined since $dim(M) + dim(\Gamma(g)) = dim(M \times M)$. In particular, for any smooth map $f : M \to M$ we may consider the intersection number of the graph of f and the identity map $I(\Gamma_f, \Gamma(id))$ of Γ_f.

Definition 5.48 *Let $f : M \to M$ be a smooth map from a compact connected smooth oriented manifold to itself. The intersection number $I(\Gamma_f, \Gamma(id))$ of Γ_f and the graph of the identity map is called the* **Lefschetz number** *of the smooth map $f : M \to M$ and is denoted by $L(f)$.*

The graph of the identity $\Gamma(id) = \{(x, x) | \ x \in M\}$ is called the **diagonal** and is often denoted by $\triangle_M$. If $f, g : M \to M$ are smooth homotopic maps, then Γ_f and Γ_g are also smooth and homotopic. Hence,

$$I(\Gamma_f, \triangle_M) = L(f) = I(\Gamma_g, \triangle_M) = L(g).$$

This shows that the Lefschetz number for a smooth map is a homotopy invariant, and since every continuous map is homotopic to a smooth map (see for instance Theorem III.2.5 of [91]) we can extend the definition of the Lefschetz number to the case where f is only assumed to be continuous.

Definition 5.49 *Given a continuous map $f : M \to M$ from a compact connected smooth oriented manifold to itself, choose a smooth map $f_1 : M \to M$ that is homotopic to f with $\Gamma_{f_1} \pitchfork \triangle_M$ and define $L(f) \equiv L(f_1)$. This number is well define because $L(f_1)$ is independent of the choice of f_1 by Theorem 5.44 and the preceeding remarks.*

A **fixed point** of $f : M \to M$ is a point $x \in M$ such that $f(x) = x$. The set of fixed points of f corresponds bijectively with intersection points of $\Gamma(f)$ and $\triangle_M$. Hence, if $\Gamma(f) \pitchfork \triangle_M$, the Lefschetz number $L(f)$ is a lower bound for the number of fixed points since,

$$L(f) = \sum_{x \in \Gamma(f) \cap \triangle_M} \text{sign}(x).$$

In particular, if $f : M \to M$ does not have a fixed point, then the intersection $\Gamma(f) \cap \triangle_M$ is empty, thus transverse, and $L(f) = 0$. We have obtained the following.

Theorem 5.50 (Smooth Lefschetz Fixed Point Theorem) *Let $f : M \to M$ be a continuous map from a compact connected smooth oriented manifold to itself. If $L(f) \neq 0$, then f has a fixed point.*

The significance of the preceeding theorem becomes apparent when one realizes that the Lefschetz number $L(f)$ is actually a **topological** invariant. In fact, $L(f)$ can be defined for any continuous map $f : X \to X$ from a topological space with bounded homology to itself, and the Lefschetz Fixed Point Theorem holds for any compact **Euclidean neighborhood retract (ENR)**, i.e. a space that is a homeomorphic to a retract of some open set in some $\mathbb{R}^m$.

Definition 5.51 *Let $f : X \to X$ be a continuous map from a topological space with bounded homology to itself. Define the **topological Lefschetz number** of f to be*

$$L(f) = \sum_{j=0}^{m} (-1)^j tr_j(f_*) \in \mathbb{Z}$$

where $H_j(X; \mathbb{Z}) = 0$ for all $j > m$, and $tr_j(f_*)$ is the trace of the induced map in homology in degree j with integer coefficients, i.e. the trace of $f_* : H_j(X; \mathbb{Z}) \to H_j(X; \mathbb{Z})$.

Remark 5.52 It can be shown that the trace of $f_* : H_j(X;\mathbb{Z}) \to H_j(X;\mathbb{Z})$ is the same as the trace of $f_* : H_j(X;\mathbb{Q}) \to H_j(X;\mathbb{Q})$ (see for instance Section IV.23 of [30]). Hence, the integer $\mathrm{tr}_j(f_*)$ can be viewed as the trace of an endomorphism of a vector space.

For a proof of the following theorem we refer the reader to Corollary IV.23.5 of [30].

Theorem 5.53 (Topological Lefschetz Fixed Point Theorem) *Let X be a compact ENR (Euclidean Neighborhood Retract) and $f : X \to X$ a continuous map. If $L(f) \neq 0$, then f has a fixed point.*

We now have two different definitions for the Lefschetz number $L(f)$ when $f : M \to M$ is a continuous map from a compact connected smooth oriented manifold to itself. Both definitions yield the same integer $L(f)$, but a proof of this fact would require discussing products and duality in cohomology. Since that would take us too far afield, we refer the reader to Section VI.12 of [30] for a proof of the following theorem.

Theorem 5.54 *Let $f : M \to M$ be a continuous map from a compact connected smooth oriented manifold to itself. Then the topological Lefschetz number agrees with the oriented intersection Lefschetz number, i.e.*

$$\sum_j (-1)^j \mathrm{tr}_j(f_*) = I(\Gamma_{f_1}, \triangle_M)$$

where f_1 is a smooth map homotopic to f that satisfies $\Gamma_{f_1} \pitchfork \triangle_M$.

We will now show how to improve the result given in Theorem 5.50 for maps $f : M \to M$ that satisfy $\Gamma_f \pitchfork \triangle_M$. A map $f : M \to M$ that satisfies this transversality condition is called a **Lefschetz map**. Note that the Homotopy Transversality Theorem (Theorem 5.17) implies that every smooth map is homotopic to a Lefschetz map. To begin, we take a closer look at the transversality condition $\Gamma(f) \pitchfork \triangle_M$.

Let $x \in M$ be a fixed point of $f : M \to M$. Then $T_x\Gamma(f) \subset T_xM \times T_xM$ is the graph of the differential of f at x, i.e.

$$T_x\Gamma(f) = \{(u,v) \in T_xM \times T_xM \mid v = df_x(u)\}.$$

For dimension reasons, the condition $\Gamma(f) \pitchfork \triangle_M$ implies that $T_xM \times T_xM$ is the direct sum $T_x\Gamma(f) \oplus T_x\triangle_M$, and hence $T_x\Gamma(f) \cap T_x\triangle_M = \{0\}$. Therefore, $df_x(u) = u$ if and only if $u = 0$, which means that df_x does not have $+1$ as an eigenvalue. That is, $df_x - \mathrm{id} : T_xM \to T_xM$ is injective and hence an isomorphism. This leads us to the following definition.

Definition 5.55 *A fixed point x of a smooth map $f : M \to M$ is called **nondegenerate** if and only if $\Gamma(f)$ meets $\triangle_M$ transversally at (x, x), or equivalently, if $\det(df_x - id) \neq 0$, i.e. df_x does not have $+1$ as an eigenvalue.*

We can now prove the following version of the Lefschetz Fixed Point Theorem, which applies only to Lefschetz maps.

Theorem 5.56 *Let $f : M \to M$ be a continuous map from a compact connected smooth oriented manifold of dimension m to itself that has only nondegenerate fixed points. Then*

$$\#fix(f) \geq L(f),$$

and if f is homotopic to the identity, then

$$\#fix(f) \geq \sum_{j=0}^{m} (-1)^j \dim H_j(M; \mathbb{Q}),$$

i.e. the number of fixed points of f is bounded below by the Euler characteristic of M.

Proof:

Using the oriented intersection definition of the Lefschetz number $L(f)$, we see that $\Gamma(f) \pitchfork \triangle_M$ implies that $L(f)$ is a lower bound for the number of fixed points since,

$$L(f) = \sum_{x \in \Gamma(f) \cap \triangle_M} \text{sign}(x).$$

When $f : M \to M$ is homotopic to the identity, the topological definition of the Lefschetz number (which is the same as the oriented intersection Lefschetz number by Theorem 5.54) shows that $L(f) = \mathcal{X}(M)$, the Euler characteristic of M. Note that the Euler characteristic of M is same for both integer and field coefficients by Remark 3.34.

$\square$

Problems

1. Let $f : M \to N$ be differentiable. Prove that

$$\{x \in M \mid f \text{ is regular at } x\}$$

is open in M.

2. Construct a smooth function $f : \mathbb{R} \to \mathbb{R}$ whose set of critical values is dense.

3. Prove that a subset $M \subseteq N$ of a smooth manifold N can be made into a m-dimensional submanifold if and only if around each point $x \in M$ there exists an open set U in N and a coordinate chart $\phi : U \to \mathbb{R}^n$ with $\phi(y) = (y_1, \ldots, y_n)$ such that $M \cap U = \{y \in N | \, y_{m+1}(x) = \cdots = y_n(x) = 0\}$.

4. Prove that an immersion between two smooth manifolds of the same dimension is an open map.

5. Suppose that $f : M \to N$ is an immersion between smooth manifolds M and N of the same dimension where M is compact and N is connected. Show that f is surjective.

6. Prove that a compact smooth manifold M of dimension m cannot be immersed in $\mathbb{R}^m$.

7. Prove the Fundamental Theorem of Algebra: Every non-constant complex polynomial must have a zero.

8. Prove the Constant Rank Theorem: Let $f : M \to N$ be a smooth map between finite dimensional smooth manifolds. Show that if for some $y \in N$ the map f is of constant rank k on a neighborhood of $f^{-1}(y)$, then $f^{-1}(y)$ is a submanifold of M of dimension $m - k$ where m is the dimension of M (or it is empty).

9. Let M and Z be smooth closed submanifolds of a finite dimensional smooth manifold N. Let $m = \dim M$, $z = \dim Z$, and $n = \dim N$. Suppose that $M \pitchfork Z$ and $n \leq m + r$. Show that for every $x \in M \cap Z$ there exists a coordinate chart $\phi : U \to \mathbb{R}^n$ where $U \subseteq N$ is an open neighborhood about x such that the first m coordinates are local coordinates for $U \cap M$ and the last z coordinates are local coordinates for $U \cap Z$.

10. Let $f : M \to N$ and $g : Z \to N$ be smooth maps where M, N, and Z are smooth manifolds. Show that $f \pitchfork g$ if and only if $f \times g$ is transverse to the diagonal $\triangle \subset N \times N$.

11. Let $f : M \to N$ and $g : Z \to N$ be smooth maps where M, N, and Z are smooth manifolds. The **fibered product** of M and Z over N is defined to be

$$M \times_N Z = \{(x, y) \in M \times Z | \, f(x) = g(y)\}.$$

Show that if $f \pitchfork g$, then $M \times_N Z$ is a smooth manifold of dimension $m + z - n$ where m, n, and z are the dimensions of the manifolds M, N, and Z respectively.

12. Suppose that M and Z are immersed submanifolds of N. Show that if $M \pitchfork Z$, then $T_x(X \cap Z) = T_x X \cap T_x Z$ for all $x \in X \cap Z$. (The tangent space of the intersection is the intersection of the tangent spaces.)

13. Suppose that M and Z are immersed submanifolds of N. Prove or disprove the following statement: If $M \cap Z$ is a submanifold of N, then $M \pitchfork Z$.

14. Suppose that M and Z are immersed submanifolds of N. Furthermore, assume that $M \cap Z$ is an immersed submanifold of dimension dim M + dim Z − dim N. Is it true that $M \pitchfork Z$?

15. Let $f : M \to N$ be a smooth map between smooth finite dimensional smooth manifolds with boundary. Say that a point $y \in N$ is a **regular value** of f if and only if df_x is surjective at every point $x \in f^{-1}(y)$ and $df_x|_{T_*\partial M}$ is surjective at every point $x \in f^{-1}(y) \cap \partial M$. Show that The Preimage Theorem (Corollary 5.9) extends to the category of smooth manifolds with boundary.

16. Show that The Inverse Image Theorem (Theorem 5.11) extends to the category of smooth manifolds with boundary.

17. Show that The Inverse Image Theorem (Theorem 5.11) extends to the category of smooth Banach manifolds with corners.

18. Let $f : M \to \partial M$ be a smooth map from a smooth manifold with boundary to its boundary. Show that f cannot leave ∂M pointwise fixed.

19. Prove the Smooth Brouwer Fixed Point Theorem: Any smooth map $f : D^n \to D^n$ must have a fixed point, i.e. a point $x \in D^n$ such that $f(x) = x$ where D^n denotes the closed unit disk in $\mathbb{R}^n$.

20. Show that the Brouwer Fixed Point Theorem also holds for continuous maps.

21. Let $f, g : M \to N$ be smoothly homotopic maps between finite dimensional smooth manifolds of the same dimension, where M is compact and without boundary. Prove the following Homotopy Lemma: If $y \in N$ is a regular value for both f and g, then $\#f^{-1}(y) \equiv \#g^{-1}(y) \mod 2$.

22. Let $f : M \to N$ be a smooth map between finite dimensional smooth manifolds of the same dimension. Show that if y and z are both regular values of f, then $\#f^{-1}(y) \equiv \#f^{-1}(z) \mod 2$.

23. Prove that the identity map of a compact finite dimensional smooth manifold without boundary is not homotopic to a constant map.

24. Let E and E' be n-dimensional vector bundles over B and let $f : E \to E'$ be a continuous map that takes each fiber of E isomorphically onto the corresponding fiber of E'. Show that f is a homeomorphism, and hence, E and E' are isomorphic.

25. Suppose that $g : E \to E'$ is a bundle map between smooth n-dimensional vector bundles, i.e. g is continuous and carries each fiber of E isomorphically onto a fiber of E', and let $\bar{g} : B \to B'$ be the corresponding map between the base spaces. Show that the pullback $\bar{g}^*(E')$ is isomorphic to E.

26. Show that an $\mathbb{R}^n$-bundle is trivial if and only if it admits n cross-sections that are nowhere dependent.

27. Suppose that $f : M \to N$ is a submersion between finite dimensional connected smooth manifolds M and N. Show that the kernels of df_x for $x \in M$ are the fibers in a smooth vector bundle over M. This bundle is called the **kernel bundle** of the submersion $f : M \to N$.

28. Let M be a closed, connected, compact finite dimensional smooth manifold, and let N be a connected finite dimensional smooth manifold. Show that if there exists a submersion $f : M \to N$, then N is closed, f is surjective, and M is a fiber bundle over N with projection f.

29. Prove that a differentiable manifold M of dimension m is orientable if and only if it admits a nowhere vanishing m-form.

30. Let M and Z be two compact oriented submanifolds of an oriented manifold N that satisfy $\dim(M) + \dim(Z) = \dim(N)$. Show that the intersection numbers satisfy

$$I(M, Z) = (-1)^{\dim(M)\dim(Z)} I(Z, M).$$

31. Let M be a compact smooth oriented m-dimensional submanifold of a $2m$-dimensional smooth oriented manifold N. Show that if m is odd, then the intersection number $I(M, M) = 0$.

32. Let $f : M \to N$ be a smooth map between smooth oriented manifolds of the same dimension m where M is compact and N is connected. If $y \in N$ is a regular value of f, then the Brouwer **degree** of f is defined to be

$$\deg(f; y) = \sum_{x \in f^{-1}(y)} \text{sign } df_x$$

where sign df_x is either $+1$ or -1 depending on whether $df_x : T_x M \to T_y M$ is orientation preserving or not. Show that the Brouwer degree does not depend on the choice of the regular value $y \in N$. Also, show that the Brouwer degree depends only on the homotopy class of the function f.

33. Show that if M is a compact smooth manifold with Euler characteristic $\mathcal{X}(M) \neq 0$, then every tangent vector field on M must have a zero.

34. Show that a map $f : M \to M$ from a compact smooth manifold to itself has only a finite number of non-degenerate fixed points.

35. Let x be a non-degenerate fixed point of a Lefschetz map $f : M \to M$. The sign of $\det(df_x - \mathrm{id})$ is called the **local Lefschetz number** of f and is denoted by $L_x(f)$. Show that if $f : M \to M$ is a Lefschetz map from a compact connected smooth oriented manifold to itself, then

$$L(f) = \sum_{x=f(x)} L_x(f).$$

36. Let M be a compact smooth oriented manifold, and let V be a tangent vector field on M. Define the **index** of an isolated zero p of V as follows: Choose a chart $\phi : U \to \mathbb{R}^m$ such that $\phi(p) = 0$, $S^{m-1} \subset \phi(U)$, and p is the only critical point in U. The vector field V induces a vector field $\tilde{V}$ on $\phi(U)$ via $\tilde{V} = d\phi_*(V)$, and the index $\mathrm{ind}_p(V)$ is defined to be the degree of the map

$$
\begin{array}{ccc}
S^{m-1} & \to & S^{m-1} \\
x & \mapsto & \dfrac{\tilde{V}(x)}{|\tilde{V}(x)|}.
\end{array}
$$

Prove the **Poincaré-Hopf Index Theorem**: Let M be a compact smooth manifold of dimension m and V a smooth tangent vector field on M with isolated zeros. Then,

$$\sum_{p \in \mathrm{zeros}(V)} \mathrm{ind}_p(V) = \sum_{j=0}^{m} (-1)^j \dim H_j(M; \mathbb{Q}),$$

i.e. the sum of the indices of the zeros of V is equal to the Euler characteristic $\mathcal{X}(M)$ of M.

Chapter 6

Morse-Smale Functions

In this chapter we introduce the Morse-Smale transversality condition for gradient vector fields, and we prove the Kupka-Smale Theorem (Theorem 6.6) which says that the space of smooth Morse-Smale gradient vector fields is a dense subspace of the space of all smooth gradient vector fields on a finite dimensional compact smooth Riemannian manifold (M, g) [92] [135]. We also prove Palis' λ-Lemma (Theorem 6.17) following [114] and [143], and we derive several important consequences of the λ-Lemma. These consequences include transitivity for Morse-Smale gradient flows (Corollary 6.21), a description of the closure of the stable and unstable manifolds of a Morse-Smale gradient flow (Corollary 6.27), and the fact that for any Morse-Smale gradient flow there are only finitely many gradient flow lines between critical points of relative index one (Corollary 6.29). In the last section of this chapter we present a couple of results due to Franks [59] that relate the stable and unstable manifolds of a Morse-Smale function $f : M \to \mathbb{R}$ to the cells and attaching maps in the CW-complex X determined by f (see Theorem 3.28).

6.1 The Morse-Smale transversality condition

In this section we introduce the Morse-Smale transversality condition, and we prove the Kupka-Smale Theorem (Theorem 6.6) which implies that the space of smooth Morse-Smale functions is a dense subspace of the space of all smooth functions on a finite dimensional compact smooth Riemannian manifold (M, g).

Definition 6.1 *A Morse function* $f : M \to \mathbb{R}$ *on a finite dimensional smooth Riemannian manifold* (M, g) *is said to satisfy the* **Morse-Smale transversality condition** *if and only if the stable and unstable manifolds of* f *intersect* *transversally, i.e.*

$$W^u(q) \pitchfork W^s(p)$$

*for all $p, q \in Cr(f)$. A Morse function that satisfies the Morse-Smale transversality condition is called a **Morse-Smale** function.*

As an immediate consequence of the Morse-Smale transversality condition we have the following.

Proposition 6.2 *Let $f : M \to \mathbb{R}$ be a Morse-Smale function on a finite dimensional compact smooth Riemannian manifold (M, g). If p and q are critical points of f such that $W^u(q) \cap W^s(p) \neq \emptyset$, then $W^u(q) \cap W^s(p)$ is an embedded submanifold of M of dimension $\lambda_q - \lambda_p$.*

Proof:

By Theorem 4.2, $W^u(q)$ and $W^s(p)$ are smooth embedded submanifolds of M of dimension λ_q and $m - \lambda_p$ respectively. By Lemma 4.20 and Corollary 5.12, $W^u(q) \cap W^s(p)$ is a smooth embedded submanifold of dimension

$$\dim W^u(q) + \dim W^s(p) - m = \lambda_q + (m - \lambda_p) - m = \lambda_q - \lambda_p.$$

$\square$

Note: We will denote $W^u(q) \cap W^s(p) = W(q, p)$.

Corollary 6.3 *If $f : M \to \mathbb{R}$ is a Morse-Smale function on a finite dimensional compact smooth Riemannian manifold (M, g), then the index of the critical points is strictly decreasing along gradient flow lines. That is, if p and q are critical points of f with $W(q, p) \neq \emptyset$, then $\lambda_q > \lambda_p$.*

Proof:

If $W(q, p) \neq \emptyset$, then $W(q, p)$ contains at least one flow line from q to p. Since a gradient flow line has dimension one we must have $\dim W(q, p) \geq 1$.

$\square$

Example 6.4 (The tilted torus) The torus T^2 resting vertically on the plane $z = 0$ in $\mathbb{R}^3$ with the standard height function $f : T^2 \to \mathbb{R}$ discussed in Example 3.6 and Example 4.25 is **not** a Morse-Smale function. This follows from Corollary 6.3 since the standard height function $f : T^2 \to \mathbb{R}$ has gradient flow lines that begin at the critical point r of index 1 and end at the critical point q of the same index.

However, according to a theorem due to Kupka and Smale (Theorem 6.6) there exists an arbitrarily small perturbation of the standard height function on T^2 which is Morse-Smale. One way to visualize such a perturbation is to use the height function on a torus that has been tilted slightly. The gradient flow lines for this perturbation are shown below.

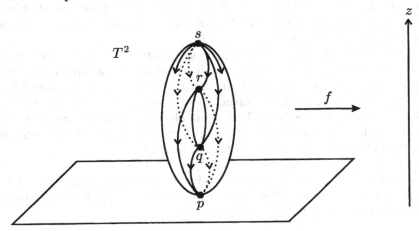

The Kupka-Smale Theorem

In 1963 both Kupka [92] and Smale [135] proved that Morse-Smale gradient vector fields are "generic". In fact, they both proved more generally that "Kupka-Smale" vector fields of class C^r are generic for all $r \geq 1$. Here we will limit ourselves to the case of gradient vector fields where the definitions of "Morse-Smale" and "Kupka-Smale" coincide. The proof of the Kupka-Smale Theorem for gradient vector fields is somewhat simpler than the proof of the general Kupka-Smale Theorem since the flow of a gradient vector field cannot have any closed orbits (see Proposition 3.18).

Definition 6.5 *A topological space X is called* **separable** *if and only if it contains a countable dense subset. A subset A of a topological space X is called* **residual** *if and only if it is a countable intersection of open dense subsets of X, i.e.*

$$A = \bigcap_{j=1}^{\infty} G_j$$

where $G_j \subseteq X$ is open and dense in X for all $j \in \mathbb{N}$. A subset of a topological space X is called **generic** *if and only if it contains a residual set, and a topological space X is called* **Baire space** *if and only if every generic subset is dense.*

Theorem 6.6 (Kupka-Smale Theorem) *If (M, g) is a finite dimensional compact smooth Riemannian manifold, then the set of Morse-Smale gradient vector fields of class C^r is a generic subset of the set of all gradient vector fields on M of class C^r for all $1 \leq r \leq \infty$.*

Remark 6.7 Note that the Riemannian metric gives a homeomorphism between the space of smooth gradient vector fields and the space of smooth functions on M modulo the relation that identifies any two functions that differ by a constant. Thus, the Kupka-Smale Theorem implies that the set of Morse-Smale functions is a generic subset of $C^\infty(M, \mathbb{R})$.

Baire's Theorem says that every complete metric space is a Baire space. That is, in a complete metric space the intersection of countably many open dense sets is dense (see for instance Theorem 7.27 of [123] or Theorem 5.6 of [125]). Since $C^\infty(M, \mathbb{R})$ is a complete metric space (see for instance Theorem 2.4.4 of [77]), the Kupka-Smale Theorem implies that the set of smooth Morse-Smale functions is dense in $C^\infty(M, \mathbb{R})$. Similarly, the set of Morse-Smale functions of class C^r is dense in $C^r(M, \mathbb{R})$ for all $r \geq 1$.

The following proof of Theorem 6.6 is based on Section 3.3 of [114] (where the general Kupka-Smale Theorem is proved). For other proofs of the Kupka-Smale Theorem see [1], [92], [135] and [143].

Lemma 6.8 *Let E be a separable Baire space and $F \subseteq E$ a dense subset. A subset $\Lambda \subseteq E$ is generic if and only if each $\xi \in F$ has an open neighborhood $\mathcal{N}_\xi$ in E such that $\Lambda \cap \mathcal{N}_\xi$ is generic in $\mathcal{N}_\xi$.*

Proof:

Let $\mathcal{N}_{\xi_1}, \mathcal{N}_{\xi_2}, \ldots$ be a countable cover of F such that $\Lambda \cap \mathcal{N}_{\xi_i}$ is generic in $\mathcal{N}_{\xi_i}$ for all $i = 1, 2, \ldots$. Then

$$\bigcap_{j=1}^{\infty} U_{ij} \subseteq \Lambda \cap \mathcal{N}_{\xi_i}$$

where U_{ij} is open and dense in $\mathcal{N}_{\xi_i}$. Let $W_{ij} = U_{ij} \cup (\mathcal{N} - \overline{\mathcal{N}}_{\xi_i})$ where $\mathcal{N} = \bigcup_{i=1}^{\infty} \mathcal{N}_{\xi_i}$. Then the W_{ij} are open and dense in E, and

$$\bigcap_{i=1}^{\infty} \bigcap_{j=1}^{\infty} W_{ij} \subseteq \Lambda.$$

Hence, Λ is generic. The converse is trivial.

□

Now let E be the set of all gradient vector fields on (M, g) of class C^r for some $1 \leq r \leq \infty$, and let F be the subset of E consisting of those gradient vector fields whose critical points are all hyperbolic. Note that F is a dense subset of E by Theorem 5.31. Moreover, it can be shown that E is a separable Baire space (see for instance Theorem 2.4.4 of [77] or Propositions 1.2.2 and

1.2.3 of [114]). To prove Theorem 6.6 we will apply the preceeding lemma with Λ equal to the space of Morse-Smale gradient vector fields on (M, g) of class C^r.

Suppose that $\xi \in F$ is a gradient vector field with hyperbolic critical points $p_1, \ldots, p_s$. For each $i = 1, \ldots, s$ let $W_0^s(p_i)$ and $W_0^u(p_i)$ be compact neighborhoods of p_i in $W^s(p_i)$ and $W^u(p_i)$ respectively such that the boundaries of $W_0^s(p_i)$ and $W_0^u(p_i)$ are tranverse to the vector field ξ. Also, for each $i = 1, \ldots, s$ let Σ_i^s be a codimension 1 submanifold of M that contains the boundary of $W_0^s(p_i)$ and is transverse to the gradient vector field ξ and the local stable manifold of p_i.

Every gradient vector field ζ in a small enough neighborhood $\mathcal{N}_\xi \subseteq E$ of ξ will be transverse to Σ_i^s for all $i = 1, \ldots, s$ and the critical points of ζ will all be hyperbolic and near the critical points of ξ (see Theorem 5.31). Thus, if $q_1, \ldots, q_s$ are the critical points of ζ, then for every $i = 1, \ldots, s$ there exists a compact neighborhood $W_0^s(q_i, \zeta)$ of q_i in $W^s(q_i, \zeta)$ whose boundary is the intersection of Σ_i^s with $W_0^s(q_i, \zeta)$. Similarly, for every $i = 1, \ldots, s$ there is a compact neighborhood $W_0^u(q_i, \zeta)$ of q_i in $W^u(q_i)$.

The definition of the map g in the proof of the Local Stable Manifold Theorem (Theorem 4.11) shows that the compact set $W_0^s(q_i, \zeta)$ depends continuously on ζ. That is, the local stable manifold is the graph of a function g which is defined using the flow of ξ, and hence, the graph of g depends continuously on ξ. Similarly, the compact set $W_0^s(q_i, \zeta)$ also depends continuously on ζ.

Remark 6.9 The statement that $\zeta \mapsto W_0^s(q_i, \zeta)$ is continuous can be made more precise using the definition of ε C^1-close found in the next section: Given any $\zeta \in \mathcal{N}_\xi$ and $\varepsilon > 0$ there exists a $\delta > 0$ such that if $\|\xi - \zeta\|_{C^1} < \delta$, then $W_0^s(p_i, \xi)$ is ε C^1-close to $W_0^s(q_i, \zeta)$. If ξ is of class C^r, then similar statements hold with respect to a metric that generates the C^r topology.

For each $n \in \mathbb{N}$ define $W_n^s(q_i, \zeta) = \varphi_{-n}(W_0^s(q_i, \zeta))$ and $W_n^u(q_i, \zeta) = \varphi_n(W_0^u(q_i, \zeta))$, where φ_t denotes the 1-parameter group of diffeomorphisms generated by ζ. The maps $\zeta \mapsto W_n^s(q_i, \zeta)$ and $\zeta \mapsto W_n^u(q_i, \zeta)$ are continuous because φ_n and φ_{-n} are diffeomorphisms that depend continuously on ζ. Moreover, $W_n^s(q_i, \zeta)$ and $W_n^u(q_i, \zeta)$ are compact submanifolds with boundary, $W^s(q_i, \zeta) = \bigcup_{n \geq 0} W_n^s(q_i, \zeta)$, and $W^u(q_i, \zeta) = \bigcup_{n \geq 0} W_n^u(q_i, \zeta)$.

Let Λ denote the set of all Morse-Smale gradient vector fields on (M, g) of class C^r, and let $\Lambda_n(\mathcal{N}_\xi)$ be the set of all vector fields $\zeta \in \mathcal{N}_\xi$ such that $W_n^s(q_i, \zeta)$ is transverse to $W_n^u(q_j, \zeta)$ for all $i, j = 1, \ldots, s$.

Lemma 6.10 *Let $\xi \in F$, and let $\mathcal{N}_\xi$ be a neighborhood of ξ as above. Then for all $n \in \mathbb{N}$ the set $\Lambda_n(\mathcal{N}_\xi)$ is open and dense in $\mathcal{N}_\xi$.*

Proof:

Let $p_1, \ldots, p_s$ be the critical points of ξ and write $\Lambda_{n,i,j}(\mathcal{N}_\xi)$ for the set of vector fields $\zeta \in \mathcal{N}_\xi$ such that $W_n^s(q_i, \zeta) \pitchfork W_n^u(q_j, \zeta)$. With this notation we have $\Lambda_n(\mathcal{N}_\xi) = \bigcap_{i,j=1}^s \Lambda_{n,i,j}(\mathcal{N}_\xi)$. Thus, to prove the lemma it is enough to show that $\Lambda_{n,i,j}(\mathcal{N}_\xi)$ is open and dense in $\mathcal{N}_\xi$ for all $i, j = 1, \ldots, s$.

Proof that $\Lambda_{n,i,j}(\mathcal{N}_\xi)$ is open in $\mathcal{N}_\xi$:

Let $\zeta \in \Lambda_{n,i,j}(\mathcal{N}_\xi)$. Since both $W_n^s(q_i, \zeta)$ and $W_n^u(q_j, \zeta)$ are compact submanifolds with boundary, $W_n^s(q_i, \zeta) \pitchfork W_n^u(q_j, \zeta)$, and the maps $\zeta \mapsto W_n^s(q_i, \zeta)$ and $\zeta \mapsto W_n^u(q_j, \zeta)$ are continuous, it follows that there exists an open neighborhood $\mathcal{N}_{\xi,\zeta}$ of ζ in $\mathcal{N}_\xi$ such that for all $\tilde{\zeta} \in \mathcal{N}_{\xi,\zeta}$ we have $W_n^s(\tilde{q}_i, \tilde{\zeta}) \pitchfork W_n^u(\tilde{q}_j, \tilde{\zeta})$ (compare with Theorem 5.16). Therefore, $\mathcal{N}_{\xi,\zeta} \subseteq \Lambda_{n,i,j}(\mathcal{N}_\xi)$, which proves that $\Lambda_{n,i,j}(\mathcal{N}_\xi)$ is open in $\mathcal{N}_\xi$.

Proof that $\Lambda_{n,i,j}(\mathcal{N}_\xi)$ is dense in $\mathcal{N}_\xi$:

Let $\zeta \in \mathcal{N}_\xi$. We will show that there exists a neighborhood $\mathcal{N}_{\xi,\zeta}$ of ζ in $\mathcal{N}_\xi$ such that $\Lambda_{n,i,j}(\mathcal{N}_\xi) \cap \mathcal{N}_{\xi,\zeta}$ is open and dense in $\mathcal{N}_{\xi,\zeta}$. In particular, there exist elements of $\Lambda_{n,i,j}(\mathcal{N}_\xi)$ arbitrarily close to ζ.

Consider the compact set $K = W_n^s(q_i, \zeta) \cap W_n^u(q_j, \zeta)$. If $x \in K$, then there exists a tubular flow $(F_x, f_{\zeta,x})$ containing x and a positive real number b_x such that $[-b_x, b_x] \times I^{m-1} \subseteq f_{\zeta,x}(F_x)$ and the vector field $(f_{\zeta,x})_*\zeta$ coincides with the unit vector field on $[-b_x, b_x] \times I^{m-1}$ (consider the Tubular Flow Theorem). Let $A_x \subseteq F_x$ be an open neighborhood of x such that the closure of A_x is contained in the interior of $(f_{\zeta,x})^{-1}([-b_x, b_x] \times I_{1/4}^{m-1})$. By shrinking F_x if necessary we may assume that $W_{2n}^s(q_i, \zeta) \cap F_x$ and $W_{2n}^u(q_j, \zeta) \cap F_x$ have only one connected component each. Let $A_1, \ldots, A_l$ be a finite cover of K by such open sets and write $(F_k, f_{\zeta,k})$ for the corresponding tubular flows. Then $(f_{\zeta,k})^{-1}([-b_k, b_k] \times I_{1/4}^{m-1})$ contains A_k.

Since the maps $\zeta \mapsto W_n^s(q_i, \zeta)$ and $\zeta \mapsto W_n^u(q_j, \zeta)$ are continuous, there exists a neighborhood $\mathcal{N}_{\xi,\zeta}$ of ζ in $\mathcal{N}_\xi$ such that for all $\tilde{\zeta} \in \mathcal{N}_{\xi,\zeta}$ we have $W_n^s(\tilde{q}_i, \tilde{\zeta}) \cap W_n^u(\tilde{q}_j, \tilde{\zeta}) \subseteq \bigcup_{k=1}^l A_k$. By shrinking $\mathcal{N}_{\xi,\zeta}$ if necessary we can also assume that for every $\tilde{\zeta} \in \mathcal{N}_{\xi,\zeta}$ and $k = 1, \ldots, l$ there is a tubular flow $(F_{\tilde{\zeta},k}, f_{\tilde{\zeta},k})$ for $\tilde{\zeta}$ such that $A_k \subseteq F_{\tilde{\zeta},k}$, $[-b_k, b_k] \times I^{m-1} \subseteq f_{\tilde{\zeta},k}(F_{\tilde{\zeta},k})$, and the interior of $(f_{\tilde{\zeta},k})^{-1}([-b_k, b_k] \times I_{1/4}^{m-1})$ contains the closure of A_k (this follows from the fact that the flow depends continuously on the vector field). Now let $\tilde{\Lambda}_k(\mathcal{N}_\xi)$ be the set of vector fields $\tilde{\zeta} \in \mathcal{N}_{\xi,\zeta}$ such that $W_n^s(\tilde{q}_i, \tilde{\zeta})$ is transverse to $W_n^u(\tilde{q}_j, \tilde{\zeta})$ at every point in $\overline{A}_k$. It is clear that $\tilde{\Lambda}_k(\mathcal{N}_\xi)$ is an open subset of $\mathcal{N}_{\xi,\zeta}$, and the lemma will be proved once we show that $\tilde{\Lambda}_k(\mathcal{N}_\xi)$ is dense in $\mathcal{N}_{\xi,\zeta}$ since $\Lambda_{n,i,j}(\mathcal{N}_\xi) \cap \mathcal{N}_{\xi,\zeta} = \bigcap_{k=1}^l \tilde{\Lambda}_k(\mathcal{N}_\xi)$.

Take a smooth vector field $\tilde{\zeta} \in \mathcal{N}_{\xi,\zeta}$ and denote

$$S_+(\tilde{\zeta}) = f_{\tilde{\zeta},k}(W_n^s(\tilde{q}_i, \tilde{\zeta}) \cap F_{\tilde{\zeta},k}) \cap (\{b_k\} \times I^{m-1})$$

and
$$U_+(\tilde{\zeta}) = f_{\tilde{\zeta},k}(W_n^u(\tilde{q}_j, \tilde{\zeta}) \cap F_{\tilde{\zeta},k}) \cap (\{b_k\} \times I^{m-1}).$$

It easy to see that if $S_+(\tilde{\zeta})$ is transverse to $U_+(\tilde{\zeta})$ in $\{b_k\} \times I_{1/4}^{m-1}$, then $W_n^s(\tilde{q}_i, \tilde{\zeta})$ is transverse to $W_n^u(\tilde{q}_j, \tilde{\zeta})$ in A_k. Also, using the Tubular Flow Theorem one can show that given any $\varepsilon > 0$ and $v \in \mathbb{R}^{m-1}$ with $\|v\|$ sufficiently small, there is a smooth vector field $\tilde{\zeta}'$ on M with $\|\tilde{\zeta} - \tilde{\zeta}'\|_{C^r} < \varepsilon$ that satisfies

1. $\tilde{\zeta}' = \tilde{\zeta}$ outside of $f_{\tilde{\zeta},k}^{-1}([-b_k, b_k] \times I^{m-1})$ and

2. $L(-b_k, y) = (b_k, y + v)$ for all $y \in I_{1/4}^{m-1}$,

where $L : \{-b_k\} \times I^{m-1} \to \{b_k\} \times I^{m-1}$ is the map that associates to each $(-b_k, y)$ the point where the orbit of $(f_{\tilde{\zeta},k})_*\tilde{\zeta}'$ through the point $(-b_k, y)$ intersects $\{b_k\} \times I^{m-1}$ (see Lemma 3.2.4 of [114]).

On the other hand, by general position arguments, it is possible to choose a small enough v so that $S_+(\tilde{\zeta})$ is transverse to $U_+(\tilde{\zeta}) + v$ (compare with Theorem 5.17 and Theorem 5.19). Then $f_{\tilde{\zeta},k}^{-1}(S_+(\tilde{\zeta}))$ is the intersection of $W_{2n}^s(\tilde{q}_i, \tilde{\zeta}')$ with the transversal section $f_{\tilde{\zeta},k}^{-1}(\{b_k\} \times I^{m-1})$, and $f_{\tilde{\zeta},k}^{-1}(U_+(\tilde{\zeta}) + v)$ is the intersection of $W_{2n}^u(\tilde{q}_j, \tilde{\zeta}')$ with the section $f_{\tilde{\zeta},k}^{-1}(\{b_k\} \times I_{1/4}^{m-1})$. Thus, these two submanifolds are transverse in $f_{\tilde{\zeta},k}^{-1}([-b_k, b_k] \times I_{1/4}^{m-1})$. Hence, $W_n^s(\tilde{q}_i', \tilde{\zeta}')$ is transverse to $W_n^u(\tilde{q}_j', \tilde{\zeta}')$ on $\overline{A}_k$. This shows that $\tilde{\zeta}' \in \tilde{\Lambda}_k(\mathcal{N}_\xi)$.

Therefore, any smooth vector field in $\mathcal{N}_{\xi,\varsigma}$ can be approximated by a vector field in $\tilde{\Lambda}_k(\mathcal{N}_\xi)$. Since any vector field in $\mathcal{N}_{\xi,\varsigma}$ can be approximated by a smooth vector field, we have shown that $\tilde{\Lambda}_k(\mathcal{N}_\xi)$ is dense in $\mathcal{N}_{\xi,\varsigma}$. This completes the proof of the lemma.

$\square$

Proof of Theorem 6.6:

Let $\xi \in F$, and let $\mathcal{N}_\xi$ be a neighborhood of ξ as above. Lemma 6.10 implies that $\Lambda_n(\mathcal{N}_\xi)$ is open and dense in $\mathcal{N}_\xi$ for all $n \in \mathbb{N}$. Thus, $\Lambda \cap \mathcal{N}_\xi = \bigcap_{n=1}^{\infty} \Lambda_n(\mathcal{N}_\xi)$ is generic in $\mathcal{N}_\xi$, and Lemma 6.8 implies that Λ is generic in E.

$\square$

Remark 6.11 In 1969 Palis proved that if M is a finite dimensional compact smooth manifold, then the set of "Morse-Smale diffeomorphisms" of class C^r on M is an open and non-empty subset of the set of all diffeomorphisms of class C^r for all $r \geq 1$ [112]. In the same paper Palis also proved that the set of all "Morse-Smale vector fields" of class C^r on M is an open and non-empty subset of the set of all vector fields of class C^r for all $r \geq 1$. Palis proved these theorems using the now famous λ-Lemma which we discuss in the next section.

It is important to note that this result is false for "Kupka-Smale vector fields" (see the Remark following the proof of Theorem 3.1 of [114]). For the precise definitions of "Kupka-Smale vector fields" and "Morse-Smale vector fields" see Section 3.3 and Section 4.4 of [114]. A "Morse-Smale vector field" is a "Kupka-Smale vector field" where the set of "non-wandering points" coincides with the critical elements of the vector field. For gradient vector fields the definitions of "Kupka-Smale" and "Morse-Smale" coincide. Thus, Palis' result implies that on a finite dimensional smooth Riemannian manifold (M, g) the space of Morse-Smale gradient vector fields of class C^r is an open subspace of the space of all gradient vector fields of class C^r for all $r \geq 1$.

Palis' results were part a program, due to Smale, to relate the Morse-Smale transversality condition to the notion of structural stability, defined by Andronov and Pontrjagin in 1937 [4]. A diffeomorphism $\varphi : M \rightarrow M$ is called C^r **structurally stable** if and only if there exists a neighborhood $U \subseteq \text{Diff}^r(M)$ (the set of diffeomorphisms of class C^r) of φ such that every $\varphi' \in U$ is topologically conjugate to φ, i.e. there exists a homeomorphism $h : M \rightarrow M$ such that $h \circ \varphi' \circ h^{-1} = \varphi$. (The definition for vector fields is similar.)

In a paper from 1962 [115] Peixoto proved that for any $r \geq 1$ a C^r vector field on a compact orientable manifold M of dimension 2 is C^r structurally stable if and only if it is Morse-Smale. In [112] Palis proved that if $\dim M \leq 3$, then both Morse-Smale diffeomorphisms and Morse-Smale vector fields are structurally stable, and in a paper from 1970 [113] Palis and Smale proved a more general and difficult result that says that every Morse-Smale diffeomorphism (or vector field) of class C^r is C^r structurally stable. However when $\dim M \geq 3$, the Anosov systems provide examples of structurally stable systems that are **not** Morse-Smale. For more details see Chapter 7 of [83], Chapter 4 of [114], and Chapter 4 of [143].

Remark 6.12 A result similar to the Kupka-Smale Theorem says that it is also possible to perturb the Riemannian metric g (instead of the Morse function f) to make the gradient flow satisfy the Morse-Smale transversality condition. That is, for any given Morse function $f : M \rightarrow \mathbb{R}$ on a finite dimensional compact smooth manifold M there is a dense set of Riemannian metrics g such that the Morse function f satisfies the Morse-Smale transversality condition with respect to g. The proof of this fact uses the "modern" techniques of Floer homology. See Chapter 2 of [131] for more details.

However, it should be noted that the preceeding result does not hold for Morse-Bott functions. That is, one can define an appropriate generalized Smale transversality condition for Morse-Bott functions, but it is **not** always possible to perturb the Riemannian metric to make a given Morse-Bott function satisfy the generalized Smale transversality condition. See Section 2 of [95] for some interesting counterexamples.

6.2   The λ-Lemma

In this section we prove one of the crucial theorems in the theory of smooth dynamical systems, the λ-Lemma. A version of this theorem, as well as several important corollaries, were announced in a paper by Smale in 1960 [136]. Smale referenced the proof of his theorem to a preprint "On structural stability" which never appeared in print. The proof of the theorem and the proofs of the corollaries from [136] did not appear in print until 1969 when Palis, one of Smale's students, published his doctoral thesis [112].

Definition 6.13 *Let d_1 be a metric that generates the C^1 topology. Let P_1, P_2, and M be finite dimensional smooth manifolds, and let $i_1 : P_1 \to M$ and $i_2 : P_2 \to M$ be smooth injective immersions with images $i_1(P_1) = W_1 \subseteq M$ and $i_2(P_2) = W_2 \subseteq M$. W_1 is said to be ε C^1-close to W_2 with respect to the metric d_1 if and only if there exists a smooth map $h : P_1 \to P_2$ such that*

$$d_1(i_1, i_2 \circ h) < \varepsilon.$$

Remark 6.14 The C^1 topology is metrizable (see for instance Theorem 2.4.4 of [77]), and d_1 in the preceeding definition can be any metric that generates the same topology as the subbasis elements in Definition 5.28. Also, note that the preceeding relationship is not symmetric. For instance, it is possible for a straight line to be ε C^1-close to a plane, but not conversely.

Example 6.15 Suppose that $M = \mathbb{R}^2$ and W_1, W_2 are the graphs of smooth functions $f_1, f_2 : \mathbb{R} \to \mathbb{R}$. If there exists a diffeomorphism $h : \mathbb{R} \to \mathbb{R}$ such that

$$|f_1(x) - (f_2 \circ h)(x)| + |f_1'(x) + (f_2 \circ h)'(x)| < \varepsilon$$

for all $x \in \mathbb{R}$, then W_1 is ε C^1-close to W_2 with respect to the standard metric on $C^1(\mathbb{R}, \mathbb{R}^2)$. For instance, if W_2 is a horizontal shift of W_1, then W_1 is ε C^1-close to W_2 for every $\varepsilon > 0$. However, if W_2 is a vertical shift of W_1 by more than ε, then W_1 is not ε C^1-close to W_2.

Definition 6.16 *Let $W \subseteq M$ be an immersed submanifold of M. A set $B \subset W$ is called an r-cell if and only if B regarded as a subset of W is diffeomorphic to a closed ball of dimension r. That is, if $i : P \to M$ is an injective immersion with $i(P) = W \subseteq M$, then $B \subseteq W$ is an r-cell if and only if $i^{-1}(B) \subseteq P$ is diffeomorphic to a closed r-ball.*

In [112] Palis proved the following theorem. For the statement of the theorem fix any metric d_1 that generates the C^1 topology.

Theorem 6.17 (λ-Lemma) *Let M be smooth manifold, and suppose that $p \in M$ is a hyperbolic fixed point of $\varphi \in \text{Diff}\,(M)$ with $\dim W^u(p) = r$ for some $0 < r < m$. If N is an injectively immersed submanifold of M, invariant under φ and having a point of transverse intersection q with $W^s(p)$, then for every r-cell neighborhood B of p in $W^u(p)$ and for every $\varepsilon > 0$, there exists an r-cell $D \subseteq N$ such that B is ε C^1-close to D. Moreover, for all $n \in \mathbb{Z}_+$ there is an r-cell $D_n \subseteq \varphi^n(D)$ such that B is ε C^1-close to D_n.*

Remark 6.18 Since the conclusion of the λ-Lemma holds for any $\varepsilon > 0$, the exact choice of the metric d_1 isn't particularly relevant.

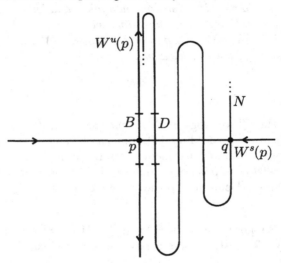

The above figure and the following proof are based on [114] and [143].

Proof:

Let q be a point of transverse intersection of N and $W^s(p)$. Then $\varphi^n(q)$ is also a point of transverse intersection of N and $W^s(p)$ for all $n \in \mathbb{Z}_+$, and since $\varphi^n(q) \to p$ as $n \to \infty$ we may assume that the point of transverse intersection q and p lie inside a common coordinate chart. Hence, it suffices to prove the theorem for the case $M = \mathbb{R}^m$.

Assume that $p = 0 \in \mathbb{R}^m$ and choose the coordinate system so that for some open neighborhood U of $0 \in \mathbb{R}^m$ we have $U \approx W^s_\mu(0) \times W^u_\mu(0)$ for some $\mu > 0$ (see Definition 4.8). In the neighborhood U we have

$$\varphi(x_s, x_u) = (Ax_s + f_s(x_s, x_u), Bx_u + f_u(x_s, x_u))$$

where $(x_s, x_u) \in W^s(0) \times W^u(0)$,

$$d\varphi(0,0) = \begin{pmatrix} A & 0 \\ 0 & B \end{pmatrix},$$

$f_s(0,0) = f_u(0,0) = df_s(0,0) = df_u(0,0) = 0$, and

$$\frac{\partial f_s}{\partial x_u}(0, x_u) = \frac{\partial f_u}{\partial x_s}(x_s, 0) = 0.$$

By an appropriate choice of inner product on $\mathbb{R}^m$ we may assume that

$$\|A\| = a < 1, \quad \text{and} \quad \|B^{-1}\| = 1/b < 1.$$

For a neighborhood $U_1 \subseteq U$ of zero define

$$k = \max \left\{ \sup_{\overline{U_1}} \left\| \frac{\partial f_s}{\partial x_s} \right\|, \sup_{\overline{U_1}} \left\| \frac{\partial f_s}{\partial x_u} \right\|, \sup_{\overline{U_1}} \left\| \frac{\partial f_u}{\partial x_s} \right\|, \sup_{\overline{U_1}} \left\| \frac{\partial f_u}{\partial x_u} \right\| \right\},$$

and choose U_1 small enough so that k satisfies the following inequalities.

$$0 < k < 1$$
$$b_1 = b - k > 1$$
$$a_1 = a + k < 1$$
$$k < \tfrac{1}{4}(b_1 - 1)^2$$

Now let B be a cell neighborhood of 0 in $W^u(0)$. For some $n \in \mathbb{Z}_+$ we have $\varphi^{-n}(B) \subset U_1$, and if for every $\varepsilon > 0$ the conclusion of the theorem holds for $\varphi^{-n}(B)$, then the theorem also holds for B. Hence, without loss of generality we may assume that B, as well as the point of transverse intersection q of N and $W^s(p)$, are contained in U_1.

Let $x \in U_1$, and let $v^0 = (v_s^0, v_u^0) \in T_x N \subset T_x U_1$ be a unit vector where v_s^0 and v_u^0 denote the components of v^0 with respect to $W^s(0)$ and $W^u(0)$ respectively. Define the **inclination** of v^0 to be

$$\lambda(v^0) = \frac{\|v_s^0\|}{\|v_u^0\|} \quad \text{whenever} \quad v_u^0 \neq 0.$$

Let $x^n = \varphi^n(x)$, $v^n = d\varphi(x^{n-1})(v^{n-1})$, and $\varphi(x) = (\varphi_s(x), \varphi_u(x)) \in W^s(0) \times W^u(0)$.

$$d\varphi(x) = \begin{pmatrix} \dfrac{\partial \varphi_s}{\partial x_s}(x) & \dfrac{\partial \varphi_s}{\partial x_u}(x) \\[2ex] \dfrac{\partial \varphi_u}{\partial x_s}(x) & \dfrac{\partial \varphi_u}{\partial x_u}(x) \end{pmatrix} = \begin{pmatrix} A + \dfrac{\partial f_s}{\partial x_s}(x) & \dfrac{\partial f_s}{\partial x_u}(x) \\[2ex] \dfrac{\partial f_u}{\partial x_s}(x) & B + \dfrac{\partial f_u}{\partial x_u}(x) \end{pmatrix}$$

and since $q^n \in W^s(0)$ for all $n \in \mathbb{Z}_+$ and $\frac{\partial f_u}{\partial x_s}(x_s, 0) = 0$ we have the following.

$$d\varphi(q^n) = \begin{pmatrix} A + \dfrac{\partial f_s}{\partial x_s}(q^n) & \dfrac{\partial f_s}{\partial x_u}(q^n) \\[2ex] 0 & B + \dfrac{\partial f_u}{\partial x_u}(q^n) \end{pmatrix}$$

For any $\varepsilon > 0$ there is a $\delta > 0$ such that

$$U_2 \overset{\text{def}}{=} B_s(\delta) \times B \subseteq U_1$$

where $B_s(\delta) \subseteq W^s(0)$ denotes the ball of radius δ in $W^s(0)$ centered at 0, and since $\frac{\partial f_s}{\partial x_u}(0, x_u) = 0$ we can choose $\delta > 0$ small enough so that

$$k_1 \overset{\text{def}}{=} \sup_{U_2} \left\| \frac{\partial f_s}{\partial x_u} \right\| < \min(\varepsilon, k).$$

As above, without loss of generality we may assume that $q \in U_2$.

Using the above inequalities we obtain the following estimate for the inclination which holds for any $v^0 \in T_x N$, $x \in N$ such that $x^j = \varphi^j(x) \in U_2$ for all $j = 0, \ldots, n$. Let $v^n = (v_s^n, v_u^n) \in W^s(0) \times W^u(0)$.

$$\begin{aligned} \lambda(v^n) &= \frac{\|d\varphi_s(x^{n-1})(v^{n-1})\|}{\|d\varphi_u(x^{n-1})(v^{n-1})\|} \\[2ex] &= \frac{\|(A + \frac{\partial f_s}{\partial x_s}(x^{n-1}))(v_s^{n-1}) + \frac{\partial f_s}{\partial x_u}(x^{n-1})(v_u^{n-1})\|}{\|\frac{\partial f_u}{\partial x_s}(x^{n-1})(v_s^{n-1}) + (B + \frac{\partial f_u}{\partial x_u}(q^{n-1}))(v_u^{n-1})\|} \\[2ex] &\leq \frac{(a+k)\|v_s^{n-1}\| + k_1\|v_u^{n-1}\|}{(b-k)\|v_u^{n-1}\| - k_1\|v_s^{n-1}\|} \\[2ex] &\leq \frac{\lambda(v^{n-1}) + k_1}{b_1 - k\lambda(v^{n-1})} \end{aligned}$$

When $x = q$ we get the following sharper inequality.

$$\begin{aligned} \lambda(v^n) &= \frac{\|(A + \frac{\partial f_s}{\partial x_s}(q^{n-1}))(v_s^{n-1}) + \frac{\partial f_s}{\partial x_u}(q^{n-1})(v_u^{n-1})\|}{\|(B + \frac{\partial f_u}{\partial x_u}(q^{n-1}))(v_u^{n-1})\|} \\[2ex] &\leq \frac{(a+k)\|v_s^{n-1}\| + k_1\|v_u^{n-1}\|}{(b-k)\|v_u^{n-1}\|} \end{aligned}$$

$$\leq \frac{\lambda(v^{n-1}) + k_1}{b_1}$$

Hence by induction we have for $v^0 \in T_q N$

$$\lambda(v^n) \leq \frac{\lambda(v^0)}{b_1^n} + k_1 \sum_{j=1}^{n} \frac{1}{b_1^j} \leq \frac{\lambda(v^0)}{b_1^n} + \frac{k_1}{b_1 - 1}.$$

Since $k_1 < k < \frac{1}{4}(b_1 - 1)^2$ and $b_1 > 1$, there exists an $n_0 \in \mathbb{Z}_+$ such that for $n > n_0$ we have

$$\lambda(v^n) \leq \frac{\lambda(v^0)}{b_1^n} + \frac{b_1 - 1}{4} < \frac{b_1 - 1}{3}.$$

The value of $\lambda(v^0)$ depends continuously on $x \in U_2$ and $v^0 \in T_x N$. Hence, by the preceeding inequality there exists a cell neighborhood $D_0 \subseteq N \cap U_2$ of $q^{n_0} \in N$ such that for any $x \in D_0$ and any $v^0 \in T_x N$ of length 1 we have

$$\lambda(v^0) \leq \frac{1}{2}(b_1 - 1).$$

We claim that if $\varphi^j(x) \in U_2$ for all $j = 0, 1, \ldots, n$, then

$$\lambda(v^n) \leq \frac{1}{2}(b_1 - 1).$$

Assume this is true for some n. Then by the above estimate for $\lambda(v^n)$ and the fact that $k_1 < k < 1$ and $k < \frac{1}{4}(b_1 - 1)^2$ we have the following,

$$\lambda(v^{n+1}) \leq \frac{\lambda(v^n) + k_1}{b_1 - k\lambda(v^n)} \leq \frac{\frac{1}{2}(b_1 - 1) + \frac{1}{4}(b_1 - 1)^2}{b_1 - \frac{1}{2}(b_1 - 1)} = \frac{1}{2}(b_1 - 1)$$

and the inequality claimed above is proved by induction.

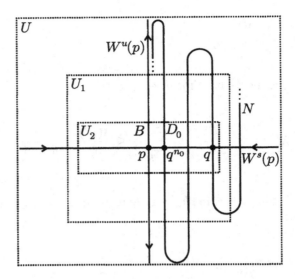

Using our first estimate for $\lambda(v^n)$ and the preceeding one we have the following estimate that holds uniformly with respect to any $x \in D_0$ such that $\varphi^j(x) \in U_2$ for all $j = 0, 1, \ldots, n$ and any unit vector $v^0 \in T_x N$.

$$
\begin{aligned}
\lambda(v^n) &\leq \frac{\lambda(v^{n-1}) + k_1}{b_1 - k\lambda(v^{n-1})} \\
&\leq \frac{\lambda(v^{n-1}) + k_1}{b_1 - \frac{1}{2}(b_1 - 1)} \\
&= \frac{\lambda(v^{n-1}) + k_1}{\frac{1}{2}(1 + b_1)} \\
&= \frac{\lambda(v^{n-1}) + k_1}{b_2}
\end{aligned}
$$

where $b_2 = \frac{1}{2}(1 + b_1) > 1$. Hence by induction we have

$$
\lambda(v^n) \leq \frac{\lambda(v^0)}{b_2^n} + k_1 \sum_{j=1}^{n} \frac{1}{b_2^j} \leq \frac{\lambda(v^0)}{b_2^n} + \frac{k_1}{b_2 - 1}
$$

and since $k_1 \leq \varepsilon$ we have

$$
\lambda(v_n) \leq \frac{\lambda(v^0)}{b_2^n} + \frac{\varepsilon}{b_2 - 1} \leq \varepsilon \left(1 + \frac{1}{b_2 - 1} \right)
$$

for n large enough, say $n \geq n_1$, for any $x \in D_0$ such that $\varphi^j(x) \in U_2$ for all $j = 0, 1, \ldots, n$, and for any unit vector $v^0 \in T_x N$.

Let $\pi_u : U \to W^u(0)$ be the projection to $W^u(0)$. Since φ is expanding on $W^u(0)$ there exists an $n_2 \geq n_1$ such that $B \subseteq \pi_u(D_{n_2})$ where D_{n_2} denotes the connected component of $\varphi^{n_2}(D_0) \cap U_2$ which contains the point q^{n_2}. If $\dim N = r$, then for every $(0, x_u) \in B$ there is exactly one point $(x_s, x_u) \in D_{n_2}$ such that $\pi_u(x_s, x_u) = (0, x_u)$; otherwise by the connectedness of D_{n_2} there would exist $x \in D_{n_2}$ such that $v_u^0 = 0$ for some $v^0 \in T_x N$, and this would contradict the bound on $\lambda(v^0) = \|v_s^0\|/\|v_u^0\|$. Define a map $h : B \to D_{n_2}$ by $h(0, x_u) = (x_s, x_u)$ where $(x_s, x_u) \in D_{n_2}$ is the unique point such that $\pi_u(x_s, x_u) = (0, x_u)$. The bound on $\lambda(v^n)$ shows that h is a map that satisfies the condition for B to be $\varepsilon(1+(b_2-1)^{-1})$ C^1-close to D_{n_2}. Moreover, for any $n \geq n_2$ there is a component of $\varphi^n(D_0) \cap U_2$ that contains a cell D_n such that B is $\varepsilon(1 + (b_2 - 1)^{-1})$ C^1-close to D_n.

If $\dim N > r$, then we can intersect N with a hyperplane H through $W^u(0)$ and the point q^{n_2}. Applying the previous case to the submanifold $N \cap H$ completes the proof of the λ-Lemma.

$\square$

6.3 Consequences of the λ-Lemma

The λ-Lemma has several corollaries that are essential for Morse homology. Many of these corollaries first appeared in one form or another in [136]. Proofs for some of the corollaries can be found in [112], [114], [133], and [143].

Corollary 6.19 *Let M be a smooth manifold, and suppose that p and q are hyperbolic fixed points of $\varphi \in Diff(M)$. If $W^u(q)$ and $W^s(p)$ have a point of transverse intersection, then*

$$\overline{W^u(q)} \supseteq W^u(p).$$

Proof:

Let $a \in W^u(p)$, and let $A, B \subset W^u(p)$ be r-cell neighborhoods of a and p respectively where $\dim W^u(p) = r$. There exists an $n \in \mathbb{Z}_+$ such that $\varphi^{-n}(A) \subseteq B$ because $\varphi^{-1}|_{W^u(p)}$ is a contraction to p. By Theorem 6.17, for any $\delta > 0$ there exists an r-cell neighborhood $D_\delta \subset W^u(q)$ such that B is δ C^1-close to D_δ, and since $\varphi^{-n}(A) \subseteq B$, $\varphi^{-n}(A)$ is also δ C^1-close to D_δ. Thus, for any $\varepsilon > 0$ we can chose a $\delta > 0$ such that A is ε C^1-close to $\varphi^n(D_\delta) \subset W^u(q)$. In particular, there exist points in $W^u(q)$ as close to $a \in W^u(p)$ as desired.

$\square$

Corollary 6.20 (Transitivity for Diffeomorphisms) *Let M be a smooth manifold, and suppose that p_1, p_2, and p_3 are hyperbolic fixed points of $\varphi \in$*

Diff (M). *If* $W^u(p_3)$ *and* $W^s(p_2)$ *have a point of transverse intersection and so do* $W^u(p_2)$ *and* $W^s(p_1)$, *then* $W^u(p_3)$ *and* $W^s(p_1)$ *also have a point of transverse intersection. Moreover, if* $W^u(p_3) \pitchfork W^s(p_2)$ *and* $W^u(p_2) \pitchfork W^s(p_1)$, *then*

$$\overline{W(p_3, p_1)} \supseteq W(p_3, p_2) \cup W(p_2, p_1) \cup \{p_1, p_2, p_3\}.$$

Proof:

Let $a \in M$ be a point of transverse intersection of $W^u(p_2)$ and $W^s(p_1)$, and let $A, B \subset W^u(p_2)$ be r-cell neighborhoods of a and p_2 respectively where $\dim W^u(p_2) = r$. There exists an $n \in \mathbb{Z}_+$ such that $\varphi^{-n}(A) \subseteq B$ because $\varphi^{-1}|_{W^u(p_2)}$ is a contraction to p_2. By Theorem 6.17, for any $\delta > 0$ there exists an r-cell neighborhood $D_\delta \subset W^u(p_3)$ such that B is δ C^1-close to D_δ, and since $\varphi^{-n}(A) \subseteq B$, $\varphi^{-n}(A)$ is also δ C^1-close to D_δ. Thus, for any $\varepsilon > 0$ we can choose a $\delta > 0$ such that A is ε C^1-close to $\varphi^n(D_\delta)$. Since A has a point of transverse intersection with $W^s(p_1)$, there exists an $\varepsilon > 0$ such that $\varphi^n(D_\delta)$ also has a point of transverse intersection with $W^s(p_1)$ by Theorem 5.16.

Now assume that $W^u(p_3) \pitchfork W^s(p_2)$ and $W^u(p_2) \pitchfork W^s(p_1)$. For any $a \in W^u(p_2) \cap W^s(p_1)$ we can choose the r-cell A used in the preceeding argument as small as desired. Hence, $\overline{W(p_3, p_1)} \supseteq W(p_2, p_1)$. To see that $\overline{W(p_3, p_1)} \supseteq W(p_3, p_2)$ repeat the above argument using φ^{-1} in place of φ. Since $p_3, p_2 \in \overline{W(p_3, p_2)}$ and $p_1 \in \overline{W(p_2, p_1)}$ we have $\overline{W(p_3, p_1)} \supseteq \{p_3, p_2, p_1\}$.

$\square$

Now assume that $f : M \to \mathbb{R}$ is a Morse-Smale function on a finite dimensional compact smooth Riemannian manifold (M, g). The gradient vector field of f generates a global 1-parameter family of diffeomorphisms φ_t, and by Lemma 4.5 and Lemma 4.19 the critical points of f are hyperbolic fixed points of the diffeomorphism $\varphi_t : M \to M$ for any fixed $t \in \mathbb{R}$. With this in mind, we can restate the preceeding corollary as follows.

Corollary 6.21 (Transitivity for Gradient Flows) *Let* p, q, *and* r *be critical points of a Morse-Smale function* $f : M \to \mathbb{R}$. *If* $W(r, q) \neq \emptyset$ *and* $W(q, p) \neq \emptyset$, *then* $W(r, p) \neq \emptyset$. *Moreover,*

$$\overline{W(r, p)} \supseteq W(r, q) \cup W(q, p) \cup \{p, q, r\}.$$

The preceeding corollary allows us to define a partial ordering on the critical points of a Morse-Smale function $f : M \to \mathbb{R}$ on a finite dimensional compact smooth Riemannian manifold (M, g) as follows.

Definition 6.22 *Let* p *and* q *be critical points of* $f : M \to \mathbb{R}$. *We say that* q *is **succeeded** by* p, $q \succeq p$, *if and only if* $W(q, p) = W^u(q) \cap W^s(p) \neq \emptyset$, *i.e. there exists a gradient flow line from q to p. The set of critical points of f, Cr(f), together with the partial ordering $\succeq$ is called the **phase diagram** of f.*

Corollary 6.19 and Corollary 6.21 imply the following.

Corollary 6.23 *For any critical point q of f : M → ℝ we have the following inclusion.*

$$\overline{W^u(q)} \supseteq \bigcup_{q \succeq p} W^u(p)$$

Moreover, if p and q are critical points of f : M → ℝ such that q $\succeq$ p, then

$$\overline{W(q,p)} \supseteq \bigcup_{q \succeq \tilde{q} \succeq \tilde{p} \succeq p} W(\tilde{q}, \tilde{p})$$

where the union is over all critical points between q and p in the phase diagram of f.

To show that the opposite inclusions also hold we will use the following series of lemmas. These lemmas can be found in a more general form in Chapter 2 of [133] where they are attributed to Palis.

Lemma 6.24 *If $\overline{W^u(q)} \cap W^u(p) \neq \emptyset$, then $p \in \overline{W^u(q)}$.*

Proof:
   By continuity, $\varphi_t(\overline{W^u(q)}) \subseteq \overline{W^u(q)}$ for any $t \in \mathbb{R}$. Hence, if $x \in \overline{W^u(q)} \cap W^u(p)$, then $p = \lim_{t \to -\infty} \varphi_t(x) \in \overline{W^u(q)}$ since $\overline{W^u(q)}$ is closed.

□

Lemma 6.25 *If $p \neq q$ is a critical point of f : M → ℝ such that $p \in \overline{W^u(q)}$, then $\overline{W^u(q)}$ intersects $W^s(p) - \{p\}$.*

Proof:
   Choose a small open neighborhood $U \subset M$ around the critical point p such that p is the only critical point in $\overline{U}$. If $W(q,p) \neq \emptyset$, then we are done. If $W(q,p) = \emptyset$, then we can choose a sequence of points $x_j \in W^u(q) \cap U$ on distinct flow lines converging to p. Since $x_j \in W^u(q)$, for every j there is a unique smallest positive integer k_j such that $\varphi_1^{-k_j}(x_j) \in W^u(q)$ is not in U. Note that $k_j \to \infty$ as $j \to \infty$ since $\varphi_1^k(p) = p \in U$ for all $k \in \mathbb{Z}$ and $x_j \to p$ as $j \to \infty$.
   Let $x \in \overline{W^u(q)}$ be a limit point of the sequence $\varphi_1^{-k_j}(x_j)$. Without loss of generality we will assume that $\varphi_1^{-k_j}(x_j) \to x$ as $j \to \infty$. For any $k > 0$

we have $\varphi_1^k(\varphi_1^{-k_j}(x_j)) \to \varphi_1^k(x)$ as $j \to \infty$, and $\varphi_1^k(\varphi_1^{-k_j}(x_j)) \in U$ for all j sufficiently large by the minimality of k_j and the fact that $k_j \to \infty$ as $j \to \infty$. Hence, $\varphi_1^k(x) \in \overline{U}$ for all $k > 0$, which implies that $x \in W^s(p)$. Moreover, $x \notin U$ since $\varphi_1^{-k_j}(x_j) \notin U$ for all j, and thus $x \neq p$.

□

Corollary 6.26 *Let q and p be critical points of $f : M \to \mathbb{R}$. If $\overline{W^u(q)} \cap W^u(p) \neq \emptyset$, then $q \succeq p$, i.e.*

$$\overline{W^u(q)} \subseteq \bigcup_{q \succeq p} W^u(p).$$

Proof:

By Lemma 6.24 we have $p \in \overline{W^u(q)}$, and by Lemma 6.25 there exists some $x \in \overline{W^u(q)} \cap (W^s(p) - \{p\})$. Then, $x \in W^u(p_2)$ for some critical point p_2 by Proposition 3.19, and applying Lemma 6.24 again we have $p_2 \in \overline{W^u(q)}$. Repeating this argument we obtain a maximal sequence of critical points $p_k \succeq p_{k-1} \succeq \cdots \succeq p_1$ with $p = p_1$ and $p_j \in \overline{W^u(q)}$ for all $j = 1, \ldots, k$. This sequence is finite since f has only finitely many critical points by Corollary 3.3 and there are no cycles by Corollary 6.3. Thus, $p_k = q$, and by Corollary 6.21 we have $q \succeq p$.

□

Combining Corollary 6.23 and Corollary 6.26 we obtain the following.

Corollary 6.27 *For any critical point q of $f : M \to \mathbb{R}$ we have*

$$\overline{W^u(q)} = \bigcup_{q \succeq p} W^u(p).$$

Similarly,

$$\overline{W^s(q)} = \bigcup_{r \succeq q} W^s(r).$$

Proof:

For the second equality note that the stable manifold of f is the unstable manifold of $-f$.

□

Corollary 6.28 *If p and q are critical points of $f : M \to \mathbb{R}$ such that $q \succeq p$, then*

$$\overline{W(q,p)} = \overline{W^u(q)} \cap \overline{W^s(p)} = \bigcup_{q \succeq \tilde{q} \succeq \tilde{p} \succeq p} W(\tilde{q}, \tilde{p})$$

where the union is over all critical points between q and p in the phase diagram of f.

Proof:

Since $W(q,p) = W^u(q) \cap W^s(p)$ we have $\overline{W(q,p)} \subseteq \overline{W^u(q)} \cap \overline{W^s(p)}$. If $x \in \overline{W^u(q)} \cap \overline{W^s(p)}$, then by the previous corollary there exists a sequence of critical points $q = p_k \succeq p_{k-1} \cdots \succeq p_1 = p$ such that either $x = p_j$ for some j or $x \in W(p_j, p_{j-1})$ for some j. Since $q \succeq p_{j-1} \succeq p$ we can apply Lemma 6.21 to conclude $\overline{W(q,p)} \supseteq W(q, p_{j-1})$, and hence, $\overline{W(q,p)} \supseteq \overline{W(q, p_{j-1})}$. Since $q \succeq p_j \succeq p_{j-1}$ we can apply Lemma 6.21 again to conclude $\overline{W(q, p_{j-1})} \supseteq W(p_j, p_{j-1})$, and hence,

$$\overline{W(p,q)} \supseteq \overline{W(q, p_{j-1})} \supseteq \overline{W(p_j, p_{j-1})}.$$

Therefore, $x \in \overline{W(q,p)}$.

$\square$

Corollary 6.29 *If p and q are critical points of relative index one, i.e. if $\lambda_q - \lambda_p = 1$, then*

$$\overline{W(q,p)} = W(q,p) \cup \{p, q\}.$$

Moreover, $W(q,p)$ has finitely many components, i.e. the number of gradient flow lines from q to p is finite.

Proof:

$W(q,p) \cup \{p, q\}$ is closed because Corollary 6.3 implies that there aren't any intermediate critical points between q and p in the phase diagram of f. Thus, $W(q,p) \cup \{p, q\} \subseteq M$ is compact since it's a closed subset of a compact space. The gradient flow lines from q to p form an open cover of $W(q,p)$, and this open cover can be extended to an open cover of $W(q,p) \cup \{p, q\}$ by taking the union of each gradient flow line with small open sets in $W(q,p) \cup \{p, q\}$ around p and q. Since every open cover of a compact space has a finite subcover, the number of gradient flow lines from q to p is finite.

$\square$

6.4   The CW-complex associated to a Morse-Smale function

In this section we prove some results that relate the unstable manifolds of a Morse-Smale function $f : M \to \mathbb{R}$ to the cells in the CW-complex determined by f. These results are proved using an additional assumption on the Riemannian metric that is not used elsewhere in this book. In particular, the proof we give in Chapter 7 of the Morse Homology Theorem (Theorem 7.4) does not use this additional assumption and is independent of the results in this section.

Let (M, g) be a finite dimensional compact smooth Riemannian manifold, and let $f : M \to \mathbb{R}$ be a Morse function on M. We proved in Theorem 3.28 that M has the homotopy type of a CW-complex X with one cell of dimension k for each critical point of index k, and we proved in Theorem 4.2 that for every critical point p of index k the unstable manifold $W^u(p) \subseteq M$ is a smoothly embedded open disk of dimension k. Moreover, in Proposition 4.22 we showed that M is a disjoint union of the unstable manifolds of f, i.e.

$$M = \coprod_{p \in \mathrm{Cr}(f)} W^u(p).$$

It is natural to ask exactly how the unstable manifolds are related to the cells in the CW-complex. That is, is there some relationship stronger than a bijection between the unstable manifolds in M and the cells in the CW-complex X? Also, can we compute the homotopy classes of the attaching maps in the CW-complex X using the stable and unstable manifolds of f?

Answers to both of these questions were given by Franks in 1979 assuming that $f : M \to \mathbb{R}$ is a Morse-Smale function and the metric g satisfies a certain generic condition near the critical points of f [59].

Definition 6.30 *A gradient vector field ∇f is said to be in **standard form** near a critical point p if and only if there exists a smooth coordinate chart around p such that in the local coordinates determined by the chart we have*

$$\nabla f = -x_1 \frac{\partial}{\partial x_1} - \cdots - x_k \frac{\partial}{\partial x_k} + x_{k+1} \frac{\partial}{\partial x_{k+1}} + \cdots + x_m \frac{\partial}{\partial x_m}.$$

*If the gradient vector field is in standard form near every critical point, then we will say that the Riemannian metric g is **compatible with the Morse charts** for the function f.*

Remark 6.31 The Morse Lemma (Lemma 3.11) says that in a neighborhood of a critical point p there exists a smooth chart $\phi : U \to \mathbb{R}^m$, where U is an open neighborhood of p and $\phi(p) = 0$, such that if $\phi(x) = (x_1, \ldots, x_m)$ for $x \in U$, then

$$(f \circ \phi^{-1})(x_1, \ldots, x_m) = f(p) - x_1^2 - x_2^2 - \cdots - x_k^2 + x_{k+1}^2 + x_{k+2}^2 + \cdots + x_m^2.$$

The coordinate chart ϕ is called a **Morse chart** for the function f. With respect to the standard metric on $\mathbb{R}^m$ the gradient of $f \circ \phi^{-1} : \phi(U) \to \mathbb{R}$ is

$$\nabla(f \circ \phi^{-1}) = -2x_1 \frac{\partial}{\partial x_1} - \cdots - 2x_k \frac{\partial}{\partial x_k} + 2x_{k+1} \frac{\partial}{\partial x_{k+1}} + \cdots + 2x_m \frac{\partial}{\partial x_m}.$$

However, this does not take into account the Riemannian metric g on M. The local formula for the gradient vector field (which depends on the Riemannian metric g) was computed in Lemma 4.3:

$$\nabla f = \sum_{i,j} g^{ij} \frac{\partial f}{\partial x_i} \frac{\partial}{\partial x_j}.$$

By choosing Morse charts on small neighborhoods of the critical points and pulling back the standard metric on $\mathbb{R}^m$ we can (using a bump function) modify a given Riemannian metric g to create a metric that gets sent to the standard metric on $\mathbb{R}^m$ by the chosen Morse charts, i.e. $g^{ij} = \delta_{ij}$ in the preceeding formula. Hence, a Riemannian metric g can always be modified on an arbitrarily small neighborhood of the critical set to create a Riemannian metric that is compatible with the Morse charts for f.

The Cell Equivalence Theorem

Theorem 3.28 says that a Morse function $f : M \to \mathbb{R}$ on a finite dimensional compact smooth Riemannian manifold (M, g) determines a CW-complex X that is homotopic to M. If we assume that f satisfies the Morse-Smale transversality condition and the Riemannian metric g is compatible with the Morse charts for f, then the Cell Equivalence Theorem (due to Franks) gives a relationship between M and X that is "stronger than homotopy equivalence and weaker than homeomorphism" ([59] p. 202). This relationship (which Franks calls "cell equivalence") basically says that two CW-complexes are "equivalent" if they can "inductively be built up by attaching corresponding cells with homotopic attaching maps" ([59] p. 202).

Franks makes this notion of "cell equivalence" more precise as follows ([59] p. 202): *If e and e' are cells of a CW-complex X, we will say that $e \geq e'$ if the closure of e contains any part of the interior of e'. If we make this relation transitive (and denote the new transitive relation by $\geq$ too), then we obtain a partial ordering on the cells of X. If S is a subset of X with the property that $e \in S$ and $e \geq e'$ implies that $e' \in S$, then the union of the cells in S forms a subcomplex of X. In particular, if e is a cell of X we define the base of e, denoted $X(e)$, to be the smallest subcomplex of X containing e. Thus, $X(e)$ is the union of all cells $e' \in X$ such that $e \geq e'$.*

Definition 6.32 ([59] p. 202) *Two finite CW-complexes X and X' are called cell equivalent if and only if there is a homotopy equivalence $h : X \to X'$ with the property that there is a bijective correspondence between cells in X and cells in X' such that if $e \subseteq X$ corresponds to $e' \subseteq X'$, then h maps $X(e)$ to $X'(e')$ and is a homotopy equivalence of these subcomplexes.*

Remark 6.33 ([59] p. 202) *It is easy to show by induction on the number of cells that cell equivalence is an equivalence relation. In fact, given a cell* *equivalence* $h : X \rightarrow X'$ *it is not difficult to construct a cell equivalence* $h' : X' \rightarrow X$ *which is a homotopy inverse of* h. *It is also easy to see that cell equivalence preserves the partial order* $\geq$.

Theorem 6.34 (Cell Equivalence Theorem) *Let* $f : M \rightarrow \mathbb{R}$ *be a Morse-Smale function on a finite dimensional compact smooth Riemannian manifold* (M, g), *and assume that the Riemannian metric* g *is compatible with the Morse charts for* f. *Then there exists a CW-complex* X, *unique up to cell equivalence, and a homotopy equivalence* $h : M \rightarrow X$ *such that for each critical point* p *of index* k, *the image* $h(W^u(p))$ *is contained in the base* $X(e)$ *of a unique* k-*cell* $e \subseteq X$.

In this way $h : M \rightarrow X$ *establishes a bijective correspondence between the critical points of* f *of index* k *and the* k-*cells of* X. *Moreover, the partial order* $\succeq$ *on the critical points of* f *corresponds to the partial order* $\geq$ *on the cells of the CW-complex* X.

Proof:

To prove this theorem we follow the proof of Theorem 3.28 and use the two additional assumptions: f satisfies the Morse-Smale transversality condition and the Riemannian metric g is compatible with the Morse charts for f.

Let $c_0 < c_1 < c_2 < \cdots < c_l$ be the critical values of $f : M \rightarrow \mathbb{R}$. The set M^t is vacuous for all $t < c_0$, and M^{t_0} is homotopic to a discrete set of points for all $c_0 < t_0 < c_1$, i.e the set of critical points $\{p \in \text{Cr}(f)|\ f(p) = c_0\}$. Suppose now for the purposes of induction that $c_{i-1} < t_{i-1} < c_i < t_i < c_{i+1}$ for some $i = 1, 2, \ldots, l$ and there is a homotopy equivalence $h_{i-1} : M^{t_{i-1}} \rightarrow X_{i-1}$ satisfying the conclusions of the theorem, where X_{i-1} is some CW-complex determined uniquely up to cell equivalence.

By Theorem 3.20, Theorem 3.25, and Remark 3.26, there exists a homotopy equivalence

$$M^{t_i} \simeq M^{t_{i-1}} \cup_{f_1} D^{\lambda_{p_1}} \cup_{f_2} \cdots \cup_{f_r} D^{\lambda_{p_r}}$$

where $p_1, \ldots, p_r$ are the critical points in $f^{-1}(c_i)$ and $f_1, \ldots, f_r$ are attaching maps. The assumption that the Riemannian metric g is compatible with the Morse charts for f implies that the cell e^k in the proof of Theorem 3.25 is a closed ball around p in $W^u(p)$. Hence, we can choose

$$D^{\lambda_{p_j}} = W^u(p_j) \cap \overline{(M - M^{t_{i-1}})}$$

and the attaching map $f_j : S^{\lambda_{p_j}-1} \rightarrow \partial M^{t_{i-1}}$ as the inclusion map for all $j = 1, \ldots, r$.

For every $j = 1, \ldots, r$, the map $h_{i-1} \circ f_j : S^{\lambda_{p_j}-1} \rightarrow X_{i-1}$ is homotopic to a map

$$\Psi_j : S^{\lambda_{p_j}-1} \rightarrow X_{i-1}^{(\lambda_{p_j}-1)}$$

by the Cellular Approximation Theorem (Theorem 2.20) where $X_{i-1}^{(\lambda_{p_j}-1)}$ denotes the $(\lambda_{p_j}-1)$-skeleton of X_{i-1}. This homotopy preserves the partial ordering $\geq$. Thus, we have a homotopy equivalence

$$h_i : M^{t_i} \to X_i \stackrel{\text{def}}{=} X_{i-1} \cup_{\Psi_1} D^{\lambda_{p_1}} \cup_{\Psi_2} \cdots \cup_{\Psi_r} D^{\lambda_{p_r}}$$

by Lemma 3.29 and Lemma 3.30. Moreover, the image $h_i(W^u(p_j))$ is contained in the base $X(e_j)$ of a unique k-cell e_j for all $j = 1, \ldots, r$ and the partial ordering $\succeq$ on M corresponds to the partial ordering $\geq$ on X_i under h_i by Corollary 6.27. The theorem now follows by induction.

$$\square$$

Remark 6.35 In [59] Franks proved the Cell Equivalence Theorem for any "gradient-like" Morse-Smale vector field on M. A **gradient-like** vector field is a vector field with no closed orbits that is in standard form near all of its critical points. Franks made use of a theorem of Smale's that implies that every Morse-Smale gradient-like vector field is the gradient vector of some self-indexing Morse function with respect to some Riemannian metric (see Theorem B of [137]).

Note that a gradient vector field is not necessarily gradient-like (the gradient vector field must be in standard form near its critical points to be gradient-like). This makes the terminology confusing.

Remark 6.36 A result due to Laudenbach (found in the appendix to [20]) says that if $f : M \to \mathbb{R}$ is a Morse-Smale function and the Riemannian metric g is compatible with the Morse charts for f, then the unstable manifolds of f give rise to the structure of a CW-complex on M. Laudenbach proved this result by using the "canonical Morse model", which is a cobordism between level sets, to show that the closure of an unstable manifold has the structure of a "submanifold with conical singularities". Laudenbach makes the following comment after Remark 3: This result can probably be extended to general Morse-Smale gradient vector fields. To do this, one needs to change the definition of a submanifold with conical singularities by delinearizing the cone construction.

The Connecting Manifold Theorem

Let $f : M \to \mathbb{R}$ be a Morse-Smale function on a finite dimensional compact smooth Riemannian manifold (M, g), and assume that the Riemannian metric g is compatible with the Morse charts for f. The function f determines a CW-complex X, and the Connecting Manifold Theorem (Theorem 6.40) says that the "connecting manifolds", which are framed submanifolds of the "unstable spheres", correspond to the relative attaching maps in X via the Thom-Pontryagin construction.

Recall that a **framing** of a submanifold $N \subset M$ is a smooth function F that assigns to each $x \in N$ a basis

$$F(x) = (V_1(x), \ldots, V_{m-n}(x))$$

for the normal space $\nu_x N \subseteq T_x M$, and a **framed submanifold** of M is a pair (N, F). Note that a submanifold $N \subseteq M$ has a framing if and only if its normal bundle is trivial. If $h : M \to S^{m-n}$ is a smooth map and $y \in S^{m-n}$ is a regular value, then Theorem 5.11 implies that $N = h^{-1}(y)$ is a submanifold of M of dimension n whose normal bundle is trivial. In fact, $dh_x : \nu_x N \to T_y S^{m-n}$ is an isomorphism for all $x \in N = h^{-1}(y)$, and so a basis for $T_y S^{m-n}$ determines a framing for N.

Conversely, if (N, F) is a framed submanifold of M, then we can choose a tubular neighborhood $E(\nu) \hookrightarrow M$ of N (see for instance Section II.11 of [30] or Section III.2 of [91]) and define a map $h : M \to S^{m-n}$ as follows. Let $\phi : E(\nu) \to N \times \mathbb{R}^{m-n}$ be the diffeomorphism determined by the framing F, and let $s : R^{m-n} \to S^{m-n}$ denote the inverse of the stereographic projection map from the north pole. Define

$$h(x) = \begin{cases} s(\pi_2((\phi(x)))) & \text{if } x \in E(\nu) \\ * & \text{if } x \notin E(\nu) \end{cases}$$

where $\pi_2 : N \times \mathbb{R}^{m-n} \to \mathbb{R}^{m-n}$ denotes projection onto the second component and $* \in S^{m-n}$ denotes the north pole. This construction, known as the **Thom-Pontryagin construction**, produces a continuous map $h : M \to S^{m-n}$ from the framed submanifold (N, F).

The Thom-Pontryagin Theorem says that the above constructions induce a bijection between the set of homotopy classes of maps $[M, S^{m-n}]$ and the framed cobordism group $\Omega^n(M)$ of n-dimensional compact framed submanifolds of M (see for instance Theorem IX.5.5 of [91] or Section 7 of [103]). In particular, if F_0 and F_1 are homotopic framings of a submanifold $N \subseteq M$, then the Thom-Pontryagin construction produces homotopic maps $h_0 : M \to S^{m-n}$ and $h_1 : M \to S^{m-n}$ (see for instance Lemma IX.5.2 of [91]). The problems at the end of this chapter discuss more results related to framed cobordism and the Thom-Pontryagin Theorem.

The following definition is a natural analogue of Definition 6.22.

Definition 6.37 *Let p and q be critical points of a Morse-Smale function $f :$ $M \to \mathbb{R}$. We will say that p is an **immediate successor** of q if and only if $q \succeq p$ and there are no **intermediate critical points** between q and p. That is, there is no critical point $\alpha \neq p, q$ such that $q \succeq \alpha$ and $\alpha \succeq p$.*

Let $f : M \to \mathbb{R}$ be a Morse-Smale function, assume that the critical point p is an immediate successor of the critical point q, and let $t \in \mathbb{R}$ be a regular value of f between $f(p)$ and $f(q)$. By the Preimage Theorem (Corollary 5.9), the level set $f^{-1}(t)$ is a submanifold of M of dimension $m - 1$, and it's clear that $f^{-1}(t)$ intersects both $W^u(q)$ and $W^s(p)$ transversally. Hence, the **unstable sphere** of q:

$$S^u(q) = W^u(q) \cap f^{-1}(t)$$

and the **stable sphere** of p:

$$S^s(p) = W^s(p) \cap f^{-1}(t).$$

are embedded submanifolds of dimension $\lambda_q - 1$ and $m - \lambda_p - 1$ respectively. Moreover, since $W^u(q)$ intersects $W^s(p)$ transversally in M, it's clear that $S^u(q)$ intersects $S^s(p)$ transversally in the level set $f^{-1}(t)$. Thus, $N(q,p) = S^u(q) \cap S^s(p) \subset f^{-1}(t)$ is an embedded submanifold of dimension $\lambda_q - \lambda_p - 1$.

The embedded disk $W^s(p)$ is contractible, and hence there exists a framing F for $W^s(p)$ in M. Restricting F to $N(q,p)$ gives a framing of $N(q,p)$ in $S^u(q)$ since $S^u(q) \pitchfork S^s(p)$.

Definition 6.38 *Let $f : M \to \mathbb{R}$ be a Morse-Smale function on a finite dimensional smooth Riemannian manifold (M, g). If the critical point p is an immediate successor of the critical point q, then a **connecting manifold** is defined to be a framed submanifold $(N(q,p), F)$ where $N(q,p) = S^u(q) \cap S^s(p)$ and the framing F of $N(q,p) \subset S^u(q)$ is obtained by restricting a framing of $W^s(p)$ in M.*

Remark 6.39 Since $W^s(p)$ is contractible there are, up to homotopy, exactly two choices for its framing in M. Hence, up to homotopy, there are exactly two choices for the framing F of a connecting manifold $N(q,p)$ in $S^u(q)$. If M is oriented and we choose an orientation for the embedded disk $W^s(p)$, then the orientation of M determines a well-defined homotopy class for the framing of $W^s(p)$ in M (see Remark 5.39).

The Connecting Manifold Theorem says that the Thom-Pontryagin construction applied to a connecting manifold $(N(q,p), F)$ produces a map

$$h : S^{\lambda_q - 1} \to S^{\lambda_p}$$

that is homotopic (up to precomposing with a representative of $\pm 1 \in \pi_j(S^j)$ where $j = \lambda_q - 1$) to the corresponding relative attaching map in the CW-complex X determined by f. Recall that the **relative attaching map** of a j-cell D^j to a k-cell D^k in a CW-complex X is defined to be the composition

$$\partial D^j = S^{j-1} \xrightarrow{f_\partial} X^{(j-1)} \longrightarrow X^{(j-1)}/(X^{(j-1)} - \mathrm{int}\, D^k) \simeq S^k$$

where $f_\partial : S^{j-1} \to X^{(j-1)}$ is the map attaching D^j to the $(j-1)$-skeleton (see Remark 2.16).

Theorem 6.40 (Connecting Manifold Theorem) *Suppose that* $f : M \to \mathbb{R}$ *is a Morse-Smale function on a finite dimensional compact smooth Riemannian manifold* (M, g), *and assume that the metric* g *is compatible with the Morse charts for* f. *Suppose that the critical point* p *is an immediate successor of the critical point* q *and let* $(N(q,p), F)$ *be a connecting manifold. Let* X *be the CW-complex associated to* f *and let* f_{qp} *be the relative attaching map of the cell in* X *corresponding to* q *to the cell corresponding to* p. *Then the Thom-Pontryagin construction applied to the framed submanifold* $(N(q,p), F)$ *produces a map that is homotopic to* f_{qp} *up to precomposing with a representative of* $\pm 1 \in \pi_j(S^j)$ *where* $j = \lambda_q - 1$.

Remark 6.41 The sign $\pm 1 \in \pi_j(S^j)$ depends on the homotopy class of the framing of the connecting manifold $(N(q,p), F)$. An orientation of $W^s(p)$ and an orientation on M will determine a homotopy class for the framing of $N(q,p)$. So, the sign $\pm 1 \in \pi_j(S^j)$ is determined by the orientation chosen for $W^s(p)$ when M is oriented.

Proof of Theorem 6.40:

Since the Riemannian metric g is compatible with the Morse charts for f, there exists local coordinates $(x_1, \ldots, x_k, y_1, \ldots, y_{m-k})$ on a neighborhood U of the critical point p such that

$$f(\vec{x}, \vec{y}) = f(p) - x_1^2 - \cdots - x_k^2 + y_1^2 + \cdots + y_{m-k}^2$$

and

$$\nabla f = -2x_1 \frac{\partial}{\partial x_1} - \cdots - 2x_k \frac{\partial}{\partial x_k} + 2y_1 \frac{\partial}{\partial y_1} + \cdots + 2y_{m-k} \frac{\partial}{\partial y_{m-k}}.$$

in the local coordinates on U. Shrink U so that

$$U = \{(\vec{x}, \vec{y}) | -\varepsilon^2 \leq -|\vec{x}|^2 + |\vec{y}|^2 \leq \varepsilon^2 \text{ and } |\vec{x}| \, |\vec{y}| \leq \delta\}$$

for some small $\varepsilon, \delta > 0$.

Let $W_0 = f^{-1}((-\infty, c-\varepsilon])$ and $W_1 = f^{-1}((-\infty, c+\varepsilon])$ where $c = f(p)$. Using Remark 3.26, Remark 3.27, and the assumption that the Riemannian metric g is compatible with the Morse charts for f we see that there is a deformation retraction

$$r : W_1 \to W_0 \cup_i W^u(p_i)$$

where $\{p_i\}$ are the critical points in the level set $f^{-1}(c)$ and $p = p_1$. Let $S^u = W^u(q) \cap f^{-1}(c+\varepsilon)$, and let $S^s = W^s(p) \cap f^{-1}(c+\varepsilon) \subset U$; then S^u

and S^s intersect transversally in a manifold N since the function f satisfies the Morse-Smale transversality condition.

The proof of Theorem 3.25 shows that the relative attaching map f_{qp} is homotopic to the composition

$$\tilde{f}_{qp} : S^u \xrightarrow{r} W_0 \cup_i W^u(p_i) \xrightarrow{\pi} S^k$$

where π collapses $W_0 \cup_{i \neq 1} W^u(p_i)$ to a point and S^k is the one point compactification of the disk $W^u(p) - W_0$. Since S^u is transverse to S^s, it is clear that $y = \tilde{f}_{qp}(p)$ is a regular value of $\tilde{f}_{qp}$ and $N(q, p) = \tilde{f}_{qp}^{-1}(y)$. Thus, the framed connecting manifold $N(q, p)$ corresponds to the homotopy class of f_{qp} via the Thom-Pontryagin construction when the framing of $N(q, p)$ is in the same homotopy class as the framing determined by pulling back a basis of $T_y S^k$. If these two framing are in opposite homotopy classes, then precomposing f_{qp} with a smooth representative of $-1 \in \pi_j(S^u)$ gives a map in the same homotopy class as the map determined by the Thom-Pontryagin construction.

$\square$

Remark 6.42 In [59] Franks proved the Connecting Manifold Theorem for any "gradient-like" Morse-Smale vector field on M. See Remark 6.35 for more details.

Remark 6.43 The signs $\pm 1 \in \pi_j(S^j)$ do not affect the homotopy type of X. See Problems 19 and 20 at the end of this chapter for more details.

Problems

1. Let $T^2 = \mathbb{R}^2/\mathbb{Z}^2$ be the 2-torus with its standard metric. Show that $f(x, y) = \cos(2\pi x) + \cos(2\pi y)$ is a Morse-Smale function. Show that the critical points and the connecting orbits can be depicted as follows.

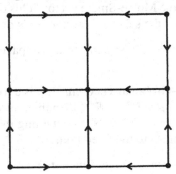

2. Consider the projective plane $\mathbb{R}P^2$ as a closed disk D^2 with opposite points on the boundary identified. Find a formula for a Morse-Smale function on $\mathbb{R}P^2$ with the gradient flow depicted below.

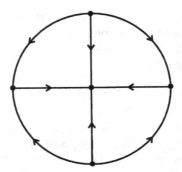

3. Show that the unstable manifolds in Example 6.4 are open cells in a CW-structure on T^2.

4. A vector field on S^1 is called Morse-Smale if and only if it has finitely many zeros and all its zeros are hyperbolic. Show that a vector field on S^1 is C^1-structurally stable if and only if it is Morse-Smale.

5. Let X be a C^r vector field on a smooth manifold M, and let $I = [0, 1]$. A **tubular flow** for X is a pair (U, f) where U is an open set in M and f is a C^r diffeomorphism of U onto the cube $I^m = I \times I^{m-1} \subset \mathbb{R} \times \mathbb{R}^{m-1}$ which takes the trajectories of X in U to the straight lines $I \times \{y\} \subset I \times I^{m-1}$. Prove the **Long Tubular Flow Theorem**: Let X be a C^r vector field on a smooth manifold M, and let $\gamma \subset M$ be an arc of a trajectory of X that is compact and not closed. Then there exists a tubular flow (U, f) of X such that $\gamma \subset F$.

6. Prove Peixoto's Theorem: A vector field of class C^r on a compact orientable 2-manifold M is C^r-structurally stable if and only if it is Morse-Smale. Moreover, the Morse-Smale vector fields of class C^r on M form an open dense subset of the set of C^r vector fields on M.

7. Prove that a Morse-Smale system on a compact manifold is structurally stable.

8. Prove the Hartman-Grobman Theorem: If p is a hyperbolic fixed point of a C^1 diffeomorphism $\varphi : \mathbb{R}^m \to \mathbb{R}^m$, then there is a neighborhood $U \subseteq \mathbb{R}^m$ of p and a neighborhood $U' \subseteq \mathbb{R}^m$ containing the origin such that $\varphi|_U$ is topologically conjugate to the fixed point $0 \in \mathbb{R}^m$ of $d\varphi_p|_{U'}$.

9. The **index filtration** of a Morse function $f : M \to \mathbb{R}$ on a finite dimensional compact smooth Riemannian manifold M is defined to be the increasing sequence of subsets

$$F_0 \subseteq F_1 \subseteq \cdots \subseteq F_m = M$$

where

$$F_k = \bigcup_{\lambda_p \le k} W^u(p)$$

for all $k = 0, \ldots, m = \dim M$. That is, the union is taken over all the unstable manifolds $W^u(p)$ where the critical point p has index less than or equal to k. Find a necessary and sufficient condition on the Morse function f such that $F_k \subseteq F_{k+1}$ is a cofibration for all $k = 0, \ldots, m - 1$.

10. Prove the local version of the Stable/Unstable Manifold Theorem (Theorem 4.2) assuming that the metric is compatible with the Morse charts for $f :$ $M \to \mathbb{R}$.

11. A bundle is **stably trivial** if its Whitney sum with a trivial bundle is trivial. Suppose that $E \to B$ is a stably trivial vector bundle over a CW-complex B. Show that if the dimension of the base B is strictly less than the fiber dimension of E, then E is trivial.

12. Two framed submanifolds (N_0, F_0) and (N_1, F_1) of a finite dimensional compact smooth manifold M are **framed cobordant** if and only if there exists a framed submanifold (W, G) of $M \times \mathbb{R}$ dimension $n + 1$ such that the part of W below $t = 0$ coincides with $(N_0 \times \mathbb{R}, F_0)$ and the part of W above $t = 1$ coincides with $(N_1 \times \mathbb{R}, F_1)$. Show that framed cobordism is an equivalence relation.

13. Let M be a finite dimensional compact smooth manifold, and let $\Omega^k(M)$ be the set of equivalence classes of compact closed framed k-dimensional submanifolds of M under the relation of framed cobordism. Show that $\Omega^k(M)$ has the structure of an abelian group when $2k < m$ where the operation of addition is given by disjoint union.

14. Suppose that F_0 and F_1 are homotopic framings of a submanifold $N \subseteq M$. Prove that the Thom-Pontryagin construction produces homotopic maps $g_0 : M \to S^{m-n}$ and $g_1 : M \to S^{m-n}$.

15. Let $g : M \to S^{m-n}$ be a smooth map, and assume that both $y \in S^{m-n}$ and $y' \in S^{m-n}$ are regular values for g. Let $(V_1, \ldots, V_{m-n})$ be a positively oriented basis for $T_y S^{m-n}$ and $(V_1', \ldots, V_{m-n}')$ a positively oriented basis for $T_{y'} S^{m-n}$. Show that the framed submanifold $(g^{-1}(y), g^*(V_1, \ldots, V_{m-n}))$ is framed cobordant to $(g^{-1}(y'), g^*(V_1', \ldots, V_{m-n}'))$.

16. Prove the Thom-Pontryagin Theorem: There is a bijection between the set $[M, S^{m-n}]$ of homotopy classes of maps from M to S^{m-n} and the set $\Omega^n(M)$ of equivalence classes of n-dimensional compact framed submanifolds of M, where the equivalence relation is given by framed cobordism.

17. Prove the Hopf Degree Theorem: If M is a finite dimensional compact connected smooth oriented manifold, then two maps $M \to S^n$ are homotopic if and only if they have the same degree.

18. Let $f : M \to \mathbb{R}$ be a Morse-Smale function on a finite dimensional compact smooth Riemannian manifold (M, g), and assume that the Riemannian metric g is compatible with the Morse charts for f. Suppose that the critical point p is an immediate successor of the critical point p, neither p nor q is local maximum or minimum of f, and the tangent bundle of M is stably trivial. Show that the framed connecting manifold $N(q, p)$ and the framed connecting manifold $N(p, q)$ determined by the gradient flow of $-f$ determine the same stable homotopy class up to sign.

19. Suppose that $f : X \to Y$ and $f' : X' \to Y'$ are continuous maps between topological spaces, $g : X \to X'$ and $h : Y \to Y'$ are homeomorphisms, and $hf = f'g$. Show that the induced map $C_f \to C_{f'}$ on the mapping cones is a homeomorphism.

$$
\begin{array}{ccccc}
X & \xrightarrow{\ f\ } & Y & \longrightarrow & C_f \\
\approx \downarrow g & & \approx \downarrow h & & \downarrow \\
X' & \xrightarrow{\ f'\ } & Y' & \longrightarrow & C_{f'}
\end{array}
$$

20. Let Y be a CW-complex, and let $f_\partial : S^{n-1} \to Y$ be the attaching map for an n-cell D^n. Show that there is a homotopy equivalence

$$Y \cup_{f_\partial} D^n \simeq Y \cup_{f_\partial \circ \sigma} D^n$$

where $\sigma : S^{n-1} \to S^{n-1}$ satisfies $[\sigma] = -1 \in \pi_{n-1}(S^{n-1})$. Note that this shows that the signs $\pm 1 \in S_j(S^j)$ in Theorem 6.40 do not affect the homotopy type of the CW-complex.

Problems on unstable manifolds versus CW-cells

The following sequence of problems, provided by Bob Wells, describes the construction of a Morse function on $S^1 \times S^2$ whose unstable manifolds **do not** form a CW-structure. The reader should compare these results with the results in Section 6.4 (which only apply to Morse-Smale functions and Riemannian metrics that are compatible with the Morse charts).

21. Let $R > 0$ and let $\Psi : \mathbb{R}^n \to \mathbb{R}^n$ be a smooth map with $\Psi(x) = x$ for all $x \in \mathbb{R}^n$ with $|x| > R$. Prove that Ψ is surjective.

22. Let $R > 0$ and let $\Psi : \mathbb{R}^n \to \mathbb{R}^n$ be a smooth map with $\Psi(x) = x$ for all $x \in \mathbb{R}^n$ with $|x| > R$. Suppose that $d\Psi(x)$ invertible for every $x \in \mathbb{R}^n$. Show that Ψ is bijective, and hence, a diffeomorphism.

23. Let $f : \mathbb{R} \to \mathbb{R}$ be defined by

$$f(x) = \begin{cases} e^{-1/x^2} & \text{for } x > 0 \\ 0 & \text{for } x \leq 0. \end{cases}$$

Show that f is smooth.

24. For $a < b$ define $\beta_{a,b}(t) = f(t-a)f(b-t)$ where f is the smooth function from the previous problem. Show that the maximum value of $\beta_{a,b}(t)$ is:

$$\max\{\beta_{a,b}(t)| \ t \in \mathbb{R}\} = e^{-(a-b)^2/2}.$$

25. For $a < b$ define the plateau function $F_{a,b}$ by setting

$$F_{a,b}(x) = \frac{1}{H(a,b)} \int_{-\infty}^{x} \beta_{a,b}(t) \, dt$$

where

$$H(a,b) = \int_a^b e^{-[(t-a)^2+(t-b)^2]} \, dt.$$

Show that $H(a,b) = H(-b,-a)$ and $F_{a,b}$ is smooth with the following graph:

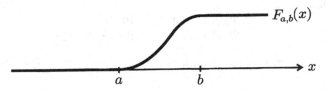

where $F_{a,b}(x) = 1$ for all $x > b$.

26. For $a < b < c < d$ define

$$F_{a,b,c,d}(x) = F_{a,b}(x)F_{-d,-c}(-x).$$

Show that the graph of $F_{a,c,b,d}$ is

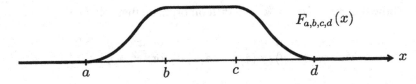

and the following inequalities hold.

$$-\frac{e^{-(c-d)^2/2}}{H(c,d)} \leq F'_{a,b,c,d}(x) \leq \frac{e^{-(a-b)^2/2}}{H(a,b)}$$

Remark: From now on let $F(x) = F_{-2,-1,1,2}(x)$ and $H = H(1,2) = H(-2,-1)$. Then by the previous problem we have the following inequality.

$$-\frac{e^{-1/2}}{H} \leq F'(x) \leq \frac{e^{-1/2}}{H}$$

27. Let $r > 0$ and let $\xi \in \mathbb{R}$. Define $\varphi_{r,\xi} : \mathbb{R} \to \mathbb{R}$ by setting

$$\varphi_{r,\xi}(x) = x + \xi F(x/r).$$

If $|\xi| < re^{1/2}H$, show that $\varphi_{r,\xi} : \mathbb{R} \to \mathbb{R}$ is a diffeomorphism that satisfies

$$\begin{aligned} \varphi_{r,\xi}(x) &= x + \xi \quad \text{for } x \in [-r, r] \\ \varphi_{r,\xi}(x) &= x \qquad \text{for } x \notin [-2r, 2r]. \end{aligned}$$

28. Let $r > 0$ and let $\xi \in \mathbb{R}^n$. Define $\Psi_{r,\xi} : \mathbb{R}^n \to \mathbb{R}^n$ by setting

$$\Psi_{r,\xi}(x) = x + F\left(\frac{|x|^2}{r^2}\right)\xi.$$

If $|\xi| < \frac{r}{4}e^{1/2}H$, show that $\Psi_{r,\xi} : \mathbb{R}^n \to \mathbb{R}^n$ is a diffeomorphism that satisfies

$$\begin{aligned} \Psi_{r,\xi}(x) &= x + \xi \quad \text{for } |x| \leq r \\ \Psi_{r,\xi}(x) &= x \qquad \text{for } |x| \geq \sqrt{2}r. \end{aligned}$$

29. Let $g : \mathbb{R} \to \mathbb{R}$ be smooth with

$$|g(x)| < \frac{r}{2\sqrt{2}}e^{1/2}H$$

for all $x \in \mathbb{R}$. Define $\gamma_{r,g} : \mathbb{R}^2 \to \mathbb{R}^2$ by setting

$$\gamma_{r,g}(x, y) = \left(x, y + F\left(\frac{x^2 + y^2}{r^2}\right)g(x)\right).$$

Show that $\gamma_{r,g} : \mathbb{R}^2 \to \mathbb{R}^2$ is a diffeomorphism that satisfies

$$\begin{aligned} \gamma_{r,g}(x, y) &= (x, y) \qquad\quad \text{for } x^2 + y^2 \geq 2r^2 \\ \gamma_{r,g}(x, y) &= (x, y + g(x)) \quad \text{for } x^2 + y^2 \leq r^2. \end{aligned}$$

Remark: One choice for g is given by

$$g(x) = \begin{cases} 0 & \text{for } x \leq 0 \\ \delta e^{-1/x^2}\sin(1/x) & \text{for } x > 0. \end{cases}$$

The parameter choice $\delta = \frac{r}{4}e^{1/2}H$ satisfies the hypothesis of the previous problem. Replacing $\mathbb{R}^2$ by $S^1 \times S^1$ it is easy to find a diffeomorphism $\gamma : S^1 \times S^1 \to S^1 \times S^1$ doing the same job less restrictively:

$$\gamma(e^{i\theta}, e^{i\phi}) = (e^{i\theta}, e^{i\phi + g(\theta)}).$$

The only restriction is that $g : \mathbb{R} \to \mathbb{R}$ be smooth and 2π-periodic.

30. Let M be a Riemannian manifold, N a smooth submanifold and $f : M \to \mathbb{R}$ a smooth function. Show that N is invariant under the gradient flow of f if and only if for all $x \in N$ the orthogonal complement of the tangent space of N at x is contained in the kernel of df_x, i.e.

$$(T_x N)^\perp \subseteq \ker df_x$$

for all $x \in N$.

31. Let M be a smooth manifold. Let g and $\tilde{g}$ be Riemannian metrics on M, and let $f : M \to \mathbb{R}$ be a smooth function. Show that the gradient flow of f with respect to g is the same as that for $\tilde{g}$ if and only if the following two conditions hold:

  i) For all $x \in M$, the orthogonal complement with respect to $\tilde{g}$ of the gradient of f at x with respect to g is in the kernel of df_x, i.e.

$$(\nabla_g f)(x) \perp_{\tilde{g}} \ker df_x$$

  for all $x \in M$.

  ii) For all $x \in M$, the magnitude with respect to g of the gradient at x of f with respect to g is the magnitude with respect to $\tilde{g}$ of the gradient at x of f with respect to $\tilde{g}$, i.e.

$$|(\nabla_g f)(x)|_g = |(\nabla_{\tilde{g}} f)(x)|_{\tilde{g}}$$

  for all $x \in M$.

32. Let M be a smooth manifold, and let $f : M \to \mathbb{R}$ and $k : M \to [0, 1]$ be smooth functions. Suppose that g and $\tilde{g}$ are Riemannian metrics producing the same gradient flow for f. Show that the Riemannian metric $kg + (1-k)\tilde{g}$ also produces the same gradient flow for f.

Remark: By a **smooth isotopy** (in N) of a diffeomorphism $\psi : N \to N$ from the identity we mean a diffeomorphism $\Psi : N \times [0, 1] \to N \times [0, 1]$ of the form $\Psi(x, t) = (\psi_t(x), t)$, with $\psi_0(x) = x$ and $\psi_1(x) = \psi(x)$ for all $x \in N$.

Recall the that **orbits** of a vector field are the unparameterized versions of the solution trajectories. Thus, proportional vector fields have the same orbits.

33. Let N be a smooth compact manifold, and let g be a Riemannian metric on $N \times [0, 1]$ such that the tangent spaces of $N \times t$ and $x \times [0, 1]$ are orthogonal for every $(x, t) \in N \times [0, 1]$. Let $f : N \times [0, 1] \to [0, 1]$ be the smooth function given by $f(x, t) = t$.

    a. Show that the vector field

$$\frac{(\nabla_g f)(x, t)}{|(\nabla_g f)(x, t)|_g^2}$$

    is the unit vector field along the fibers $x \times [0, 1]$. So, its flow φ_s is given by $\varphi_s(x, t) = (x, t + s)$.

    b. Let $\Phi : N \times [0, 1] \to N \times [0, 1]$ be an isotopy $(x, t) \mapsto (\psi_t(x), t)$ such that $\psi_0(x) = x$ for all $x \in N$. Define a new Riemannian metric $\tilde{g}$ on $N \times [0, 1]$ by requiring for all $(x, t) \in N \times [0, 1]$

      i) $\tilde{g}$ and g agree on $T_x N \times t$

      ii) $(T_x N \times t) \perp_{\tilde{g}} \left(\frac{d\psi_t(x)}{dt}, \frac{\partial}{\partial t} \right)$

      iii) $\tilde{g}(\left(\frac{d\psi_t(x)}{dt}, \frac{\partial}{\partial t} \right), \left(\frac{d\psi_t(x)}{dt}, \frac{\partial}{\partial t} \right)) = 1$.

    Show that the flow generated by the vector field

$$\frac{\nabla_{\tilde{g}} f}{|\nabla_{\tilde{g}} f|_{\tilde{g}}^2}$$

    is given by $\varphi_s(\psi_t(x), t) = (\psi_{t+s}(x), t + s)$.

34. Suppose that we have the following situation:

    i) M is compact smooth manifold.

    ii) g is a Riemannian metric on M.

    iii) $f : M \to \mathbb{R}$ is a Morse function.

    iv) c is a regular value of f.

    v) $q \in M$ is a critical point of f such that $f(q) > c$ and there are no critical values of f in the interval $[c, f(q)]$.

    vi) $S^u(q, c, g) = W_g^u(q) \cap f^{-1}(c)$ is the unstable sphere with respect to g in $f^{-1}(c)$.

    vii) $\psi : f^{-1}(c) \to f^{-1}(c)$ is a diffeomorphism that is smoothly isotopic in $f^{-1}(c)$ to the identity of $f^{-1}(c)$.

Show that there exists a Riemannian metric $\tilde{g}$ on M such that the unstable sphere $S^u(q, c, \tilde{g}) = W^u_{\tilde{g}}(q) \cap f^{-1}(c)$ satisfies

$$S^u(q, c, \tilde{g}) = \psi(S^u(q, c, g)).$$

Moreover, show that for any $\varepsilon > 0$ such that $c < c+\varepsilon < f(q)$ the Riemannian metric $\tilde{g}$ can be chosen so that it agrees with g on $M - f^{-1}([c, c+\varepsilon])$.

35. Let x, y be the usual coordinates in $\mathbb{R}^2$ and u,v,w the usual coordinates in $\mathbb{R}^3$. We may regard $S^1 \times S^2 \subset \mathbb{R}^2 \times \mathbb{R}^3$ as the common solution set of the two equations $x^2 + y^2 = 1$ and $u^2 + v^2 + w^2 = 1$.

a. Show that $(1, 1)$ is a regular value of the map $(x, y, u, v, w) \mapsto (x^2 + y^2, u^2 + v^2 + w^2)$. Use Corollary 5.9 to conclude that $S^1 \times S^2$ is a smoothly embedded submanifold of $\mathbb{R}^5 = \mathbb{R}^2 \times \mathbb{R}^3$.

b. Define smooth functions $W, Y : \mathbb{R}^5 \to \mathbb{R}$ by $W(x, y, u, v, w) = w$ and $Y(x, y, u, v, w) = y$. Show that the restriction of $Y + 2W$ to $S^1 \times S^2$ is a Morse function $f : S^1 \times S^2 \to \mathbb{R}$ with the following critical points, critical values, and Morse indices.

Critical Point p	Critical Value $f(p)$	Morse Index λ_p
$(0, +1, 0, 0, +1)$	$+3$	3
$(0, -1, 0, 0, +1)$	$+1$	2
$(0, +1, 0, 0, -1)$	-1	1
$(0, -1, 0, 0, -1)$	-3	0

c. Show that the map $\alpha : S^1 \times S^1 \to f^{-1}(0)$ given by

$$\alpha(x, y, x', y') = \left(x, y, x'\sqrt{1 - \frac{1}{4}y^2}, y'\sqrt{1 - \frac{1}{4}y^2}, -\frac{1}{2}y\right)$$

is a diffeomorphism (where S^1 is identified as the solution set of $x^2 + y^2 = 1$ in $\mathbb{R}^2$), i.e. α is bijective, smooth, and α^{-1} is also smooth.

d. Let g be the Riemannian metric on $S^1 \times S^2$ given by restricting the standard Riemannian metric on $\mathbb{R}^5$. Show that

$$W^u_g((0, -1, 0, 0, +1)) \cap f^{-1}(0) = \alpha((0, -1) \times S^1)$$

and

$$W^s_g((0, +1, 0, 0, -1)) \cap f^{-1}(0) = \alpha((0, -1) \times S^1)$$

where $(0, -1) \in S^1 \subset \mathbb{R}^2$. Thus, the above stable and unstable manifolds do not intersect transversally.

e. Use Problem 29 to define a diffeomorphism $\psi' : S^1 \times S^1 \to S^1 \times S^1$ so that $(0, -1) \times S^1$ and $\psi'((0, -1) \times S^1)$ meet transversally.

For example:

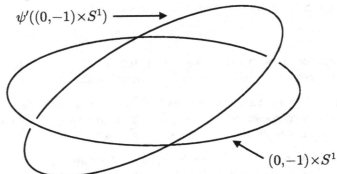

$\psi'((0,-1)\times S^1)$

$(0,-1)\times S^1$

Show that ψ' can be chosen so that the diffeomorphism $\psi : \alpha \circ \psi' \circ \alpha^{-1} :$ $f^{-1}(0) \to f^{-1}(0)$ is isotopic to the identity in $f^{-1}(0) \approx S^1 \times S^1$. Observe then that f is a Morse-Smale function with respect to the new Riemannian metric produced by Problem 34.

f. Use Problem 29 to define a diffeomorphism $\vartheta' : S^1 \times S^1 \to S^1 \times S^1$ that satisfies the following two conditions:

   i) $\vartheta'((0,-1) \times S^1)$ and $(0,-1) \times S^1$ are disjoint except in the image of the coordinate map $\rho : (-\frac{\pi}{2}, \frac{\pi}{2}) \times (-\frac{\pi}{2}, \frac{\pi}{2}) \to S^1 \times S^1$ given by

$$\rho(\theta, \phi) = (e^{i(\theta + \frac{3\pi}{2})}, e^{i\phi}),$$

   where $\rho^{-1}((0,-1) \times S^1) = 0 \times (-\frac{\pi}{2}, \frac{\pi}{2})$.

   ii) $\rho^{-1}(\vartheta'((0,-1) \times S^1)) \cap (-\frac{\pi}{4}, \frac{\pi}{4}) \times (-\frac{\pi}{2}, \frac{\pi}{2}) =$ $\{(\theta, \phi)|\ |\theta| < \frac{\pi}{4}$ and $\theta = e^{-1/\phi^2} \sin(\frac{1}{\phi})$ or $(\theta, \phi) = (0,0)\}$.

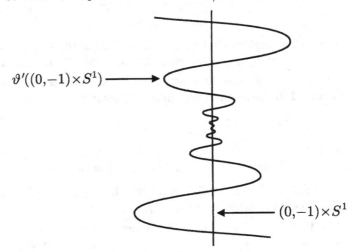

$\vartheta'((0,-1)\times S^1)$

$(0,-1)\times S^1$

Show that ϑ' can be chosen so that $\vartheta = \alpha \circ \vartheta' \circ \alpha^{-1}$ is isotopic to the identity in $f^{-1}(0) \approx S^1 \times S^1$. Hence, Problem 34 can be used to produce a Riemannian metric $\tilde{g}$ on $S^1 \times S^2$ such that

$$S^u((0, -1, 0, 0, +1), 0, \tilde{g}) = \alpha(\vartheta'((0, -1) \times S^1)).$$

g. Let B be the infinite bouquet of decreasing circles:

Show that for $x \in W_{\tilde{g}}^u((0, +1, 0, 0, -1))$ with $x \neq (0, +1, 0, 0, -1)$ a neighborhood of x in

$$W_{\tilde{g}}^u((0, -1, 0, 0, +1)) \cup W_{\tilde{g}}^u((0, +1, 0, 0, -1)) \cup \{(0, -1, 0, 0, -1)\}$$

is homeomorphic to $(-1, 1) \times B$, where $\tilde{g}$ is the Riemannian metric produced in part f. Conclude that the union of these unstable manifolds is **not** locally contractible, and hence, **not** a CW-complex. Note that with respect to the index filtration of f:

$$F_0 \subseteq F_1 \subseteq F_2 \subseteq F_3$$

the above says that the F_2 term is **not** a CW-complex and the inclusion $F_2 \subseteq F_3$ is **not** a cofibration.

36. Continuing from the previous problem, we will denote the set

$$A_n = (\alpha \circ \rho) \left(\left\{ (\theta, \phi) \mid \theta = e^{-1/\phi^2} \sin\left(\frac{1}{\phi}\right) \text{ and } \frac{\pi}{n} \geq \phi \geq \frac{\pi}{n+1} \right\} \right)$$

for $n = \pm 1, \pm 2, \ldots$. Show that $W_{\tilde{g}}^s((0, +1, 0, 0, -1)) \cap f^{-1}((-\infty, 0])$ is equal to

$$\{(0, +1, u, v, w) \mid u^2 + v^2 + w^2 = 1 \text{ and } w \leq -\frac{1}{2}\}$$

and $W_{\tilde{g}}^u((0, +1, 0, 0, -1))$ is equal to

$$\{(x, y, 0, 0, -1) \mid x^2 + y^2 = 1 \text{ and } y \neq -1\} \subset f^{-1}((-\infty, 0])$$

where $\tilde{g}$ denotes the Riemannian metric produced in part f. Using whatever spiritual exercise you wish, convince yourself that the closure of the orbit of A_n under the gradient flow of $-f$ looks something like this:

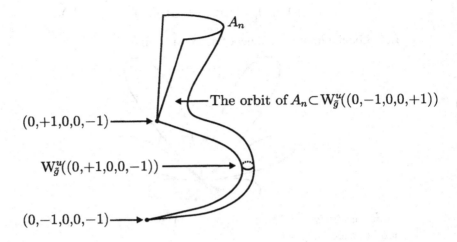

Chapter 7

The Morse Homology Theorem

In this chapter we construct the Morse-Smale-Witten chain complex and prove that its homology coincides with the singular homology. For an interesting history of this complex we refer the reader to Bott's colorful paper [26]. The story started with a Comptes Rendus Note of the French Academy of Sciences by René Thom in 1949 [145] and culminated with Witten's paper in 1982 [153], where the boundary operator described here was explicitly written. The way Witten arrived at this boundary operator was through supersymmetric mechanics. Before, Morse and Smale's work had already found the ideas required to make rigorous Witten's "physicist's" proof: these ideas have been explained in the preceeding chapters. This is why we call the resulting complex the Thom-Morse-Smale-Witten chain complex, or simply, the Morse-Smale-Witten chain complex. In the literature it is sometimes called the Witten complex [26].

The Morse-Smale-Witten boundary operator is expressed in terms of intersection numbers of the "unstable spheres" of critical points of index k and the "stable spheres" of critical points of index $k - 1$. This was essentially proved by Milnor [102], but his proof did not attract attention. Floer proved Witten's boundary formula using the Conley index [35], but acknowledged Milnor's unnoticed contribution [56]. We give here an alternative proof using the Conley index as well which is due to Salamon [128].

We first discuss orientations of stable/unstable manifolds, and then we define the Morse-Smale-Witten boundary operator. We give some examples of computing homology using the Morse-Smale-Witten chain complex, and then we introduce the Conley index and describe examples of index pairs which are pertinent to the proof of the main theorem. Using the fact that the Conley index is independent of the choice of the index pairs representing it, we prove that the Morse-Smale-Witten boundary operator coincides with the boundary operator in the CW-Homology Theorem (Theorem 2.15). The last part of this chapter is devoted to proving that the Conley index is independent of the choice of index pairs representing it [35].

7.1   The Morse-Smale-Witten boundary operator

Let (M, g) be a finite dimensional compact smooth oriented Riemannian manifold, and let $f : M \to \mathbb{R}$ be a Morse-Smale function. Recall that this means that for all critical points p and q of f, the unstable manifold $W^u(q)$ and the stable manifold $W^s(p)$ intersect transversally. We denote $W(q,p) = W^u(q) \cap W^s(p)$. This is either empty or a smooth manifold of dimension $\lambda_q - \lambda_p$ by Proposition 6.2, where λ_q is the index of q and λ_p is the index of p. In the latter case we write $q \succeq p$. We will denote by $\mathrm{Cr}(f)$ the set of all critical points of f and by $\mathrm{Cr}_k(f)$ those critical points q with $\lambda_q = k$.

Orientation conventions

For each $p \in \mathrm{Cr}(f)$, we choose a basis B_p^u of $T_p^u M = T_p W^u(p)$ giving the orientation of $T_p W^u(p)$. This orientation of $T_p^u M$ determines an orientation of $T_p^s M = T_p W^s(p)$ since $T_p M = T_p^s M \oplus T_p^u M$ (see Section 4.1 and Section 5.6). That is, the embedded submanifolds $W^u(p)$ and $W^s(p)$ have orientations compatible with the orientation of M at p. These orientations also determine orientations of $T_v T_p^u M \approx T_p^u M$ for all $v \in T_p^u M$ and $T_v T_p^s M \approx T_p^s M$ for all $v \in T_p^s M$. Hence, there is an orientation determined on $T_x W^u(p)$ for all $x \in W^u(p)$ by the embedding $E^u : T_p^u M \to W^u(p)$ defined by the Stable/Unstable Manifold Theorem (Theorem 4.2). Similarly, there is an orientation determined on $T_x W^s(p)$ for all $x \in W^s(p)$ by the embedding $E^s : T_p^s M \to W^s(p)$.

Counting flow lines with sign – the definition of $n(q,p)$

Consider now two critical points p, q of index $\lambda_p = k - 1$ and $\lambda_q = k$ respectively, and assume that $q \succeq p$. Let $\gamma : \mathbb{R} \to M$ be a gradient flow line from q to p:

$$\frac{d}{dt}\gamma(t) = -(\nabla f)(\gamma(t)), \quad \lim_{t \to -\infty} \gamma(t) = q, \quad \lim_{t \to \infty} \gamma(t) = p.$$

At any point $x \in \gamma(\mathbb{R}) \subset W(q,p)$ we can complete $-(\nabla f)(x)$ to a positive basis $(-(\nabla f)(x), \hat{B}_x^u)$ of $T_x W^u(q)$ (providing the orientation of $W^u(q)$ at x). If we pick any positive basis B_x^s of $T_x W^s(p)$ (providing the orientation of $W^s(p)$ at x), then $(B_x^s, \hat{B}_x^u)$ is a basis for $T_x M$. If $(B_x^s, \hat{B}_x^u)$ is a positive orientation for $T_x M$, then we assign $+1$ to the flow γ. Otherwise we assign -1 to the flow. Since the orientations on $W^u(q)$ and $W^s(p)$ are defined so that

E^s and E^u are orientation preserving, it's clear that this assignment does not depend on the point $x \in \gamma(\mathbb{R})$.

Recall that if $\lambda_q - \lambda_p = 1$, then $W(q, p) \cup \{q, p\}$ is a compact 1-dimensional manifold (Corollary 6.29), with an action of $\mathbb{R}$ given by flowing for time $t \in \mathbb{R}$. Hence, $\mathcal{M}(q, p) = W(q, p)/\mathbb{R}$ is a compact zero dimensional manifold, i.e. it consists of a finite number of elements, and the number of elements in $\mathcal{M}(q, p)$ is precisely the number of flows γ from q to p. To each flow γ from q to p we have assigned a number $+1$ or -1 using the orientations. The integer $n(q, p) \in \mathbb{Z}$ is defined to be the sum of these numbers.

Remark 7.1 There are a lot of choices involved in assigning $+1$ or -1 to a gradient flow line. For instance, we could complete $-(\nabla f)(x)$ to a positive basis $(\hat{B}_x^u, -(\nabla f)(x))$ instead of $(-(\nabla f)(x), \hat{B}_x^u)$. Similarly, we could use $(B_q^u, \hat{B}_p^s)$ instead of $(B_p^s, \hat{B}_q^u)$ as the basis for $T_x M$. It is easy to see that changing one of these conventions only changes the sign of $n(q, p)$, and hence these choices only affect the sign of the Morse-Smale-Witten boundary operator.

An alternate definition of $n(q, p)$ using intersection numbers

Let c be a regular value in the open interval (a, b) where $f(p) = a$ and $f(q) = b$. Consider the **unstable sphere** of q:

$$S^u(q) = W^u(q) \cap f^{-1}(c)$$

and the **stable sphere** of p:

$$S^s(p) = W^s(p) \cap f^{-1}(c)$$

inside the level set $f^{-1}(c)$. By Corollary 5.9, $f^{-1}(c)$ is an $(m-1)$-dimensional manifold which we orient as follows: for any $x \in f^{-1}(c)$ a basis $v_1, \ldots, v_{m-1}$ of $T_x f^{-1}(c)$ is positive if and only if $-(\nabla f)(x), v_1, \ldots, v_{m-1}$ is a positive basis for $T_x M$. By Corollary 5.12, $S^u(q)$ is a $(k-1)$-dimensional manifold and $S^s(p)$ is an $(m-k)$-dimensional manifold. We orient these manifolds using the same convention just described for $f^{-1}(c)$.

Since the manifolds $S^u(q)$ and $S^s(p)$ intersect transversally in the submanifold $f^{-1}(c)$, Corollary 5.12 implies that $S^u(q) \cap S^s(p)$ is a 0-dimensional manifold, were each point corresponds to a connecting orbit in $W(q, p)$, i.e.

$$S^u(q) \cap S^s(p) \approx \mathcal{M}(q, p).$$

Recall that this set is finite by Corollary 6.29. The integer $n(q, p) \in \mathbb{Z}$ can also be defined as the intersection number of the oriented manifolds $S^u(q)$ and $S^s(p)$ inside the oriented manifold $f^{-1}(c)$ (see Section 5.6).

The Morse Homology Theorem

Definition 7.2 (Morse-Smale-Witten Chain Complex) *Let* $f : M \to \mathbb{R}$ *be a Morse-Smale function on a compact smooth oriented Riemannian manifold M of dimension $m < \infty$, and assume that orientations for the unstable manifolds of f have been chosen. Let $C_k(f)$ be the free abelian group generated by the critical points of index k, and let*

$$C_*(f) = \bigoplus_{k=0}^{m} C_k(f).$$

The homomorphism $\partial_k : C_k(f) \to C_{k-1}(f)$ defined by

$$\partial_k(q) = \sum_{p \in Cr_{k-1}(f)} n(q,p)p$$

*is called the **Morse-Smale-Witten** boundary operator, and the pair $(C_*(f), \partial_*)$ is called the **Morse-Smale-Witten chain complex** of f.*

Remark 7.3 The integer $n(q,p) \in \mathbb{Z}$ in the preceeding definition is well defined by Corollary 6.29 of the λ-Lemma (Theorem 6.17).

Theorem 7.4 (Morse Homology Theorem) *The pair $(C_*(f), \partial_*)$ is a chain complex, and its homology is isomorphic to the singular homology $H_*(M; \mathbb{Z})$.*

Remark 7.5 See the beginning of this chapter for the origin of the appellation "Morse-Smale-Witten". Also, note that the preceeding theorem implies that the homology of the Morse-Smale-Witten chain complex is independent of the orientations chosen for the unstable manifolds.

Remark 7.6 The Morse-Smale-Witten chain complex can be defined with coefficients in any commutative ring with unit Λ by taking the tensor product $C_k(f) \otimes \Lambda$ for all k. Since we have an isomorphism for homology with integer coefficients, the Universal Coefficient Theorem asserts that we have isomorphisms with coefficients in Λ (see for instance Theorem V.7.4 of [30]).

Motivation for the proof of Theorem 7.4:

We proved in Chapter 3 that if M is a compact manifold and $f : M \to \mathbb{R}$ is a Morse function, then M is homotopy equivalent to a CW-complex X, where the k-skeleton $X^{(k)}$ is obtained from the $(k-1)$-skeleton by attaching one k-cell for each critical point of index k. The CW-homology of X with integer coefficients is the homology of the chain complex $(\underline{C}_*(X), \underline{\partial}_*)$ where the free abelian group $\underline{C}_k(X)$ is generated by the k-cells of X for all $k \in \mathbb{Z}_+$ and $\underline{\partial}_k : \underline{C}_k(X) \to \underline{C}_{k-1}(X)$ is induced from the homology exact sequence of the triple $(X^{(k)}, X^{(k-1)}, X^{(k-2)})$. That is, the CW-boundary operator $\underline{\partial}_k : \underline{C}_k(X) \to \underline{C}_{k-1}(X)$ is defined as the composite of the following maps:

$$\underline{C}_k(X) \xrightarrow{\Psi_k} H_k(X^{(k)}, X^{(k-1)}; \mathbb{Z}) \xrightarrow{\delta_*} H_{k-1}(X^{(k-1)}, X^{(k-2)}; \mathbb{Z}) \xrightarrow{\Phi_{k-1}} \underline{C}_{k-1}(X)$$

where the map δ_* is the connecting homomorphism, the map $\Psi_k : \underline{C}_k(X) \to H_k(X^{(k)}, X^{(k-1)}; \mathbb{Z})$ is the isomorphism given by Lemma 2.11, and $\Phi_{k-1} : H_{k-1}(X^{(k-1)}, X^{(k-2)}; \mathbb{Z}) \to \underline{C}_{k-1}(X)$ is the inverse of Ψ_{k-1} described explicitly in Proposition 2.14.

So far the Riemannian metric has played a minimal role. Now assume that the Morse function $f : M \to \mathbb{R}$ satisfies the Morse-Smale transversality condition with respect to the Riemannian metric. For any $k \in \mathbb{Z}_+$ we can identify $\underline{C}_k(X)$ and $C_k(f)$ since they are both free abelian groups whose generators are indexed by the critical points of index k. If we could prove that under this identification the following diagram commutes:

$$\cdots \xrightarrow{\partial_{k+2}} \underline{C}_{k+1}(X) \xrightarrow{\partial_{k+1}} \underline{C}_k(X) \xrightarrow{\partial_k} \underline{C}_{k-1}(X) \xrightarrow{\partial_{k-1}} \cdots$$
$$\Big\updownarrow{\approx} \qquad\qquad \Big\updownarrow{\approx} \qquad\qquad \Big\updownarrow{\approx}$$
$$\cdots \xrightarrow{\partial_{k+2}} C_{k+1}(f) \xrightarrow{\partial_{k+1}} C_k(f) \xrightarrow{\partial_k} C_{k-1}(f) \xrightarrow{\partial_{k-1}} \cdots$$

then the proof of Theorem 7.4 would be complete since

$$H_*(C_*(f), \partial_*) \approx H_*(\underline{C}_*(X), \underline{\partial}_*) \approx H_*(X; \mathbb{Z}) \approx H_*(M; \mathbb{Z})$$

where the second isomorphism comes from the CW-Homology Theorem (Theorem 2.15), and the last isomorphism comes from the fact that X is homotopy equivalent to M by Theorem 3.28.

To prove that the above diagram commutes we would need to compare the boundary operator $\underline{\partial}_k$ in the CW-chain complex with the Morse-Smale-Witten boundary operator ∂_k. The CW-Homology Theorem (Theorem 2.15) says that the boundary operator $\underline{\partial}_k$ can be computed in terms of the degrees of the relative attaching maps, and the alternate definition of $n(q, p)$ given above describes ∂_k in terms of the oriented intersection numbers $S^u(q) \cap S^s(p)$. However, the relative attaching maps are maps of spheres $S^{k-1} \to S^{k-1}$, whereas the intersection $S^u(q) \cap S^s(p)$ is always a 0-dimensional manifold for any value of k. These notions sound similar when $k = 1$, but for $k > 1$ it's not readily apparent that these numbers are the same.

Another difficulty that appears when one tries to compare ∂_k directly to $\underline{\partial}_k$ is this: The Morse-Smale-Witten boundary operator ∂_k is defined in terms of gradient flows in M, whereas the boundary operator $\underline{\partial}_k$ in the CW-chain complex is defined in terms of subsets of a CW-complex X that is **homotopic** to M. Moreover, the proof that X is homotopic to M (Theorem 3.28) relied on the Cellular Approximation Theorem (Theorem 2.20), and hence the attaching maps in X were only well-defined up to homotopy. This makes it difficult to directly compare the two boundary operators ∂_k and $\underline{\partial}_k$.

Outline of the proof of Theorem 7.4

In Section 7.4 we avoid both the topological and analytical difficulties inherent in the approach suggested above by using the Conley index and Conley's connection matrix. The Conley index was introduced by Conley in [35] and further refined by Salamon in [127]. It is a pointed homotopy class associated to an isolated compact invariant set of a dynamical system that remains the same under small perturbations of the system. The Conley index captures the change in the homotopy type of $M^t = \{x \in M \mid f(x) \le t\}$ when $t \in \mathbb{R}$ crosses a critical value and was used by Goresky and MacPherson in their study of stratified Morse theory (see Section 3.3 of [65]).

The proof we give in Section 7.4 of the Morse-Homology Theorem is due to Salamon [128], and it describes the Morse-Smale-Witten boundary operator as a special case of Conley's connection matrix [35] [60] [61]. The CW-complex $X \simeq M$ is replaced by a filtration of index pairs

$$\emptyset = N_{-1} \subseteq N_0 \subseteq N_1 \subseteq \cdots \subseteq N_m = M$$

such that N_k/N_{k-1} is the Conley index of $\mathrm{Cr}_k(f)$ for all $k = 0, \ldots, m$. This filtration satisfies $H_j(N_k, N_{k-1}; \mathbb{Z}) = 0$ for all $j \ne k$, and hence we can use the same arguments we used to prove Lemma 2.12 to show that the above filtration satisfies conditions similar to the skeletal filtration

$$\emptyset = X^{-1} \subseteq X^{(0)} \subseteq X^{(1)} \subseteq \cdots \subseteq X^{(m)} = X.$$

In particular, $H_j(N_k; \mathbb{Z}) \approx H_j(M; \mathbb{Z})$ for all $j < k$, and $H_j(N_k; \mathbb{Z}) = \{0\}$ for all $j > k$. The advantages of working with a filtration of M rather than a CW-complex homotopic to M become apparent in the proof of Lemma 7.21 where we show how to relate the Morse-Smale-Witten boundary operator to a connecting homomorphism.

Before getting into the technical details of the Conley index and the connection matrix we present in Section 7.2 examples of computing the homology of some well known spaces using the Morse-Smale-Witten chain complex. In Section 7.3 we define isolated compact invariant sets, index pairs, and the Conley index, and we describe examples of index pairs which are pertinent to the proof of the main theorem. In Section 7.4 we show how to construct the above filtration of index pairs, and we relate the Morse-Smale-Witten boundary operator

$$\partial_k : C_k(f) \to C_{k-1}(f)$$

to the connecting homomorphism

$$\delta_* : H_k(N_k, N_{k-1}) \to H_{k-1}(N_{k-2}, N_{k-2}).$$

using the fact that we can choose "nice" index pairs for the critical points and the Conley index is independent of the choice of the index pairs representing it (Lemma 7.21). Once we prove Lemma 7.21 (the main technical lemma) the proof of the Morse Homology Theorem follows easily from arguments similar to those used to prove the CW-homology Theorem (Theorem 2.15). Section 7.5 is devoted to proving that the Conley index is independent of the choice of index pairs representing it [35] [127].

7.2 Examples using the Morse Homology Theorem

A nice feature of the Morse-Smale-Witten chain complex is that the boundary operator is described geometrically. So, if we can draw the space and the Morse-Smale gradient flows, then it is often easy to compute homology using the Morse Homology Theorem (Theorem 7.4). In the following pictures the arrows indicate our choices for the **orientations** of the unstable manifolds and the induced orientations on the stable manifolds. Unlike the pictures in Chapters 4 and 6, the arrows in the following pictures **do not** indicate the direction of the gradient flow lines.

Example 7.7 (A circle) Consider $M = S^1$ oriented clockwise and the Morse-Smale function $f : M \to \mathbb{R}$ given by the height function in the following picture, where we have chosen to orient $T_q^u M$ from left to right as indicated.

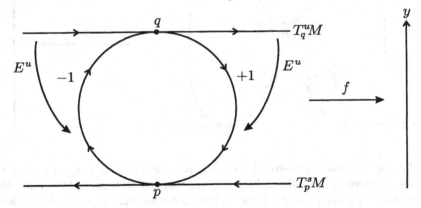

The gradient flow is downward, and hence $-(\nabla f)(x)$ agrees with the orientation of $T_x W^u(q)$ when x is on the right side and disagrees with the orientation of $T_x W^u(q)$ when x is on the left side. Thus, $\hat{B}^u_x = +1$ when x is on the right, and $\hat{B}^u_x = -1$ when x is on the left. If we give $T^u_p S^1 = \{0\}$ the orientation $+1$, then the induced orientation on $W^s(p)$ agrees with the orientation of S^1, and $(B^s_x, \hat{B}^u_x)$ is a positive orientation of $T_x S^1$ when x is on the right and a negative orientation of $T_x S^1$ when x is on the left. Hence, the flow on the right is assigned $+1$ while the flow on the left is assigned -1. (Note that if we chose to give $T^u_p S^1$ the orientation -1, then the signs would be reversed.)

Thus $n(q,p) = 0$, and the Morse-Smale-Witten chain complex

$$
\begin{array}{ccccc}
C_1(f) & \xrightarrow{\partial_1 = 0} & C_0(f) & \longrightarrow & 0 \\
\Big\uparrow{\scriptstyle \approx} & & \Big\uparrow{\scriptstyle \approx} & & \\
<q> & \xrightarrow{\partial_1 = 0} & <p> & \longrightarrow & 0
\end{array}
$$

yields

$$
H_k((C_*(f), \partial_*)) = \begin{cases} \mathbb{Z} & \text{if } k = 0, 1 \\ 0 & \text{otherwise} \end{cases}
$$

as expected.

Example 7.8 (A deformed circle) Consider $M = S^1$ oriented clockwise and the Morse-Smale function $f : M \to \mathbb{R}$ given by the height function in the following picture.

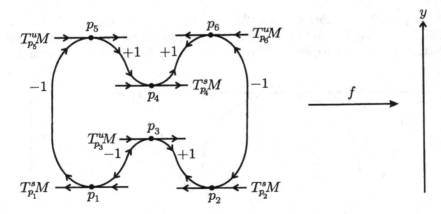

In the above picture we have chosen to orient the unstable tangent spaces at p_3 and p_5 from left to right and the tangent space at p_6 from right to left. We have chosen the orientation $+1$ for the unstable manifolds at p_1, p_2 and p_4, and hence the stable tangent spaces at p_1, p_2 and p_4 all have the same orientation as the tangent space of S^1 at these points. Thus, the sign associated to a gradient flow line in the above picture is $+1$ when the orientation of the unstable manifold points downward and -1 when the orientation of the unstable manifold points upward.

The Morse-Smale-Witten chain complex is

$$C_1(f) \xrightarrow{\partial_1} C_0(f) \longrightarrow 0$$

$$\uparrow \approx \qquad\qquad \uparrow \approx$$

$$< p_3, p_5, p_6 > \xrightarrow{\partial_1} < p_1, p_2, p_4 > \longrightarrow 0$$

where the coefficients $n(p_i, p_j)$ for the boundary operator ∂_1 are given in the following table.

$$n(p_i, p_j)$$

	p_1	p_2	p_3	p_4	p_5	p_6
p_1	0	0	0	0	0	0
p_2	0	0	0	0	0	0
p_3	-1	$+1$	0	0	0	0
p_4	0	0	0	0	0	0
p_5	-1	0	0	$+1$	0	0
p_6	0	-1	0	$+1$	0	0

It is easy to see that $H_1((C_*(f), \partial_*)) = \ker \partial_1 =< p_3 - p_5 + p_6 > \approx \mathbb{Z}$ since

$$\partial_1(p_3 - p_5 + p_6) = (p_2 - p_1) - (p_4 - p_1) + (p_4 - p_2) = 0,$$

and $\ker \partial_0 =< p_1, p_2, p_4 > \approx \mathbb{Z} \oplus \mathbb{Z} \oplus \mathbb{Z}$. The image of ∂_1 is the free abelian group generated by $\partial_1(p_3) = p_2 - p_1$, $\partial_1(p_5) = p_4 - p_1$, and $\partial_1(p_6) = p_4 - p_2$. Hence,

$$
\begin{aligned}
H_0((C_*(f), \partial_*)) &= \ker \partial_0 / \mathrm{im}\, \partial_1 \\
&\approx < p_1, p_2, p_4 > / < p_2 - p_1, p_4 - p_1, p_4 - p_2 > \\
&\approx < p_1, p_2, p_4; \ p_1 = p_2 = p_4 > \\
&\approx \mathbb{Z},
\end{aligned}
$$

and we have

$$H_k((C_*(f), \partial_*)) = \begin{cases} \mathbb{Z} & \text{if } k = 0, 1 \\ 0 & \text{otherwise} \end{cases}$$

as expected.

Example 7.9 (The n-sphere) Let

$$M = S^n = \{(x_1, \ldots, x_{n+1}) \in \mathbb{R}^{n+1} | \ x_1^2 + \cdots + x_{n+1}^2 = 1\}$$

be the n-sphere, and define $f : S^n \to \mathbb{R}$ by $f(x_1, \ldots, x_{n+1}) = x_{n+1}$. The function f is a Morse-Smale function on S^n with two critical points, the north pole $N = (0, \ldots, 0, +1)$ (the maximum) and the south pole $S = (0, \ldots, 0, -1)$ (the minimum).

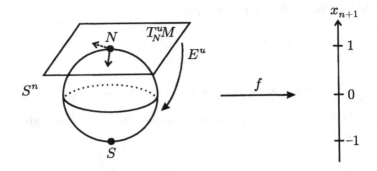

The Morse-Smale-Witten chain complex is as follows.

$$
\begin{array}{ccccccccc}
C_n(f) & \xrightarrow{\partial_n} & C_{n-1}(f) & \xrightarrow{\partial_{n-1}} & \cdots & \xrightarrow{\partial_2} & C_1(f) & \xrightarrow{\partial_1} & C_0(f) \longrightarrow 0 \\
\Big\uparrow{\scriptstyle\approx} & & \Big\uparrow{\scriptstyle\approx} & & & & \Big\uparrow{\scriptstyle\approx} & & \Big\uparrow{\scriptstyle\approx} \\
<N> & \xrightarrow{\partial_n} & 0 & \xrightarrow{\partial_{n-1}} & \cdots & \xrightarrow{\partial_2} & 0 & \xrightarrow{\partial_1} & <S> \longrightarrow 0
\end{array}
$$

When $n = 0$ it is clear that the boundary operators are all zero, in Example 7.7 we showed that $\partial_1 = 0$ when $n = 1$, and when $n > 1$ there are no critical points of relative index 1. Hence, for any $n \geq 0$ the boundary operators in the Morse-Smale-Witten chain complex of the height function on S^n are all zero. We have

$$
H_k((C_*(f), \partial_*)) = \begin{cases} \mathbb{Z} \oplus \mathbb{Z} & \text{if } k = 0 \\ 0 & \text{otherwise} \end{cases}
$$

when $n = 0$ and

$$
H_k((C_*(f), \partial_*)) = \begin{cases} \mathbb{Z} & \text{if } k = 0, n \\ 0 & \text{otherwise} \end{cases}
$$

for all $n > 0$ as expected.

Example 7.10 (A deformed 2-sphere) Consider $M = S^2$ and the Morse-Smale function $f : M \to \mathbb{R}$ given by the height function in the following picture, where the orientation chosen for S^2 is indicated by a solid vector followed by a dashed vector.

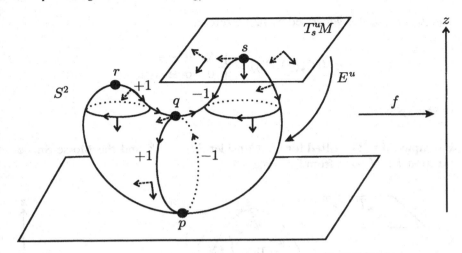

There are 4 critical points p, q, r, and s with the following indices:

$$\begin{aligned} \lambda_r = \lambda_s &= 2 \\ \lambda_q &= 1 \\ \lambda_p &= 0, \end{aligned}$$

and the Morse-Smale-Witten chain complex is as follows.

$$\begin{array}{ccccccc} C_2(f) & \xrightarrow{\partial_2} & C_1(f) & \xrightarrow{\partial_1} & C_0(f) & \longrightarrow & 0 \\ \uparrow{\scriptstyle\approx} & & \uparrow{\scriptstyle\approx} & & \uparrow{\scriptstyle\approx} & & \\ <r,s> & \xrightarrow{\partial_2} & <q> & \xrightarrow{\partial_1} & <p> & \longrightarrow & 0 \end{array}$$

From the picture we see that when $x \in W(r,q)$, the vector $-(\nabla f)(x)$ points in the direction of the solid vector giving the first element of the orientation of $T_x W^u(r)$, and hence $\hat{B}_x^u$ can be taken as the dashed vector pointing outward. In the picture we have chosen to orient $W^u(q)$ in the direction of the dashed vector pointing from back to front, and hence the induced orientation on $W^s(q)$ is from left to right because of our orientation convention $T_q M = T_q^s M \oplus T_q^u M$. Therefore, we can take $B_x^s = -(\nabla f)(x)$ and $(B_x^s, \hat{B}_x^u)$ agrees with the orientation of S^2 when $x \in W(r,q)$. Thus, $\partial_2(r) = q$.

Similarly, the orientations in the picture show that $\partial_2(s) = -q$ because $\hat{B}_x^u$ points inward when $x \in W(s,q)$. Hence, $\partial_2(r+s) = 0$, and the homology group $H_2((C_*(f), \partial_*)) \approx < r+s > \approx \mathbb{Z}$. Since $\partial_2(r) = q$, we see that $H_1((C_*(f), \partial_*)) = 0$ because the generator of $C_1(f)$ is in the image of ∂_2. Since $\partial_1(q) = \partial^2 r = 0$ we see that p is not in the image of ∂_1, and hence,

$H_0((C_*(f), \partial_*))$ is generated by p. Therefore,

$$H_k((C_*(f), \partial_*)) = \begin{cases} \mathbb{Z} & \text{if } k = 0, 2 \\ 0 & \text{otherwise} \end{cases}$$

as expected.

Example 7.11 (The tilted torus) Consider $M = T^2$ and the Morse-Smale function $f : M \to \mathbb{R}$ from Example 6.4.

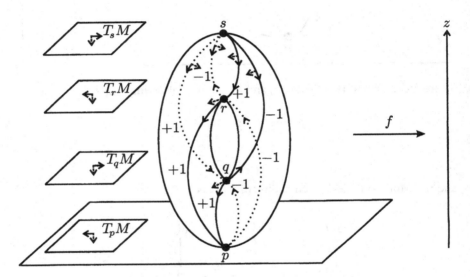

In the preceeding picture the orientation chosen for T^2 is indicated by a solid vector followed by a dashed vector, and the orientation chosen for $T_q^u M$ and $T_r^u M$ are indicated by the dashed vectors pointing outward. The induced orientations on $W^s(q)$ and $W^s(r)$ coming from our orientation convention $T_q M = T_q^s M \oplus T_q^u M$ are indicated by the arrows, and the orientation of $T_p W^s(p)$ agrees with the orientation of $T_p M$ if we choose the orientation $+1$ for $T_p^u M = \{0\}$. The reader should verify that these choices for the orientations induce the signs for the gradient flow lines shown in the picture.

There are 4 critical points p, q, r, and s with the following indices:

$$\lambda_s = 2$$
$$\lambda_q = \lambda_r = 1$$
$$\lambda_p = 0,$$

and the Morse-Smale-Witten chain complex is as follows.

$$C_2(f) \xrightarrow{\partial_2=0} C_1(f) \xrightarrow{\partial_1=0} C_0(f) \longrightarrow 0$$

$$\Big\downarrow\approx \qquad \Big\downarrow\approx \qquad \Big\downarrow\approx$$

$$<s> \xrightarrow{\partial_2=0} <q,r> \xrightarrow{\partial_1=0} <p> \longrightarrow 0$$

Therefore,

$$H_k((C_*(f), \partial_*)) = \begin{cases} \mathbb{Z} & \text{if } k = 0, 2 \\ \mathbb{Z} \oplus \mathbb{Z} & \text{if } k = 1 \\ 0 & \text{otherwise} \end{cases}$$

as expected.

7.3 The Conley index

Basic definitions

Let $\varphi_t : M \to M$ be a flow on a locally compact metric space M, i.e. a continuous action of the real numbers $\mathbb{R}$ on M:

$$\begin{aligned} \mathbb{R} \times M &\to M \\ (t, x) &\mapsto \varphi_t(x) \end{aligned}$$

that satisfies $\varphi_t \circ \varphi_s = \varphi_{t+s}$ for all $s, t \in \mathbb{R}$ and $\varphi_0 = \text{id}$. We will also write $\varphi_t(x) = x \cdot t$. A subset $S \subseteq M$ is called an **invariant subset** if and only if $\varphi_t(S) = S$ for all $t \in \mathbb{R}$. Given any subset $N \subseteq M$ we will consider the **maximal invariant subset**

$$\begin{aligned} I(N) &= \{x \in N \,|\, \varphi_t(x) \in N \text{ for all } t \in \mathbb{R}\} \\ &= \bigcap_{t \in \mathbb{R}} N \cdot t. \end{aligned}$$

A compact invariant subset S is said to be **isolated** if and only if there exists a compact neighborhood N of S such that $I(N) = S$. Such a neighborhood N is called an **isolating neighborhood**.

Definition 7.12 *An **index pair** (N, L) for an isolated compact invariant subset S is a pair of compact sets $L \subset N$ such that:*

i) $S = I(cl(N - L)) \subseteq int(N - L)$.

ii) L is **positively invariant** in N. That is,

$$x \in L \text{ and } x \cdot [0, t] \subseteq N \Rightarrow x \cdot [0, t] \subseteq L.$$

*iii) Each orbit that leaves N must go through L first. That is, if $x \in N$ and
$x \cdot t \notin N$ for $t > 0$, then there exists some $t' \in [0, t]$ such that $x \cdot [0, t'] \subseteq N$
and $x \cdot t' \in L$.*

*Because of property iii), L is called the **exit set** for the flow.*

Example 7.13 The following pictures show examples of isolated compact invariant sets and index pairs.

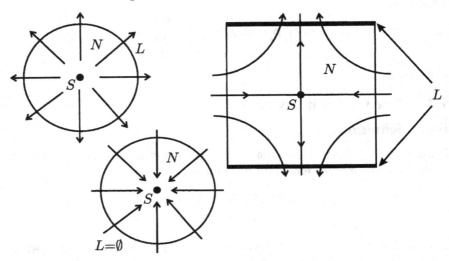

The following results were proved by Conley in [35] and by Salamon in
[127] using new and simplified definitions and proofs. The definitions given
above follow Salamon [127].

Theorem 7.14 *Every isolated compact invariant set admits an index pair.*

Theorem 7.15 *If S is an isolated compact invariant set and (N, L) and $(\tilde{N}, \tilde{L})$
are two index pairs for S, then N/L and $\tilde{N}/\tilde{L}$ are homotopy equivalent as
pointed spaces via maps that are induced by the flow.*

Definition 7.16 *The **Conley index** of an isolated compact invariant set S is
the homotopy type of N/L where (N, L) is an index pair for S.*

We will say that an index pair (N, L) is **regular** if and only if the inclusion map
$L \subset N$ is a cofibration. An index pair (N, L) can always be modified to create
an index pair $(N, \tilde{L})$ that is regular (see Section 5.1 of [127]). Hence, Theorem
7.15 and Corollary 2.31 imply that $H_*(N, L; \mathbb{Z})$ is an invariant associated to
an isolated compact invariant set S where (N, L) is any regular index pair for
S.

The proof of Theorem 7.15 will be given in Section 7.5. We will not prove Theorem 7.14 or show how to modify a given index pair to create a regular index pair since we will explicitly describe regular index pairs for all the isolated compact invariant sets considered in the proof of the Morse Homology Theorem. For a proof of Theorem 7.14 see Section III.4 of [35] or Section 4.1 of [127].

Examples of isolated compact invariant sets and index pairs

Example 7.17 (Critical points as isolated compact invariant sets)

Let $f : M \rightarrow \mathbb{R}$ be a Morse function on a finite dimensional smooth Riemannian manifold (M, g). We showed in Section 4.1 that the critical points of f are hyperbolic fixed points of the diffeomorphism φ_t determined by $-\nabla f$ for any $t \neq 0$. Moreover, for any $q \in \mathrm{Cr}_k(f)$ there is a splitting $T_q M = T_q^s M \oplus T_q^u M$ where $T_q^s M = T_q W^s(q)$ and $T_q^u M = T_q W^u(q)$. In particular, $W^s(q)$ and $W^u(q)$ intersect transversally at the point q. Hence, there exists a coordinate chart $\phi : U \rightarrow T_q M$ around $q \in M$ that maps $W^s(q) \cap U$ into $T_q^s M$ and $W^u(q) \cap U$ into $T_q^u M$ (see Problem 9 of Chapter 5). For any $\varepsilon > 0$ let

$$D_\varepsilon^s = \{v \in T_q^s M|\ \|v\| \leq \varepsilon\} \text{ and}$$
$$D_\varepsilon^u = \{v \in T_q^u M|\ \|v\| \leq \varepsilon\}.$$

Let $N_q = \phi_q^{-1}(D_\varepsilon^s \times D_\varepsilon^u)$ and $L_q = \phi_q^{-1}(D_\varepsilon^s \times \partial D_\varepsilon^u)$. We easily see that N_q is an isolating neighborhood for the compact invariant set $S = \{q\}$ and (N_q, L_q) is an index pair:

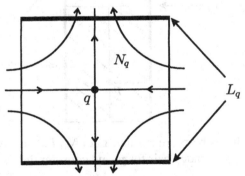

Hence, $I(N_q) = \{q\}$ and L_q is the exit set.

Note that the index pair (N_q, L_q) of q is regular. In fact,

$$(N_q, L_q) \approx (D_\varepsilon^s \times D_\varepsilon^u, D_\varepsilon^s \times \partial D_\varepsilon^u) \simeq (D_\varepsilon^u, \partial D_\varepsilon^u),$$

and thus $N_q/L_q \simeq D_\varepsilon^u/\partial D_\varepsilon^u$ is a pointed λ_q-dimensional sphere. This shows that

$$H_j(N_q, L_q; \mathbb{Z}) \approx H_j(D_\varepsilon^u, \partial D_\varepsilon^u; \mathbb{Z})$$

$$\approx \quad H_j(D^u_\varepsilon/\partial D^u_\varepsilon, *; \mathbb{Z})$$
$$\approx \quad H_j(N_q/L_q, *; \mathbb{Z})$$

for all $j \in \mathbb{Z}_+$. Moreover, an orientation of $T^u_q M$ determines a generator of the homology group $H_k(N_q, L_q; \mathbb{Z}) \approx H_k(D^u_\varepsilon, \partial D^u_\varepsilon; \mathbb{Z}) \approx \mathbb{Z}$. Hence, we can identify

$$C_k(f) \approx \bigoplus_{q \in \mathrm{Cr}_k(f)} H_k(N_q, L_q; \mathbb{Z}).$$

Example 7.18 (Gradient flow lines as isolated compact invariant sets)
Let $f : M \to \mathbb{R}$ be a Morse-Smale function on a finite dimensional compact smooth Riemannian manifold (M, g). Let $q \in \mathrm{Cr}_k(f)$ and $p \in \mathrm{Cr}_{k-1}(f)$. The results in Section 6.3 imply that $S = W(q, p) \cup \{p, q\}$ is an isolated compact invariant set. Hence, by Theorem 7.14 there exists an index pair (N_2, N_0) for S (a regular index pair is described explicitly in the proof of Lemma 7.21 in the next section). Set $N_1 = N_0 \cup (N_2 \cap M^c)$ where $f(p) < c < f(q)$ (recall that $M^c = \{x \in M \mid f(x) \le c\}$). Then (N_2, N_1) is an index pair for q and (N_1, N_0) is an index pair for p.

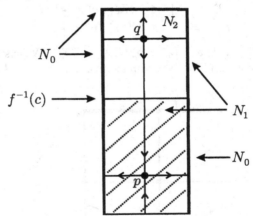

Assume that these index pairs are regular. Let (N_q, L_q) be a regular index pair for q, (N_p, L_p) a regular index pair for p, and consider the homomorphism

$$\Delta_k(q, p) : H_k(N_q, L_q; \mathbb{Z}) \to H_{k-1}(N_p, L_p; \mathbb{Z})$$

which is defined as the composition of the following maps:

$$H_k(N_q, L_q) \xrightarrow[(1)]{\approx} H_k(N_2, N_1) \xrightarrow[(2)]{\delta_*} H_{k-1}(N_1, N_0) \xrightarrow[(3)]{\approx} H_{k-1}(N_p, L_p)$$

where (1) and (3) are induced by the homotopy equivalence of Theorem 7.15 and δ_* is the connecting homomorphism in the homology exact sequence of the triple (N_2, N_1, N_0). This determines a homomorphism

$$\Delta_k : \bigoplus_{q \in \mathrm{Cr}_k(f)} H_k(N_q, L_q; \mathbb{Z}) \longrightarrow \bigoplus_{p \in \mathrm{Cr}_{k-1}(f)} H_{k-1}(N_p, L_p; \mathbb{Z}).$$

Remark 7.19 Recall that the **connecting homomorphism** δ_* in the homology exact sequence of the triple $N_0 \subseteq N_1 \subseteq N_2$:

$$\delta_* : H_k(N_2, N_1) \to H_{k-1}(N_1, N_0)$$

can be defined as follows. Consider the following commutative diagram with exact rows.

$$\begin{array}{ccccccccc}
0 & \longrightarrow & C_k(N_1, N_0) & \xrightarrow{i} & C_k(N_2, N_0) & \xrightarrow{j} & C_k(N_2, N_1) & \longrightarrow & 0 \\
& & \downarrow & & \downarrow{\scriptstyle \bar{\partial}_k} & & \downarrow & & \\
0 & \longrightarrow & C_{k-1}(N_1, N_0) & \xrightarrow{i} & C_{k-1}(N_2, N_0) & \xrightarrow{j} & C_{k-1}(N_2, N_1) & \longrightarrow & 0
\end{array}$$

Let $a \in C_k(N_2, N_1)$ be a k-cycle, and lift it to $j^{-1}(a) = \bar{a} \in C_k(N_2, N_0)$. Then $\bar{\partial}_k(\bar{a})$ is a cycle in $C_{k-1}(N_2, N_0)$ which comes from a unique cycle $\hat{a} \in C_{k-1}(N_1, N_0)$, and we define $\delta_*[a] = [\hat{a}]$. Note that this definition is consistent with the discussion in Section 2.1. In particular, the above definition of δ_* satisfies $\delta_* = j_* \circ \delta_k$ where δ_k is the connecting homomorphism in the homology exact sequence of the pair (N_2, N_1) and $j : (N_1, \emptyset) \to (N_1, N_0)$ is the inclusion.

Also, recall that the connecting homomorphism δ_* is a **natural transformation**, i.e. if $g : (N_2, N_1, N_0) \to (\tilde{N}_2, \tilde{N}_1, \tilde{N}_0)$ is a map of triples, then the following diagram commutes for all k.

$$\begin{array}{ccc}
H_k(N_2, N_1) & \xrightarrow{\delta_*} & H_{k-1}(N_1, N_0) \\
\downarrow{\scriptstyle g_*} & & \downarrow{\scriptstyle g_*} \\
H_k(\tilde{N}_2, \tilde{N}_1) & \xrightarrow{\delta_*} & H_{k-1}(\tilde{N}_1, \tilde{N}_0)
\end{array}$$

Since the homotopy equivalences in Theorem 7.15 are induced by the flow, the above diagram commutes for regular index pairs.

7.4 Proof of the Morse Homology Theorem

To prove Theorem 7.4 we first show how to construct a filtration of index pairs. Next we prove the main technical lemma (Lemma 7.21) which relates the Morse-Smale-Witten boundary operator to a connecting homomorphism. Theorem 7.4 is then easily proved using arguments similar to those used to prove the CW-Homology Theorem (Theorem 2.15).

A filtration of index pairs

Let $f : M \to \mathbb{R}$ be a Morse-Smale function on a finite dimensional compact smooth Riemannian manifold (M, g) of dimension m, and let $\varphi_t : M \to M$ be the flow determined by $-\nabla f$. For any $0 \le j \le k \le m$ define

$$W(k, j) = \bigcup_{j \le \lambda_p \le \lambda_q \le k} W(q, p).$$

These spaces are compact by Corollary 6.3 and Corollary 6.28. If N is a compact neighborhood of $W(k, j)$ such that $\text{Cr}(f) \cap N = \text{Cr}(f) \cap W(k, j)$, then we have

$$I(N) = \{x \in N | \varphi_t(x) \in N \text{ for all } t \in \mathbb{R}\} = W(k, j)$$

by Proposition 3.19. Therefore, $W(k, j)$ is an isolated compact invariant set for all $0 \le j \le k \le m$.

Note that by Corollary 6.27 the following sets are compact:

$$W_j^s = \bigcup_{j \le \lambda_p} W^s(p)$$

$$W_j^u = \bigcup_{\lambda_p \le j} W^u(p)$$

for all $j = 0, \ldots, m$, and let $N_m = M$. Choose a cofibered compact neighborhood N_{m-1} of W_{m-1}^u in M which is positively invariant in N_m and satisfies $N_{m-1} \cap W_m^s = \emptyset$. For instance, we could choose

$$N_{m-1} = N_m - \bigcup_{q \in \text{Cr}_m(f)} \text{int } N_q.$$

Then (N_m, N_{m-1}) is a regular index pair for $\text{Cr}_m(f)$. Next, choose a cofibered compact neighborhood N_{m-2} of W_{m-2}^u in N_{m-1} which is positively invariant in N_{m-1} and satisfies $N_{m-2} \cap W_{m-1}^s = \emptyset$. For instance, we could choose N_{m-2} as N_{m-1} minus the interior of an appropriate tubular neighborhood of $W^s(p)$ for all $p \in \text{Cr}_{m-1}(f)$. Then (N_{m-1}, N_{m-2}) is an index pair for $\text{Cr}_{m-1}(f)$ and (N_m, N_{m-2}) is an index pair for $W(m, m-1)$. Iterating this procedure establishes the existence of a filtration

$$\emptyset = N_{-1} \subseteq N_0 \subseteq N_1 \subseteq \cdots \subseteq N_m = M$$

such that (N_k, N_{j-1}) is a regular index pair for $W(k, j)$ for all $0 \le j \le k \le m$.

Remark 7.20 The existence of the preceeding filtration of index pairs was proved in a more general context by Conley in [35]. The above procedure for constructing the filtration follows Corollary 4.4 of [127].

Lemma 7.21 *The following diagram commutes*

$$
\begin{array}{ccc}
C_k(f) & \xrightarrow{\;\;\partial_k\;\;} & C_{k-1}(f) \\
\Big\uparrow{\scriptstyle\approx} & & \Big\uparrow{\scriptstyle\approx} \\
\displaystyle\bigoplus_{q\,\in\,Cr_k(f)} H_k(N_q, L_q; \mathbb{Z}) \xrightarrow{\;\Delta_k\;} & \displaystyle\bigoplus_{p\,\in\,Cr_{k-1}(f)} H_{k-1}(N_p, L_p; \mathbb{Z}) \\
\Big\uparrow{\scriptstyle\approx} & & \Big\uparrow{\scriptstyle\approx} \\
H_k(N_k, N_{k-1}) & \xrightarrow{\;\;\delta_*\;\;} & H_{k-1}(N_{k-1}, N_{k-2})
\end{array}
$$

where (N_q, L_q) is a regular index pair for $q \in Cr_k(f)$, (N_p, L_p) is a regular index pair for $p \in Cr_{k-1}(f)$, ∂_k is the Morse-Smale-Witten boundary operator, Δ_k is the homomorphism defined in Example 7.18, and δ_* is a connecting homomorphism.

Proof:

The bottom square commutes since the connecting homomorphism is a natural transformation and the vertical isomorphisms are induced by the flow. To show that the top square commutes we restrict to the case there $q \in Cr_k(f)$ and $p \in Cr_{k-1}(f)$ are the only critical points in $f^{-1}([a, b])$, where $a = f(p)$ and $b = f(q)$. The general case follows by an alteration of f outside an isolating neighborhood of $S = W(q, p) \cup \{p, q\}$, which does not affect either the homomorphism Δ_k or the Morse-Smale-Witten boundary operator. See the problems at the end of this chapter for further details concerning this alteration.

We are going to take advantage of the freedom to choose the index pairs. For any $t \in \mathbb{R}$ let $M^t = \{x \in M \mid f(x) \le t\}$ and let $M_t = \{x \in M \mid f(x) \ge t\}$. For $c \in (a, b)$, $\varepsilon > 0$ very small, and $T > 0$ very large, define the following index pairs:

$$
\begin{aligned}
N_q &= \{x \in M_c \mid f(\varphi_{-T}(x)) \le b + \varepsilon\} \\
L_q &= \{x \in N_q \mid f(x) = c\} \\
N_p &= \{x \in M^c \mid f(\varphi_T(x)) \ge a - \varepsilon\} \\
L_p &= \{x \in N_p \mid f(\varphi_T(x)) = a - \varepsilon\} \\
C &= N_p \cup N_q, \quad B = N_p \cup L_q, \quad A = L_p \cup \mathrm{cl}(L_q - N_p).
\end{aligned}
$$

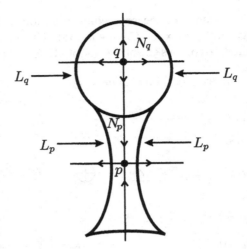

Then (N_q, L_q) and (C, B) are index pairs for q, (N_p, L_p) and (B, A) are index pairs for p, and (C, A) is an index pair for $S = W(q, p) \cup \{p, q\}$. These assertions are easy to verify.

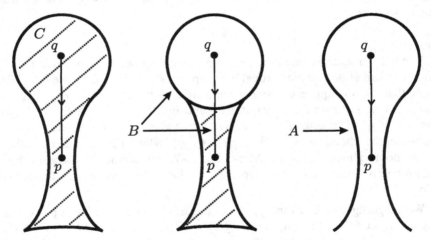

Note that the unstable closed disk $W^u(q) \cap M_c$ is contained in N_q since for every $x \in W^u(q)$ we have $f(\varphi_{-T}(x)) \le f(q) = b$ for all T. Moreover, with $\varepsilon > 0$ fixed, the set $N_q = \{x \in M_c \mid f(\varphi_{-T}(x)) \le b + \varepsilon\}$ contracts to $W^u(q) \cap M_c$ when $T \to \infty$. Hence, (N_q, L_q) contracts to

$$(W^u(q) \cap M_c, W^u(q) \cap f^{-1}(c)) \approx (D^k, \partial D^k).$$

This shows that (N_q, L_q) is a regular index pair.

Likewise, N_p is a tubular neighborhood of the stable closed disk $W^s(p) \cap M^c$ with width tending to zero as $T \to \infty$. Since $W^s(p)$ is diffeomorphic to an open disk of dimension $m - k + 1$, we have

$$(N_p, L_p) \approx (D^{k-1} \times D^{m-k+1}, \partial D^{k-1} \times D^{m-k+1}),$$

and we see that (N_p, L_p) is a regular index pair.

Observe that N_p is a thickening of the stable closed disk $W^s(p) \cap M^c$, and recall that $S^u(q) = W^u(q) \cap f^{-1}(c) \subseteq L_q$ is the unstable sphere of q. Since $W(q, p) = W^u(q) \cap W^s(p)$ has finitely many components (Corollary 6.29), $N_p \cap S^u(q)$ consists of finitely many components $V_1, \ldots, V_n$ with a single point $x_j \in W(q, p) \cap V_j$ for all $j = 1, \ldots, n$.

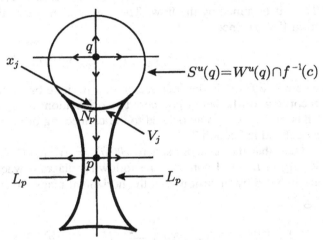

Since N_p is a tubular neighborhood of the closed stable disk $W^s(p) \cap M^c \approx D^{m-k+1}$, there is a diffeomorphism

$$\Psi_p : N_p \to D^{k-1} \times D^{m-k+1}$$

with

$$
\begin{aligned}
\Psi_p(L_p) &= \partial D^{k-1} \times D^{m-k+1} \\
\Psi_p(N_p \cap W^s(p)) &= \{0\} \times D^{m-k+1} \\
\Psi_p(V_j) &= D^{k-1} \times \{\theta_j\} \text{ where } \theta_j \in \partial D^{m-k+1}.
\end{aligned}
$$

Thus, V_j is a $(k-1)$-dimensional manifold diffeomorphic to D^{k-1} (and its boundary $\partial V_j = V_j \cap L_p$ is diffeomorphic to ∂D^{k-1}) via the maps $\Psi_j = \pi_1 \circ \Psi_p|_{V_j}$:

$$V_j \xrightarrow{\Psi_p|_{V_j}} D^{k-1} \times D^{m-k+1} \xrightarrow{\pi_1} D^{k-1}.$$

The map $\Psi_{p,1} = \pi_1 \circ \Psi_p : N_p \to D^{k-1}$ takes L_p into ∂D^{k-1} and induces an isomorphism

$$H_{k-1}(N_p, L_p) \overset{(\Psi_{p,1})_*}{\underset{\approx}{}} H_{k-1}(D^{k-1}, \partial D^{k-1}) \overset{(\Psi_j^{-1})_*}{\underset{\approx}{}} H_{k-1}(V_j, \partial V_j) \approx \mathbb{Z}.$$

The orientation of $T_p^u M$ determines a generator α of $H_{k-1}(N_p, L_p) \approx \mathbb{Z}$. A chain representing α will also determine a generator for $H_{k-1}(B, A) \approx \mathbb{Z}$, and thus we can identify $H_{k-1}(N_p, L_p) = H_{k-1}(B, A)$. The generator $\alpha \in H_{k-1}(N_p, L_p)$ is mapped onto a generator α_j of $H_{k-1}(V_j, \partial V_j)$ by $(\Psi_j^{-1})_* \circ (\Psi_{p,1})_*$. In other words, the homology class α_j is determined by the orientation of $T_{x_j} V_j$, inherited from the orientation of $T_p^u M$ via the isomorphism $T_{x_j} V_j \to T_p^u M$ determined by the flow. There is another orientation of $T_{x_j} V_j$ coming from $W^u(q)$ since

$$T_{x_j} V_j = (-\nabla f)(x_j)^{\perp} \cap T_{x_j} W^u(q) \subseteq T_{x_j} W^u(q)$$

where $-\nabla f(x_j)$ is the first vector in a positive basis. Let n_j be $+1$ or -1 according to whether or not these two orientations of $T_{x_j} V_j$ agree or disagree. Thus, n_j is the sign associated to the connecting orbit in $W(q, p)$ containing x_j defined in Section 7.1.

Note that the unstable sphere $S^u(q) = W^u(q) \cap f^{-1}(c)$ coincides with $W^u(q) \cap L_q$, and consider the following diagram where the vertical maps are induced by inclusions and the horizontal maps are connecting homomorphisms.

$$
\begin{array}{ccc}
H_k(W^u(q) \cap N_q, S^u(q)) & \overset{\delta_*}{\longrightarrow} & H_{k-1}(S^u(q), \mathrm{cl}(S^u(q) - \coprod_{j=1}^n V_j)) \\
\downarrow \approx & & \downarrow s_* \\
H_k(C, B) & \overset{\delta_*}{\longrightarrow} & H_{k-1}(B, A)
\end{array}
$$

This diagram commutes since the connecting homomorphism is a natural transformation. Note that

$$H_{k-1}(S^u(q), \mathrm{cl}(S^u(q) - \coprod_{j=1}^n V_j)) \approx \bigoplus_{j=1}^n H_{k-1}(S^u(q), \mathrm{cl}(S^u(q) - V_j)) \approx \bigoplus_{j=1}^n \mathbb{Z}$$

and define $\delta_{*,j}$ to be the jth component of the connecting homomorphism

$$\delta_* : H_k(W^u(q) \cap N_q, S^u(q)) \to H_{k-1}(S^u(q), \mathrm{cl}(S^u(q) - \coprod_{j=1}^n V_j))$$

under this identification. Thus, for any $\beta \in H_k(W^u(q) \cap N_q, S^u(q))$ we have $\delta_*(\beta) = \delta_{*,1}(\beta) + \cdots + \delta_{*,n}(\beta)$ and

$$s_*(\delta_*(\beta)) = \sum_{j=1}^{n} s_*(\delta_{*,j}(\beta)) \in H_{k-1}(B, A) \approx \mathbb{Z}.$$

Since N_q is a thickening of the unstable closed disk $W^u(q) \cap M_c = W^u(q) \cap N_q$ we have

$$(N_q, L_q) \simeq (W^u(q) \cap N_q, W^u(q) \cap L_q) = (W^u(q) \cap N_q, S^u(q)).$$

Thus, $H_k(W^u(q) \cap N_q, S^u(q)) \approx H_k(N_q, L_q)$. We get a generator β for this group as follows. Start with triangulations of the $(k-1)$-dimensional closed disks $V_j \subseteq S^u(q)$ for all $j = 1, \ldots, n$, and extend these triangulations to a triangulation of the k-dimensional manifold with boundary $W^u(q) \cap N_q \approx D^k$. This triangulation, together with the orientation of $W^u(q)$, determines the generator β. Note that β also determines a generator for $H_k(C, B) \approx H_k(N_q, L_q)$, and hence we can identify

$$H_k(C, B) = H_k(N_q, L_q) = H_k(W^u(q) \cap N_q, S^u(q)).$$

By excision we have $H_{k-1}(S^u(q), \mathrm{cl}(S^u(q) - V_j)) \approx H_{k-1}(V_j, \partial V_j) \approx \mathbb{Z}$ for all $j = 1, \ldots, n$. Under this isomorphism the image of the homology class $\delta_{*,j}(\beta)$ is represented by the original triangulation of V_j together with the original orientation, hence it corresponds to $n_j \alpha_j$. We know that $\alpha_j \in H_{k-1}(V_j, \partial V_j)$ corresponds to α under the isomorphism $H_{k-1}(N_p, L_p) \approx H_{k-1}(V_j, \partial V_j)$, and thus we obtain

$$s_*(\delta_*(\beta)) = \sum_{j=1}^{n} n_j \alpha = n(q, p)\alpha \in H_{k-1}(B, A) = H_{k-1}(N_p, L_p).$$

With the identification $\beta = q$ and $\alpha = p$ we have the formula

$$\Delta_k(q) = \sum_{p \in \mathrm{Cr}_{k-1}(f)} n(q, p)p = \partial_k(q).$$

$\square$

Proof of Theorem 7.4:

By Lemma 7.21 there is a commutative diagram

$$
\begin{array}{ccccc}
C_{k+1}(f) & \xrightarrow{\partial_{k+1}} & C_k(f) & \xrightarrow{\partial_k} & C_{k-1}(f) \\
\Big\uparrow{\scriptstyle\approx} & & \Big\uparrow{\scriptstyle\approx} & & \Big\uparrow{\scriptstyle\approx} \\
H_{k+1}(N_{k+1}, N_k) & \xrightarrow{\delta_*} & H_k(N_k, N_{k-1}) & \xrightarrow{\delta_*} & H_{k-1}(N_{k-1}, N_{k-2})
\end{array}
$$

where the maps on the bottom row are connecting homomorphisms. Therefore, the Morse-Smale-Witten boundary operator satisfies

$$\partial_k \circ \partial_{k+1} = 0.$$

(See the paragraph preceeding Theorem 2.15.)

Since $W(k, k) = \mathrm{Cr}_k(f)$ for all $0 \le k \le m$, Theorem 7.15 and Example 7.17 imply that

$$N_k/N_{k-1} \simeq \bigcup_{q \in \mathrm{Cr}_k(f)} N_q \Bigg/ \bigcup_{q \in \mathrm{Cr}_k(f)} L_q \simeq \bigvee_{q \in \mathrm{Cr}_k(f)} S_q^k.$$

Thus, $H_j(N_k, N_{k-1}) = 0$ for all $j \ne k$, and the homology exact sequence of the pair (N_k, N_{k-1}):

$$\cdots \to H_{j+1}(N_k, N_{k-1}) \to H_j(N_{k-1}) \to H_j(N_k) \to H_j(N_k, N_{k-1}) \to \cdots$$

implies that the inclusion map $H_j(N_{k-1}) \to H_j(N_k)$ is an isomorphism for all $j \ne k, k-1$. Hence, $H_j(N_k; \mathbb{Z}) \to H_j(M; \mathbb{Z})$ is an isomorphism for all $j < k$.

Using arguments similar to those used in the proof of Lemma 2.12 we can also show that $H_j(N_k; \mathbb{Z}) = \{0\}$ for all $j > k$: We will use induction on k. When $k = 0$ it is clear that $H_j(N_0) = 0$ for all $j > 0$ since $(N_0, \emptyset)$ is an index pair for $\mathrm{Cr}_0(f)$. Now assume that $H_j(N_{k-1}) = 0$ for all $j > k-1$ and consider the homology exact sequence of the pair (N_k, N_{k-1}):

$$\cdots \to H_j(N_{k-1}) \to H_j(N_k) \to H_j(N_k, N_{k-1}) \to \cdots$$

For all $j > k$, $H_j(N_k; \mathbb{Z}) = 0$ by the induction hypothesis and the fact that $H_j(N_{k-1}, N_k) = 0$.

This shows that we have the following commutative diagram where the horizontal and vertical sequences are exact:

$$
\begin{array}{ccccccc}
 & & & & & & H_{k-1}(N_{k-2}) = 0 \\
 & & & & & & \downarrow \\
0 = H_k(N_{k-1}) & \longrightarrow & H_k(N_k) & \longrightarrow & H_k(N_k, N_{k-1}) & \xrightarrow{\ \delta_k\ } & H_{k-1}(N_{k-1}) \\
 & & & & & \searrow^{\delta_*} & \downarrow^{j_*} \\
 & & & & & & H_{k-1}(N_{k-1}, N_{k-2})
\end{array}
$$

and $j : (N_{k-1}, \emptyset) \to (N_{k-1}, N_{k-2})$ is an inclusion map of pairs. In particular,

$$j_* : H_{k-1}(N_{k-1}) \to H_{k-1}(N_{k-1}, N_{k-2})$$

is injective, and hence, the kernels of the two connecting homomorphisms agree. They are both isomorphic to $H_k(N_k)$, and we have

$$H_k(N_k; \mathbb{Z}) \approx \ker \delta_* \approx \ker \partial_k$$

by the first commutative diagram. Thus, the following diagram commutes:

$$
\begin{array}{ccccccc}
H_{k+1}(N_{k+1}, N_k; \mathbb{Z}) & \xrightarrow{\delta_{k+1}} & H_k(N_k; \mathbb{Z}) & \longrightarrow & H_k(N_{k+1}; \mathbb{Z}) & \longrightarrow & 0 \\
\Big\uparrow{\approx} & & \Big\uparrow{\approx} & & \Big\uparrow{\approx} & & \\
C_{k+1}(f) & \xrightarrow{\partial_{k+1}} & \ker \partial_k & \longrightarrow & H_k(M; \mathbb{Z}) & \longrightarrow & 0
\end{array}
$$

where the first row comes from the homology exact sequence of (N_{k+1}, N_k). Therefore, the bottom row is also exact and we have

$$H_k(M; \mathbb{Z}) \approx \ker \partial_k / \operatorname{im} \partial_{k+1}.$$

$\square$

7.5 Independence of the choice of the index pairs

The goal of this section is to prove Theorem 7.15, i.e. if (N, L) and $(\tilde{N}, \tilde{L})$ are two index pairs for an isolated compact invariant set S, then N/L and $\tilde{N}/\tilde{L}$ are homotopy equivalent as pointed spaces via maps that are induced by the flow. Throughout this section we will assume that $\varphi_t : M \to M$ is a flow on a locally compact metric space M.

The proof consists of the following three lemmas from Section 4 of [127].

Lemma 7.22 *Let N be an isolating neighborhood for the isolated compact invariant set S, and let U be a neighborhood of S in M. Then there exists a $t > 0$ such that $x \cdot [-t, t] \subseteq N$ implies that $x \in U$.*

Proof:
   Suppose that there exist sequences $x_k \in M - U$ and $t_k \geq 0$ such that $t_k \to \infty$ as $k \to \infty$ and $x_k \cdot [-t_k, t_k] \subseteq N$. Then any limit point x of the sequence x_k would satisfy $x \in \operatorname{cl}(M - U)$ and $x \cdot \mathbb{R} \subseteq N$. This would imply that $x \in S \cap \operatorname{cl}(M - U) = \emptyset$, which is a contradiction.

$\square$

   The next lemma shows how the flow induces a family of maps from N/L to $\tilde{N}/\tilde{L}$ (and vice versa) for any two index pairs (N, L) and $(\tilde{N}, \tilde{L})$ of S. Note that the following parameter $T \geq 0$ always exists by the previous lemma.

Lemma 7.23 *Let (N, L) and $(\tilde{N}, \tilde{L})$ be index pairs for the isolated compact invariant set S and choose $T \geq 0$ such that the following implications hold for $t \geq T$:*

$$x \cdot [-t, t] \subseteq N - L \quad \text{implies} \quad x \in \tilde{N} - \tilde{L} \tag{7.1}$$
$$x \cdot [-t, t] \subseteq \tilde{N} - \tilde{L} \quad \text{implies} \quad x \in N - L. \tag{7.2}$$

Then the map $h : N/L \times [T, \infty) \to \tilde{N}/\tilde{L}$ defined by

$$h([x], t) = \begin{cases} [x \cdot 3t] & \text{if } x \cdot [0, 2t] \subseteq N - L \text{ and } x \cdot [t, 3t] \subseteq \tilde{N} - \tilde{L} \\ [\tilde{L}] & \text{otherwise} \end{cases}$$

for $x \in N$ and $t \geq T$ is continuous.

Proof:

First Case: $x \cdot [t, 3t] \not\subseteq \text{cl}(\tilde{N} - \tilde{L})$. In this case $x \cdot t^* \notin \text{cl}(\tilde{N} - \tilde{L})$ for some t^* with $t < t^* < 3t$. Hence, there exists a neighborhood U of $x \cdot t^*$ in M such that $U \cap \text{cl}(\tilde{N} - \tilde{L}) = \emptyset$. Since the flow is continuous this implies that there exists a neighborhood W of (x, t) in $M \times [T, \infty)$ such that $(x', t') \in W$ implies $x' \cdot t^* \in U$ and $t' < t^* < 3t'$. Thus, $x' \cdot [t', 3t'] \not\subseteq \tilde{N} - \tilde{L}$, and hence $h([x'], t') = [\tilde{L}]$ for every $(x', t') \in W$ with $x' \in N$.

The case $x \cdot [0, 2t] \not\subseteq \text{cl}(N - L)$ can be treated the same. Hence, from now on we may assume that

$$x \cdot [0, 2t] \subseteq \text{cl}(N - L) \quad \text{and} \quad x \cdot [t, 3t] \subseteq \text{cl}(\tilde{N} - \tilde{L}). \tag{7.3}$$

Second Case: $x \cdot [t, 3t] \cap \tilde{L} = \emptyset$. In this case it follows from (7.3) that $x \cdot [t, 3t] \subseteq \tilde{N} - \tilde{L}$. By (7.2) this implies that $x \cdot 2t \in N - L$, and hence $x \cdot [0, 2t] \subseteq N - L$. Therefore, $h([x], t) = x \cdot 3t \in \tilde{N} - \tilde{L}$. Now let U be a neighborhood of $x \cdot 3t$ in M. Since the flow is continuous there exists a neighborhood W of (x, t) in $M \times [T, \infty)$ such that whenever $(x', t') \in W$ we have

$$x' \cdot [0, 2t'] \cap L = \emptyset, \quad x' \cdot [t', 3t'] \cap \tilde{L} = \emptyset, \quad \text{and} \quad x' \cdot 3t' \in U.$$

If $x' \in N$, then $x' \cdot [0, 2t] \subseteq N - L$ and (7.1) implies that $x' \cdot t' \in \tilde{N} - \tilde{L}$, and hence $x' \cdot [t', 3t'] \subseteq \tilde{N} - \tilde{L}$. Therefore, $h([x'], t') = [x' \cdot 3t'] = x' \cdot 3t' \in U$ for every $(x', t') \in W$ with $x' \in N$.

Third Case: $x \cdot [t, 3t] \cap \tilde{L} \neq \emptyset$. In this case it follows from (7.3) that $x \cdot 3t \in \tilde{L}$. Let $[U]$ be a neighborhood of $h([x], t) = [\tilde{L}]$ in $\tilde{N}/\tilde{L}$ and define

$$U = (\pi^{-1}([U]) \cap \tilde{N} - \tilde{L}) \cup (M - \tilde{N}) \cup \tilde{L}$$

where $\pi : \tilde{N} \to \tilde{N}/\tilde{L}$ denotes the quotient map. Then U is a neighborhood of $\tilde{L}$ in M and

$$[U] = \pi(U \cap \tilde{N} - \tilde{L}) \cup [\tilde{L}].$$

By the continuity of the flow, there exists a neighborhood W of (x, t) in $M \times [T, \infty)$ such that, whenever $(x', t') \in W$ we have $x' \cdot 3t' \in U$. This implies that

$$h([x'], t') \in \{[x' \cdot 3t'], [\tilde{L}]\} \subseteq \pi(U \cap \tilde{N} - \tilde{L}) \cup [\tilde{L}] = [U]$$

for every $(x', t') \in W$ with $x' \in N$.

$\square$

Lemma 7.24 *Let* (N, L), $(\tilde{N}, \tilde{L})$ *and* (N', L') *be index pairs for* S. *Choose* $T \geq 0$ *such that* (7.1) *and* (7.2) *are satisfied for* $t \geq T$ *and suppose that the following implications hold for* $t \geq \tilde{T}$:

$$x \cdot [-t, t] \subseteq \tilde{N} - \tilde{L} \quad implies \quad x \in N' - L' \tag{7.4}$$
$$x \cdot [-t, t] \subseteq N' - L' \quad implies \quad x \in \tilde{N} - \tilde{L}. \tag{7.5}$$

Let $h : N/L \times [T, \infty) \to \tilde{N}/\tilde{L}$ *be defined as in Lemma 7.23, and define* $\tilde{h} : \tilde{N}/\tilde{L} \times [T, \infty) \to N'/L'$ *analogously. Then the following equation holds for* $t \geq \max \{T, \tilde{T}\}$:

$$\tilde{h}(h([x], t), t) = \begin{cases} [x \cdot 6t] & \text{if } x \cdot [0, 4t] \subseteq N - L \text{ and } x \cdot [2t, 6t] \subseteq N' - L' \\ [L'] & \text{otherwise.} \end{cases}$$

Proof:
  Observe that (7.1), (7.2), (7.4), and (7.5) imply that

$$x \cdot [0, 2t] \subseteq N - L, \quad x \cdot [t, 5t] \subseteq \tilde{N} - \tilde{L}, \quad x \cdot [4t, 6t] \subseteq N' - L'$$

is equivalent to

$$x \cdot [0, 4t] \subseteq N - L, \quad x \cdot [2t, 6t] \subseteq N' - L'.$$

$\square$

Proof of Theorem 7.15:
  Let $h_t : N/L \to \tilde{N}/\tilde{L}$ and $g_t : \tilde{N}/\tilde{L} \to N/L$ be the families given by Lemma 7.23. By Lemma 7.24 the composition $g_t \circ h_t : N/L \to N/L$ is homotopic to another family of maps of the same form as in Lemma 7.23. Taking $\tilde{N} = N$, $\tilde{L} = L$ and $T = 0$ in Lemma 7.23 we see that the identity map is also of this same form. It's clear that any two families of the form given in Lemma 7.23 are homotopic, and hence $g_t \circ h_t$ is homotopic to the identity map. Similarly, $h_t \circ g_t$ is also homotopic to the identity. Therefore,

$$N/L \simeq \tilde{N}/\tilde{L}.$$

$\square$

Problems

1. Compute the homology of the Morse-Smale-Witten chain complex of $f :$ $T^2 \to \mathbb{R}$ where $T^2 = \mathbb{R}^2/\mathbb{Z}^2$ and $f(x, y) = \cos(2\pi x) + \cos(2\pi y)$.

2. Compute the integral homology of $\mathbb{R}P^2$ using the Morse-Smale-Witten chain complex of an appropriate Morse-Smale function.

3. Compute the homology of the Morse-Smale-Witten chain complex in Example 7.8 where the orientation of $T^u_{p_1}S^1$ and $T^u_{p_4}S^1$ is chosen to be -1.

4. Compute the homology of the Morse-Smale-Witten chain complex in Example 7.10 with the opposite orientation on $T^u_q S^2$.

5. Let $f : M \to \mathbb{R}$ be a Morse function on a finite dimensional smooth manifold, and let c be a critical value of f. Show that

$$M^b/M^a \simeq \bigcup_{p \in \operatorname{Cr}(f) \cap f^{-1}(c)} S^{\lambda_p}$$

where a and b are regular values of f and c is the only critical value of f in the interval $[a, b]$.

6. Derive the Morse inequalities using Theorem 7.15 instead of Theorem 3.28.

7. Let $f : M \to \mathbb{R}$ be a Morse-Smale function on a finite dimensional compact smooth Riemannian manifold (M, g). Let $q \in \operatorname{Cr}_k(f)$ and $p \in \operatorname{Cr}_{k-1}(f)$. Show that it is possible to alter f outside of an isolating neighborhood of the isolated compact invariant set $S = W(q, p) \cup \{q, p\}$ so that q and p are the only critical points in $f^{-1}([a, b])$, where $a = f(p)$ and $b = f(q)$. Show that it is possible to make this alteration without changing the homomorphisms under consideration in the proof of Lemma 7.21.

8. Let $N_0 \subseteq N_1 \subseteq N_2$ and $\tilde{N}_0 \subseteq \tilde{N}_1 \subseteq \tilde{N}_2$ be triples such that (N_1, N_0) and $(\tilde{N}_1, \tilde{N}_0)$ are regular index pairs for an isolated compact invariant set S_1 and (N_2, N_1) and $(\tilde{N}_2, \tilde{N}_1)$ are regular index pairs for an isolated compact invariant set S_2. Prove that the following diagram commutes for all k:

$$
\begin{array}{ccc}
H_k(N_2, N_1) & \xrightarrow{\delta_*} & H_{k-1}(N_1, N_0) \\
\uparrow{\scriptstyle \approx} & & \uparrow{\scriptstyle \approx} \\
H_k(\tilde{N}_2, \tilde{N}_1) & \xrightarrow{\delta_*} & H_{k-1}(\tilde{N}_1, \tilde{N}_0)
\end{array}
$$

where δ_* denotes a connecting homomorphism.

9. A **connected simple system** consists of a collection I_0 of pointed spaces along with a collection I_m of homotopy classes of maps between these spaces such that (i) $\hom(X, Y) = \{[f] \in [X : Y] \mid [f] \in I_m\}$ consists of exactly one element for each pair of spaces $X, Y \in I_0$, (ii) if $X, Y, Z \in I_0$, $[f] \in \hom(X, Y)$, and $[g] \in \hom(Y, Z)$, then $[g \circ f] \in \hom(X, Z)$, and (iii) $\hom(X, X) = \{[\mathrm{id}_X]\}$ for all $X \in I_0$. Let $\varphi_t : M \to M$ be a flow on a locally compact metric space M, and let S be an isolated compact invariant set in M. Prove that (I_0, I_m) is a connected simple system where

$$I_0 = \{N/L \mid (N, L) \text{ is an index pair for } S \text{ in } M\}$$

and

$$I_m = \{[h_t] \mid h_t = h(-, t) \text{ is the map defined in Lemma 7.23}\}.$$

10. Use the filtration of index pairs described in Section 7.4 to show that every finite dimensional compact smooth manifold has the homotopy type of a CW-complex.

11. Give an alternate proof of the Morse Homology Theorem (Theorem 7.4) assuming that the gradient vector field is in standard form around every critical point.

12. Extend Theorem 7.4 to homology with coefficients in an abelian group G using the chain group $C_k(f, G) = C_k(f) \otimes G$ for all $k \in \mathbb{Z}_+$.

13. Extend Theorem 7.4 to the case where M is not orientable by using the orientation double cover.

14. Let $\varphi_t : M \to M$ be a flow on a locally compact metric space M. The ω-**limit sets** of a set $Y \subseteq M$ are defined to be

$$\omega(Y) \quad = I(\mathrm{cl}(Y \cdot [0, \infty))) \quad = \bigcap_{t>0} \mathrm{cl}(Y \cdot [t, \infty)),$$

$$\omega^*(Y) \quad = I(\mathrm{cl}(Y \cdot (-\infty, 0])) \quad = \bigcap_{t>0} \mathrm{cl}(Y \cdot (-\infty, -t]).$$

Let S be a compact invariant set. A compact invariant set $A \subseteq S$ is said to be an **attractor** in S if and only if there exists a neighborhood U of A in S such that $\omega(U) = A$. Similarly, a compact invariant set $A^* \subseteq S$ is said to be a **repeller** in S if and only there exists a neighborhood U of A^* in S such that $A^* = \omega^*(U)$. Prove that a compact invariant set $A \subseteq S$ is an attractor in S if and only if there exists a neighborhood U of A in S such that $x \cdot (-\infty, 0] \not\subseteq U$ for all $x \in U - A$. State and prove an analogous result for repellers.

15. Let $\varphi_t : M \to M$ be a flow on a locally compact metric space M, and let S be a compact invariant set. A finite collection $\{M(k) \mid k \in P\}$ of compact invariant sets in S is said to be a **Morse decomposition** of S if and only if there exists an ordering $k_1, \ldots, k_n$ of P such that for every $x \in S - \bigcup_{k \in P} M(k)$ there exist indices $i, j \in \{1, \ldots, n\}$ such that $i < j$ and

$$\omega(x) \subseteq M(k_i) \quad \text{and} \quad \omega^*(x) \subseteq M(k_j).$$

Every ordering of P with this property is said to be **admissible**. The sets $M(k)$ are called **Morse sets**. A partial ordering is defined on the indexing set P as follows: for $k, k^* \in P$ we define $k < k^*$ if $k \neq k^*$ and k comes before k^* in every admissible ordering of P. Prove that $k < k^*$ if and only if there exists sequences $k = k_0, k_1, \ldots, k_j = k^* \in P$ and $x_1, \ldots, x_j \in S - \bigcup \{M(k) \mid k \in P\}$ such that

$$\omega(x_i) \subseteq M(k_{i-1}) \quad \text{and} \quad \omega^*(x_i) \subseteq M(k_i) \quad \text{for all } i = 1, \ldots, j.$$

16. Let $\varphi_t : M \to M$ denote the gradient flow of a Morse-Smale function $f : M \to \mathbb{R}$ on a finite dimensional compact smooth Riemannian manifold M of dimension m. Prove that $\{\mathrm{Cr}_k(f) \mid k = 0, \ldots, m\}$ is a Morse decomposition of M.

17. Let $\varphi_t : M \to M$ be a flow on a locally compact metric space M, and let N_1 and N_2 be isolating neighborhoods for isolated compact invariant sets S_1 and S_2 respectively. Decide whether the following statements are true or false.

    a) $S_1 \cap S_2$ is an isolated compact invariant set with isolating neighborhood $N_1 \cap N_2$.

    b) $S_1 \cup S_2$ is an isolated compact invariant set with isolating neighborhood $N_1 \cup N_2$.

18. Prove Theorem 7.14: If $\varphi_t : M \to M$ is a flow on a locally compact metric space M and S is an isolated compact invariant set, then S admits an index pair (N, L).

19. Prove the following Theorem: Let $\varphi_t : M \to M$ be a flow on a locally compact metric space M. Let S be an isolated compact invariant set, $\{M(k) \mid k \in P\}$ a Morse decomposition of S with an admissible ordering $k_1, \ldots, k_n$ of P, and (N_n, N_0) an index pair for S. Then there exists a filtration $N_0 \subset N_1 \subset \cdots \subset N_n$ of compact sets such that (N_i, N_{j-1}) is an index pair for

$$M_{ij} = \left\{ x \in S \,\middle|\, \omega(x) \cup \omega^*(x) \subset \bigcup_{l=j}^{i} M(k_l) \right\}$$

whenever $1 \leq j \leq i \leq n$.

20. Let $\varphi_t : M \to M$ be a flow on a locally compact metric space M, and let S be an isolated compact invariant set. Prove that it is always possible to modify an index pair (N, L) for S to create a regular index pair $(N, \tilde{L})$.

21. Give a direct proof that $(C_*(f), \partial_*)$ is a chain complex. That is, prove that the Morse-Smale-Witten boundary operator satisfies $\partial_k \circ \partial_{k+1} = 0$ without using Lemma 7.21.

22. Give an alternate proof of the Morse Homology Theorem (Theorem 7.4) using the results in Section 6 of Milnor's book [102].

Chapter 8

Morse Theory On Grassmann Manifolds

In this chapter we show how Bott's perfect Morse functions (discussed in Example 3.7) are examples of a more general class of Morse-Smale functions defined on the complex Grassmann manifolds. The Morse-Smale functions, $f_A : G_{n,n+k}(\mathbb{C}) \to \mathbb{R}$, are defined analogous to the Morse functions constructed in Theorem 3.8. More specifically, we fix an embedding of $G_{n,n+k}(\mathbb{C})$ into the Lie algebra of skew-Hermitian matrices $\psi : G_{n,n+k}(\mathbb{C}) \hookrightarrow \mathfrak{u}(n+k) \approx \mathbb{R}^{(n+k)^2}$, and then we take the inner product of $x \in G_{n,n+k}(\mathbb{C}) \subset \mathfrak{u}(n+k)$ with a matrix $A \in \mathfrak{u}(n+k)$ using the negative of the trace form, i.e.

$$f_A(x) \overset{\text{def}}{=} < \psi(x), A > = -\text{trace}(\psi(x)A)$$

where $\psi(x)A$ denotes matrix multiplication.

Theorem 3.8 shows that for almost every matrix $A \in \mathfrak{u}(n+k)$ the function $f_A(x) = < \psi(x), A >$ is a Morse function. In Section 8.2 of this chapter we show, more explicitly, that whenever A has distinct eigenvalues f_A is a Morse function (Theorem 8.12). In Section 8.3 we describe an almost complex structure on $G_{n,n+k}(\mathbb{C}) \subset \mathfrak{u}(n+k)$ in terms of the Lie bracket on $\mathfrak{u}(n+k)$, and we compute the gradient vector field of the function f_A in terms of the almost complex structure (Theorem 8.28).

In Section 8.4 we specialize to a specific $A \in \mathfrak{u}(n+k)$ with distinct eigenvalues, and we analyze the critical points and gradient flow lines of f_A. We give a simple formula for computing the index of the critical points of f_A (Theorem 8.32) which shows that the critical points of f_A are all of even index, and we prove that there is a bijection between the critical points of f_A of index λ and the Schubert cells of $G_{n,n+k}(\mathbb{C})$ of dimension λ for all $\lambda \in \mathbb{Z}_+$ (Corollary 8.33). In fact, in Section 8.6 we show that the unstable manifolds of f_A are the Schubert cells of $G_{n,n+k}(\mathbb{C})$ (Theorem 8.40), and we use this fact to show that the function f_A satisfies the Morse-Smale transversality condition (Theorem 8.45).

Theorem 8.32 shows that f_A is a perfect Morse function, i.e. the boundary operators in the Morse-Smale-Witten chain complex are all trivial. Thus, the homology of $G_{n,n+k}(\mathbb{C})$ can be computed using either the Morse Homology Theorem (Theorem 7.4) or the CW-Homology Theorem (Theorem 2.15) and Theorem 3.28. We discuss this in Section 8.7.

In the last section of this chapter we briefly outline further generalizations and applications to the theory of Lie groups and symplectic geometry.

Much of the following discussion can be carried out for general Lie groups. In Section 8.1 we state and prove the preliminary lemmas and claims for any compact Lie group G whose Lie algebra $\mathfrak{g}$ possesses an associative inner product, and in Section 8.2 we specialize to the case where G is a unitary group and the inner product on $\mathfrak{g}$ is the negative of the trace form. The reader who is unfamiliar with the theory of Lie groups may want to consider in Section 8.1 the case of the unitary group

$$G = U(n) = \left\{ A \in M_{n \times n}(\mathbb{C}) \,|\, {}^t\bar{A}A = I_{n \times n} \right\}$$

and the Lie algebra of skew-Hermitian matrices

$$\mathfrak{g} = \mathfrak{u}(n) = \left\{ A \in M_{n \times n}(\mathbb{C}) \,|\, {}^t\bar{A} + A = 0 \right\}$$

where the Lie bracket is the usual commutator for matrices

$$[A, B] = AB - BA,$$

as that case is sufficient for most of the results this chapter.

Background required for this chapter

The results in this chapter require significantly more background than those in the preceeding chapters. For background on Lie groups, Lie algebras, the exponential map for Lie groups, and the adjoint action, we refer the reader to Chapter I of [31], Chapters II and III of [75], or Sections 3.1 through 3.52 of [148]. For background on transformation groups we refer the reader to Chapters 3 and 4 of [87] or Sections 3.58 through 3.65 of [148], and for background on Grassmann manifolds and Schubert cells we refer the reader to Section 1.5 of [67] or Sections 5 and 6 of [104].

We would like to thank Ralph Cohen, Paul Norbury, and Catalin Zara for several helpful discussions concerning the material in this chapter.

8.1 Morse theory on the adjoint orbit of a Lie group

Recall that a Lie algebra $\mathfrak{g}$ over $\mathbb{R}$ is a real vector space $\mathfrak{g}$ together with a bilinear operator $[\,,\,] : \mathfrak{g} \times \mathfrak{g} \to \mathfrak{g}$, called the **Lie bracket** that satisfies the following two properties for all $X, Y, Z \in \mathfrak{g}$.

(1) $[X,Y] = -[Y,X]$ (skew-symmetry)
(2) $[[X,Y],Z] + [[Y,Z],X] + [[Z,X],Y] = 0$ (Jacobi identity)

If G is a compact Lie group, then the tangent space at the identity T_eG can be identified with the set of left invariant vector fields on G, and under this identification $T_eG = \mathfrak{g}$ forms a Lie algebra under the Lie bracket operation on vector fields (see for instance Proposition 3.7 of [148]).

There is a smooth representation of a Lie group G on its Lie algebra $\mathfrak{g}$ called the **adjoint representation**

$$\mathrm{Ad} : G \to \mathrm{Aut}(\mathfrak{g})$$

whose differential at the identity of G gives a smooth map

$$\mathrm{ad} : \mathfrak{g} \to \mathrm{End}(\mathfrak{g})$$

which makes the following diagram commute

$$
\begin{array}{ccc}
\mathfrak{g} & \xrightarrow{\;\mathrm{ad}\;} & \mathrm{End}(\mathfrak{g}) \\
{\scriptstyle\exp}\downarrow & & \downarrow{\scriptstyle\exp} \\
G & \xrightarrow{\;\mathrm{Ad}\;} & \mathrm{Aut}(\mathfrak{g})
\end{array}
$$

where $\exp(X)$ is defined to be the value at 1 of the unique 1-parameter subgroup $\alpha : \mathbb{R} \to G$ whose tangent vector at zero is $X \in \mathfrak{g}$. Moreover, for any $X, Y \in \mathfrak{g}$ we have $\mathrm{ad}(X)(Y) = [X,Y]$ (see for instance Section 1.2 and 1.3 of [31] or Section 3.46 and Proposition 3.47 of [148]).

For any $x_0 \in \mathfrak{g}$, let G_{x_0} denote the **isotropy group** of the adjoint representation at x_0, i.e.

$$G_{x_0} = \{g \in G|\, \mathrm{Ad}(g)(x_0) = x_0\} \subseteq G,$$

and let $G \cdot x_0$ denote the **orbit** of the adjoint representation at x_0, i.e.

$$G \cdot x_0 = \{\mathrm{Ad}(g)(x_0)|\, g \in G\} \subseteq \mathfrak{g}.$$

There is an induced smooth structure on the homogeneous space G/G_{x_0} (see for instance Theorem 3.37 of [87] or Theorem 3.58 of [148]) and a smooth structure on the orbit $G \cdot x_0$ such that the map $h : G/G_{x_0} \to G \cdot x_0$ given by $h([g]) = \mathrm{Ad}(g)(x_0)$ is a G-equivariant diffeomorphism (see for instance Corollary 4.4 of [87] or Theorem 3.62 of [148]).

Definition 8.1 *Let $x_0 \in \mathfrak{g}$, where $\mathfrak{g}$ is the Lie algebra of a compact Lie group G. For any $A \in \mathfrak{g}$ we define the function $f_A : G \cdot x_0 \to \mathbb{R}$ by*

$$f_A(x) = <x, A>$$

for all $x \in G \cdot x_0$, where $<, >: \mathfrak{g} \times \mathfrak{g} \to \mathbb{R}$ is an inner product on $\mathfrak{g}$.

Note that the function f_A depends on the choice of the inner product on the real vector space $\mathfrak{g}$. By Theorem 3.8 we have the following.

Theorem 8.2 *For almost all $A \in \mathfrak{g}$, the function $f_A : G \cdot x_0 \to \mathbb{R}$ is a Morse function.*

Although this theorem is a step in the right direction, it does not suffice for our purposes. We need to be able to choose a specific inner product $<, >$ and a specific $A \in \mathfrak{g}$ such that f_A is a Morse function. To do this we will relate the Morse theory of $f_A : G \cdot x_0 \to \mathbb{R}$ to the Lie algebra structure of $\mathfrak{g}$. This will require an inner product that is compatible with the Lie algebra structure of $\mathfrak{g}$.

Definition 8.3 *A bilinear form $B : \mathfrak{g} \times \mathfrak{g} \to \mathbb{R}$ on a Lie algebra $\mathfrak{g}$ is said to be* **associative** *if and only if it satisfies*

$$B([X,Y],Z) = B(X,[Y,Z])$$

for all $X, Y, Z \in \mathfrak{g}$, where $[\,,\,]$ denotes the Lie bracket on $\mathfrak{g}$.

Example 8.4 (Matrix groups and the trace form)   A Lie subgroup of the general linear group $GL_n(\mathbb{C})$ or $GL_n(\mathbb{R})$ is called a **matrix group**. Examples of matrix groups include the unitary group $U(n)$, the special unitary group $SU(n)$, the special orthogonal group $SO(n)$, and the symplectic group $Sp(n)$. (For descriptions of these groups and their Lie algebras see the problems at the end of this chapter.)

For any matrix group G we have the following

$$
\begin{aligned}
\mathrm{Ad}(g)(X) &= gXg^{-1} \\
\mathrm{ad}(X)(Y) &= [X,Y] = XY - YX \\
\exp(X) &= e^X = \sum_{k=0}^{\infty} \frac{X^k}{k!} \\
\det e^X &= e^{\mathrm{trace}(X)}
\end{aligned}
$$

for all $g \in G$ and for all $X, Y \in \mathfrak{g}$ (see for instance Sections 3.29 through 3.47 of [148]). Moreover, using the properties of the trace it is easy to show that the trace form

$$B(X,Y) \overset{\text{def}}{=} \operatorname{Re} \operatorname{trace}(XY)$$

is an associative bilinear form, i.e.

$$\text{Re trace}([X, Y]Z) = \text{Re trace}(X[Y, Z])$$

for all $X, Y, Z \in \mathfrak{g}$, that is symmetric and invariant under the adjoint representation.

Lemma 8.5 *Suppose that G is a matrix group with Lie algebra $\mathfrak{g}$ that satisfies the following condition: if $X \in \mathfrak{g}$, then ${}^t\bar{X} \in \mathfrak{g}$. Then the trace form is nondegenerate on $\mathfrak{g}$. Moreover, if $G = U(n)$, $SU(n)$, or $SO(n)$, then the trace form is negative definite on the Lie algebra $\mathfrak{g}$.*

Proof:
    Suppose that $X \in \mathfrak{g}$ satisfies $B(X, Y) = 0$ for all $Y \in \mathfrak{g}$. Then taking $Y = {}^t\bar{X}$ we have

$$\begin{aligned}
0 &= B(X, {}^t\bar{X}) \\
&= \text{Re trace}(X^t\bar{X}) \\
&= \text{Re} \sum_{i,j} X_{ij}\bar{X}_{ij} \\
&= \sum_{ij} |X_{ij}|^2,
\end{aligned}$$

and we see that $X = 0$. For $G = U(n)$, $SU(n)$, or $SO(n)$, we have $X = -{}^t\bar{X}$ for any $X \in \mathfrak{g}$ (see the problems at the end of this chapter), and hence,

$$B(X, X) = - \sum_{ij} |X_{ij}|^2.$$

$\square$

Hence, for $G = U(n)$, $SU(n)$, and $SO(n)$, the negative of the trace form is an associative inner product on the Lie algebra $\mathfrak{g}$.

Example 8.6 (Semisimple Lie groups and the Killing form) The trace form can be extended to general Lie algebras using a representation of the Lie algebra on a vector space. The most important example of this is the **Killing form** $B : \mathfrak{g} \times \mathfrak{g} \to \mathbb{R}$ which is the trace form determined by the representation $\text{ad} : \mathfrak{g} \to \text{End}(\mathfrak{g})$:

$$B(X, Y) = \text{trace}(\text{ad}(X) \circ \text{ad}(Y))$$

for all $X, Y \in \mathfrak{g}$. Using the properties of the trace one can show that the Killing form is an associative bilinear form that is symmetric and invariant under the adjoint representation (see for instance Section II.6 of [75]). Moreover, it is possible to show that the Killing form of a compact connected Lie group G is negative definite if and only if G is **semisimple** (see for instance Proposition V.5.13 of [31] or Section II.6 of [75]). Hence, for any compact connected semisimple Lie group G the negative of the Killing form is an associative inner product on the Lie algebra $\mathfrak{g}$.

From now on we will assume that the Lie algebra $\mathfrak{g}$ has an associative inner product. Our next step is to describe the tangent and normal spaces of the orbit $G \cdot x_0$ in terms of the Lie algebra structure.

Lemma 8.7 *Let $\eta : G \times M \to M$ be a transitive smooth action of a compact Lie group G on a smooth manifold M, and let $x \in M$. For all $X \in T_x M$ there exists a $Y \in \mathfrak{g}$ such that*

$$\frac{d}{dt}\eta(\exp(tY), x)\Big|_{t=0} = X.$$

Proof:

Define $\eta_x : G \to M$ by $\eta_x(g) = \eta(g, x)$ for all $g \in G$. Then the following diagram commutes

$$
\begin{array}{ccc}
 & G & \\
{\scriptstyle \pi}\Big\downarrow & & \searrow{\scriptstyle \eta_x} \\
G/G_x & \xrightarrow[\beta]{} & M
\end{array}
$$

where β is a diffeomorphism (see for instance Corollary 4.4 of [87] or Theorem 3.62 of [148]) and π is a submersion (see for instance Corollary 3.38 of [87] or Theorem 3.58 of [148]). Hence, η_x is a submersion and, in particular, $d\eta_x|_e :$ $\mathfrak{g} \to T_x M$ is surjective where e is the identity of G. Given $X \in T_x M$ choose a $Y \in \mathfrak{g}$ that satisfies $d\eta_x(Y) = X$. Then,

$$
\begin{aligned}
\frac{d}{dt}\eta(\exp(tY), x)\Big|_{t=0} &= \frac{d}{dt}\eta_x(\exp(tY))\Big|_{t=0} \\
&= d\eta_x(Y) \\
&= X
\end{aligned}
$$

where the second equality follows for instance from Remark 3.36 of [148].

$\square$

Proposition 8.8 *The tangent space at $x \in G \cdot x_0$ is given by*

$$T_x(G \cdot x_0) = [\mathfrak{g}, x] = [x, \mathfrak{g}].$$

Proof:

If we let $X \in T_x(G \cdot x_0)$, then the preceeding lemma says that there exists a $Y \in \mathfrak{g}$ such that

$$\frac{d}{dt}(\mathrm{Ad}(\exp(tY))(x))\Big|_{t=0} = X,$$

where the left hand side is just $[Y, x]$ (see for instance Proposition 3.47 of [148]). Therefore, $T_x(G \cdot x_0) \subseteq [\mathfrak{g}, x]$.

On the other hand, for any $Y \in \mathfrak{g}$ the equality

$$[Y, x] = \frac{d}{dt}(\mathrm{Ad}(\exp(tY))(x))\Big|_{t=0}$$

shows that $[Y, x]$ is the derivative of a path in $G \cdot x_0$ passing through x at $t = 0$. Therefore, $[\mathfrak{g}, x] \subseteq T_x(G \cdot x_0)$, and hence,

$$T_x(G \cdot x_0) = [\mathfrak{g}, x] = [x, \mathfrak{g}]$$

where the second equality follows from the fact that a Lie bracket is skew-symmetric.

$\square$

Proposition 8.9 *The normal space at* $x \in G \cdot x_0 \subseteq \mathfrak{g}$ *with respect to an associative inner product on* $\mathfrak{g}$ *is given by*

$$N_x(G \cdot x_0) = \{Z \in \mathfrak{g} |\ [Z, x] = 0\}.$$

Proof:

The vector $Z \in \mathfrak{g}$ is orthogonal to the tangent space at $x \in G \cdot x_0$ if and only if one (and hence all) of the following equivalent conditions hold

$$
\begin{aligned}
& < Z, X >= 0 && \text{for all } X \in T_x(G \cdot x_0) \\
\Leftrightarrow\quad & < Z, [x, Y] >= 0 && \text{for all } Y \in \mathfrak{g} \\
\Leftrightarrow\quad & < [Z, x], Y >= 0 && \text{for all } Y \in \mathfrak{g} \\
\Leftrightarrow\quad & \phantom{<} [Z, x] = 0 &&
\end{aligned}
$$

where the second equivalence follows from associativity, and the last equivalence follows from the fact that an inner product is nondegenerate.

$\square$

We will now show how to compute the critical points of f_A using the Lie algebra structure.

Lemma 8.10 *For any $x \in G \cdot x_0$ and for any $X \in T_x(G \cdot x_0)$ the directional derivative of $f_A : G \cdot x_0 \to \mathbb{R}$ in the direction of X is*

$$D_X f_A = <X, A>.$$

Proof:

First note that $f_A : G \cdot x_0 \to \mathbb{R}$ has an obvious extension $\tilde{f}_A : \mathfrak{g} \to \mathbb{R}$, and in any direction tangent to $G \cdot x_0$ the directional derivative of f_A and $\tilde{f}_A$ are identical. Hence, we can compute the partial derivative of f_A in the direction $X \in T_x(G \cdot x_0)$ as follows.

$$
\begin{aligned}
D_X f_A &= \lim_{t \to 0} \frac{1}{t} \left[\tilde{f}_A(x + tX) - \tilde{f}_A(x) \right] \\
&= \lim_{t \to 0} \frac{1}{t} \left[<x + tX, A> - <x, A> \right] \\
&= \lim_{t \to 0} \frac{1}{t} \left[t <X, A> \right] \\
&= <X, A>
\end{aligned}
$$

$\square$

Proposition 8.11 *A point $p \in G \cdot x_0$ is a critical point of $f_A : G \cdot x_0 \to \mathbb{R}$ if and only if*

$$[p, A] = 0.$$

Proof:

The point $p \in G \cdot x_0$ is a critical point of f_A is and only if one (and hence all) of the following equivalent conditions hold

$$
\begin{aligned}
& D_X f_A = 0 && \text{for all } X \in T_p(G \cdot x_0) \\
\Leftrightarrow \quad & <X, A> = 0 && \text{for all } X \in T_p(G \cdot x_0) \\
\Leftrightarrow \quad & <[Z, p], A> = 0 && \text{for all } Z \in \mathfrak{g} \\
\Leftrightarrow \quad & <Z, [p, A]> = 0 && \text{for all } Z \in \mathfrak{g} \\
\Leftrightarrow \quad & [p, A] = 0
\end{aligned}
$$

where the third equivalence follows from associativity, and the last equivalence follows from the fact that an inner product is nondegenerate.

$\square$

Note that the associativity of the inner product was only used in the proofs of Propositions 8.9 and 8.11.

8.2 A Morse function on an adjoint orbit of the unitary group

In the previous section we described the tangent and normal spaces of an orbit $G \cdot x_0$ and the critical points of the function $f_A : G \cdot x_0 \to \mathbb{R}$ in terms of the Lie algebra structure of $\mathfrak{g}$ for any compact Lie group G whose Lie algebra $\mathfrak{g}$ possesses an associative inner product.

We now restrict to the case of the unitary group $G = U(n + k)$ and the Lie algebra of skew-Hermitian matrices $\mathfrak{g} = \mathfrak{u}(n + k)$ where the inner product $< , >$ is the negative of the trace form, i.e.

$$< A, B >= -\text{trace}(AB)$$

for all $A, B \in \mathfrak{u}(n + k)$. We also fix the point

$$x_0 = \left(\begin{array}{cc} \left. \begin{array}{ccc} i & & 0 \\ & \ddots & \\ 0 & & i \end{array} \right\} n & 0 \\ 0 & k \left\{ \begin{array}{ccc} 0 & \cdots & 0 \\ \vdots & \ddots & \vdots \\ 0 & \cdots & 0 \end{array} \right. \end{array} \right) \in \mathfrak{u}(n + k),$$

and we denote the adjoint action

$$U(n + k) \times \mathfrak{u}(n + k) \to \mathfrak{u}(n + k)$$

by

$$g \cdot x \overset{\text{def}}{=} \text{Ad}(g)(x) = gxg^{-1}$$

for all $g \in U(n + k)$ and $x \in \mathfrak{u}(n + k)$. In Section 8.5 we will show that with these choices the orbit $U(n + k) \cdot x_0$ is diffeomorphic to $G_{n,n+k}(\mathbb{C})$, the complex Grassmann manifold consisting of the set of n-dimensional complex planes in $\mathbb{C}^{n+k}$. We do this by showing that both $G_{n,n+k}(\mathbb{C})$ and $U(n+k) \cdot x_0$ are diffeomorphic to $U(n + k)/(U(n) \times U(k))$, which is a smooth manifold of real dimension $2nk$.

The rest of this section is devoted to the proof of the following theorem. For versions of this theorem that apply to more general Lie groups see [17] and the references therein.

Theorem 8.12 *If $A \in \mathfrak{u}(n + k)$ has distinct eigenvalues, then the function $f_A : U(n + k) \cdot x_0 \to \mathbb{R}$ given by $f_A(x) =< x, A >$ is Morse function.*

Our proof of this theorem is based on a lemma from Section 6 of [100] that allows us to turn a problem involving the second order partial derivatives of f_A

into a problem involving the first order partial derivatives of another function E defined on the total space of the normal bundle of $U(n+k) \cdot x_0 \subset u(n+k)$. We begin with the following observation.

Lemma 8.13 *The function $f_A : U(n+k) \cdot x_0 \to \mathbb{R}$ given by $f_A(x) =< x, A >$ satisfies*

$$f_A(x) = -\frac{1}{2}(g_A(x) - C)$$

for some constant $C \in \mathbb{R}$ where $g_A(x) \overset{def}{=} \|x - A\|^2$ for all $x \in U(n+k) \cdot x_0$. Hence, the functions f_A and g_A have the same critical points, and a critical point p is degenerate for f_A if and only if it is degenerate for g_A. Moreover, a non-degenerate critical p of index λ_p for the function f_A is a non-degenerate critical point of index $2nk - \lambda_p$ for the function g_A.

Proof:
    Since the trace form is invariant under the adjoint representation,

$$< g \cdot x_0, g \cdot x_0 >=< x_0, x_0 >$$

for all $g \in U(n + k)$. Hence, $\|x\|^2 = \|x_0\|^2$ for all $x \in U(n + k) \cdot x_0$, i.e. the orbit $U(n + k) \cdot x_0$ lies in the sphere of radius $\|x_0\|$ in $u(n + k)$. Therefore,

$$
\begin{aligned}
g_A(x) &= \|x - A\|^2 \\
&= < x - A, x - A > \\
&= < x, x > -2 < x, A > - < A, A > \\
&= -2 < x, A > +C
\end{aligned}
$$

for all $x \in U(n + k) \cdot x_0$, where $C = \|x_0\|^2 - \|A\|^2$.

$\square$

Manifolds in Euclidean space

We now recall some definitions from Section 6 of [100]. Let M be a manifold embedded in some Euclidean space $\mathbb{R}^r$. Define a function $E : N \to \mathbb{R}^r$ by $E(x, \vec{v}) = x + \vec{v}$ where N is the total space of the normal bundle of M in $\mathbb{R}^r$, i.e.

$$N = \{(x, \vec{v}) \in \mathbb{R}^r \times \mathbb{R}^r \mid x \in M \text{ and } \vec{v} \in N_x(M)\}.$$

Definition 8.14 *A point $e \in \mathbb{R}^r$ is called a **focal point** of $x \in M$ with **multiplicity** μ if and only if $E(x, \vec{v}) = e$ for some $\vec{v}$ with $(x, \vec{v}) \in N$ and the Jacobian of $E : N \to \mathbb{R}^r$ at $(x, \vec{v})$ has nullity $\mu > 0$.*

Remark 8.15 Intuitively, a focal point of M is a point in $\mathbb{R}^r$ where nearby normals intersect.

The following is Lemma 6.5 of [100].

Lemma 8.16 *A point $p \in M$ is a degenerate critical point of the function $g_A : M \to \mathbb{R}$ given by $g_A(x) = \|x - A\|^2$ if and only if A is a focal point of $p \in M$. The nullity of p as a critical point of g_A of is equal to the multiplicity of A as a focal point of $p \in M$.*

Remark 8.17 If $f : M \to \mathbb{R}$ is a smooth function, then the **nullity** of a critical point $p \in M$ of f is the dimension of the null space of the Hessian $H_p(f)$. In particular, p is a degenerate critical point of f if and only if its nullity is positive.

To prove this lemma we will use the following notation.

Definition 8.18 *Let $u_1, \ldots, u_m$ be local coordinates on M. The inclusion of M into $\mathbb{R}^r$ determines r smooth functions*

$$x_1(u_1, \ldots, u_m), \ldots, x_r(u_1, \ldots, u_m)$$

*given by projecting onto the axes in $\mathbb{R}^r$. We will denote $\vec{x}(u_1, \ldots, u_m) = (x_1, \ldots, x_r)$. The **first fundamental form** associated to this coordinate system is the following symmetric $m \times m$ matrix of real valued functions:*

$$(g_{ij}) = \left(\frac{\partial \vec{x}}{\partial u_i} \cdot \frac{\partial \vec{x}}{\partial u_j} \right).$$

*The **second fundamental form** is the symmetric $m \times m$ matrix of vector valued functions $(\vec{\ell}_{ij})$ where*

$$\vec{\ell}_{ij} \overset{def}{=} \text{normal component of } \frac{\partial^2 \vec{x}}{\partial u_i \partial u_j}.$$

The proof of the following lemma follows that of Lemma 6.3 of [100].

Lemma 8.19 *The nullity of the Jacobian of E at $(p, t\vec{v}) \in N$ equals the nullity of an $r \times r$ matrix of the form*

$$\begin{pmatrix} \left. \frac{\partial \vec{x}}{\partial u_i} \right|_p \cdot \left. \frac{\partial \vec{x}}{\partial u_j} \right|_p - t\vec{v} \cdot \vec{\ell}_{ij} & 0 \\ * & I_{(r-m) \times (r-m)} \end{pmatrix}$$

where $I_{(r-m)\times(r-m)}$ denotes the $(r-m)\times(r-m)$ identity matrix. Hence, $p+t\vec{v}$ is a focal point of $p \in M$ with multiplicity μ if and only if the upper left $m \times m$ minor of the above matrix is singular with nullity μ.

Proof:

Locally, choose $r-m$ orthonormal vector fields

$$\vec{w}_1(u_1,\ldots,u_m),\ldots,\vec{w}_{r-m}(u_1,\ldots,u_m)$$

spanning the normal bundle of $M \subseteq \mathbb{R}^r$, and introduce local coordinates $(u_1,\ldots,u_m,t_1,\ldots,t_{r-m})$ on the total space of the normal bundle N as follows: Let $(u_1,\ldots,u_m,t_1,\ldots,t_{r-m})$ correspond to the point

$$\left(\vec{x}(u_1,\ldots,u_k),\sum_{\alpha=1}^{r-m}t_\alpha\vec{w}_\alpha(u_1,\ldots,u_m)\right) \in N.$$

In these coordinates the function $E: N \to \mathbb{R}^r$ becomes

$$\vec{e}(u_1,\ldots,u_m,t_1,\ldots,t_{r-m}) = \vec{x}(u_1,\ldots,u_k) + \sum_{\alpha=1}^{r-m}t_\alpha\vec{w}_\alpha(u_1,\ldots,u_m),$$

and the partial derivatives of $\vec{e}$ are

$$\frac{\partial\vec{e}}{\partial u_i} = \frac{\partial\vec{x}}{\partial u_i} + \sum_{\alpha=1}^{r-m}t_\alpha\frac{\partial\vec{w}_\alpha}{\partial u_i}$$

$$\frac{\partial\vec{e}}{\partial t_\alpha} = \vec{w}_\alpha.$$

Multiplying the Jacobian of $\vec{e}$ on the left by the $r \times r$ nonsingular matrix whose rows are the linearly independent vectors $\frac{\partial\vec{x}}{\partial u_1},\ldots,\frac{\partial\vec{x}}{\partial u_m},\vec{w}_1,\ldots,\vec{w}_{r-m}$, we get an $r \times r$ matrix whose nullity equals the nullity of the Jacobian of E. This $r \times r$ matrix has the following form:

$$\begin{pmatrix} \left(\frac{\partial\vec{x}}{\partial u_i}\cdot\frac{\partial\vec{x}}{\partial u_j} + \sum_\alpha t_\alpha\frac{\vec{w}_\alpha}{\partial u_i}\cdot\frac{\partial\vec{x}}{\partial u_j}\right) & 0 \\ \left(\sum_\alpha t_\alpha\frac{\partial\vec{w}_\alpha}{\partial u_i}\cdot\vec{w}_\beta\right) & I_{(r-m)\times(r-m)} \end{pmatrix}.$$

Using the identity

$$0 = \frac{\partial}{\partial u_i}\left(\vec{w}_\alpha\cdot\frac{\partial\vec{x}}{\partial u_j}\right) = \frac{\partial\vec{w}_\alpha}{\partial u_i}\cdot\frac{\partial\vec{x}}{\partial u_j} + \vec{w}_\alpha\cdot\frac{\partial^2\vec{x}}{\partial u_i\partial u_j}$$

we see that the upper left $m \times m$ minor of the above matrix is

$$\left(\frac{\partial\vec{x}}{\partial u_i}\cdot\frac{\partial\vec{x}}{\partial u_j} - \sum_{\alpha=1}^{r-m}t_\alpha\vec{w}_\alpha\cdot\vec{\ell}_{ij}\right)$$

where $\displaystyle\sum_{\alpha=1}^{r-m} t_\alpha \vec{w}_\alpha$ is some vector $t\vec{v}$ that is normal to M.

□

Lemma 8.20 *The point $p \in M \subseteq \mathbb{R}^r$ is a critical point of $g_A(x) = \|x - A\|^2$ if and only if $A - p$ is normal to M at p. Moreover, if $\vec{v} = A - p$ is normal to M at p, then the Hessian of g_A at the critical point p satisfies*

$$H_p(g_A) = 2 \left(\frac{\partial \vec{x}}{\partial u_i}\bigg|_p \cdot \frac{\partial \vec{x}}{\partial u_j}\bigg|_p - \vec{v} \cdot \vec{\ell}_{ij} \right).$$

Proof:

Using the notation in Definition 8.18 we have

$$g_A(\vec{x}(u_1, \ldots, u_m)) = \|\vec{x} - A\|^2 = \vec{x} \cdot \vec{x} - 2\vec{x} \cdot A + A \cdot A.$$

Therefore,

$$\frac{\partial g_A}{\partial u_i} = 2 \frac{\partial \vec{x}}{\partial u_i} \cdot (\vec{x} - A),$$

and we see that g_A has a critical point at p if and only if $p - A$ (and hence $A - p$) is normal to M at p. The second partial derivatives of g_A are

$$\frac{\partial^2 g_A}{\partial u_i \partial u_j} = 2 \left(\frac{\partial \vec{x}}{\partial u_i} \cdot \frac{\partial \vec{x}}{\partial u_j} + \frac{\partial^2 \vec{x}}{\partial u_i \partial u_j} \cdot (\vec{x} - A) \right).$$

So, if $\vec{v} = A - p$ is normal to M at p, then p is a critical point of g_A and the Hessian of g_A at p is given by

$$H_p(g_A) = 2 \left(\frac{\partial \vec{x}}{\partial u_i}\bigg|_p \cdot \frac{\partial \vec{x}}{\partial u_j}\bigg|_p - \vec{v} \cdot \vec{\ell}_{ij} \right).$$

□

Proof of Lemma 8.16:

If p is a critical point of $g_A(x) = \|x - A\|^2$, then by Lemma 8.20, $\vec{v} = A - p$ is normal to M at p. Thus, $E(p, \vec{v}) = A$ where $(p, \vec{v}) \in N$. Moreover, Lemmas 8.19 and 8.20 show that the nullity of the Hessian $H_p(g_A)$ and the nullity of the Jacobian of E at $(p, \vec{v})$ are the same.

□

The following is Lemma 6.9 of [100]. We prove it here for future reference.

Lemma 8.21 *The index of $g_A = \|x - A\|^2$ at a non-degenerate critical point $p \in M$ is equal to the number of focal points of $p \in M$ that lie on the line segment from p to A; each focal point being counted with multiplicity.*

Proof:

The index of the matrix

$$H_p(g_A) = 2 \left(\left. \frac{\partial \vec{x}}{\partial u_i} \right|_p \cdot \left. \frac{\partial \vec{x}}{\partial u_j} \right|_p - \vec{v} \cdot \vec{\ell}_{ij} \right)$$

is equal to the number of its negative eigenvalues. If we choose local coordinates such that $\left(\left. \frac{\partial \vec{x}}{\partial u_i} \right|_p \cdot \left. \frac{\partial \vec{x}}{\partial u_j} \right|_p \right)$ is the identity matrix, then this is equal to the number of eigenvalues of $(\vec{v} \cdot \vec{\ell}_{ij})$ that are greater than 1.

For any eigenvalue $\lambda > 1$, the point $p + \frac{1}{\lambda}\vec{v}$ is a focal point of $p \in M$ by Lemma 8.19. Moreover, the multiplicity of the focal point $p + \frac{1}{\lambda}\vec{v}$ is equal to the multiplicity of λ as an eigenvalue. Similarly, if $p + t\vec{v}$ is a focal point of $p \in M$ of multiplicity μ with $0 < t < 1$, then Lemma 8.19 implies that $\frac{1}{t} > 1$ is an eigenvalue of $(\vec{v} \cdot \vec{\ell}_{ij})$ of multiplicity μ.

$\square$

Remark 8.22 If $\vec{v}$ is a unit vector that is normal to M at p, then the matrix

$$\left(\vec{v} \cdot \vec{\ell}_{ij} \right)$$

is called the **second fundamental form** of M at p in the direction $\vec{v}$. The eigenvalues of this matrix are called the **principal curvatures** of M at p in the normal direction $\vec{v}$, and the reciprocals of these eigenvalues are called the **principal radii of curvature**.

The adjoint orbit

We now apply the preceeding results to the manifold $M = U(n + k) \cdot x_0 \subset \mathfrak{u}(n + k)$. Note that

$$N = \{(x, Z) \in \mathfrak{u}(n + k) \times \mathfrak{u}(n + k) \mid x \in U(n + k) \cdot x_0 \text{ and } [Z, x] = 0\}$$

by Proposition 8.9.

To compute the Jacobian of $E : N \to \mathfrak{u}(n + k)$ we will need a basis for the tangent space of N.

Lemma 8.23 *Let $N \subset \mathfrak{u}(n + k) \times \mathfrak{u}(n + k)$ be the total space of the normal bundle of $U(n+k) \cdot x_0$, let $x \in U(n+k) \cdot x_0$, and let $Z \in N_x(U(n+k) \cdot x_0)$. If*

$[Y_1, x], \ldots, [Y_{2nk}, x]$ *is a basis for* $T_x(U(n+k) \cdot x_0)$ *and* $Z_{2nk+1}, \ldots, Z_{(n+k)^2}$ *is a basis for* $N_x(U(n+k) \cdot x_0)$, *then*

$$
\begin{aligned}
X_1 &= ([Y_1, x], [Y_1, Z]) \\
&\vdots \qquad \vdots \\
X_{2nk} &= ([Y_{2nk}, x], [Y_{2nk}, Z]) \\
X_{2nk+1} &= (\vec{0}, Z_{2nk+1}) \\
&\vdots \qquad \vdots \\
X_{(n+k)^2} &= (\vec{0}, Z_{(n+k)^2}).
\end{aligned}
$$

is a basis for $T_{(x,Z)}N$.

Proof:

For $j = 1, \ldots, 2nk$ define paths $\gamma_j : \mathbb{R} \to U(n+k) \cdot x_0$ by $\gamma_j(t) = (\exp tY_j) \cdot x$. The γ_j's are paths through x that satisfy $\gamma_j'(0) = [Y_j, x]$ for all $j = 1, \ldots, 2nk$ (see for instance Proposition 3.47 of [148]). We claim that $\gamma_j^\perp(t) = (\exp tY_j) \cdot Z \in N_{\gamma_j(t)}(U(n+k) \cdot x)$ for all $j = 1, \ldots, 2nk$. To see this, note that

$$
\begin{aligned}
[\gamma_j(t), \gamma_j^\perp(t)] &= [(\exp tY_j) \cdot x, (\exp tY_j) \cdot Z] \\
&= (\exp tY_j) \cdot [x, Z] \\
&= (\exp tY_j) \cdot 0 \\
&= 0.
\end{aligned}
$$

So, we have $p_j(t) = (\gamma_j(t), \gamma_j^\perp(t)) \in N$ for all $t \in \mathbb{R}$ and $p_j(0) = (x, Z)$. Therefore, $p_j'(0) = ([Y_j, x], [Y_j, Z]) \in T_{(x,Z)}N$ for all $j = 1, \ldots, 2nk$.

Now define $p_j(t) = (x, Z + tZ_j)$ for all $j = 2nk+1, \ldots, (n+k)^2$. Clearly $p_j(t) \in N$ for all $t \in \mathbb{R}$ and $p_j(0) = (x, Z)$. So, the derivative $p_j'(0) = (\vec{0}, Z_j) \in T_{(x,Z)}N$ for all $j = 2nk+1, \ldots, (n+k)^2$. Linear independence is obvious from the choice of the Y_j's and Z_j's.

$\square$

Using the same notation as in the previous lemma we compute the partial derivatives of E as follows.

Lemma 8.24 *Let* $(x, Z) \in N$. *For all* $j = 1, \ldots, 2nk$ *we have,*

$$
D_{X_j}E|_{(x,Z)} = [Y_j, x] + [Y_j, Z],
$$

and for all $j = 2nk+1, \ldots, (n+k)^2$ *we have,*

$$
D_{X_j}E|_{(x,Z)} = Z_j.
$$

Proof:
For any $j = 1, \ldots, 2nk$ we have,

$$
\begin{aligned}
D_{X_j} E\big|_{(x,Z)} &= \lim_{t \to 0} \frac{1}{t}\left(E((x,Z) + tX_j) - E(x,Z) \right) \\
&= \lim_{t \to 0} \frac{1}{t}\left(E\left(x + t[Y_j, x], Z + t[Y_j, Z]\right) - (x + Z) \right) \\
&= \lim_{t \to 0} \frac{1}{t}\left(x + t[Y_j, x] + Z + t[Y_j, Z] - x - Z \right) \\
&= [Y_j, x] + [Y_j, Z].
\end{aligned}
$$

For any $j = 2nk + 1, \ldots, (n+k)^2$ we have,

$$
\begin{aligned}
D_{X_j} E\big|_{(x,Z)} &= \lim_{t \to 0} \frac{1}{t}\left(E((x,Z) + tX_j) - E(x,Z) \right) \\
&= \lim_{t \to 0} \frac{1}{t}\left(E(x, Z + tZ_j) - (x + Z) \right) \\
&= \lim_{t \to 0} \frac{1}{t}\left(x + B + tZ_j - x - B \right) \\
&= Z_j.
\end{aligned}
$$

$\square$

Proof of Theorem 8.12:
Let $A \in \mathfrak{u}(n + k)$ be a matrix with distinct eigenvalues, and let $p \in U(n + k) \cdot x_0$ be a critical point of f_A. Since $p \in U(n + k) \cdot x_0$ there exists some $g \in U(n + k)$ such that $g \cdot p = x_0$, and since the inner product $<\,,\,>$ is invariant under the adjoint representation we have

$$
f_A(x) = <x, A> = <g \cdot x, g \cdot A> = f_{g \cdot A}(g \cdot x)
$$

for all $x \in U(n + k) \cdot x_0$. Hence, p is a non-degenerate critical point of f_A if and only if x_0 is a non-degenerate critical point of the function $f_{g \cdot A}$. Since A and $g \cdot A = gAg^{-1}$ have the same eigenvalues, this shows that it suffices to prove the theorem in the case where the critical point $p = x_0$.

By Proposition 8.11, x_0 is a critical point of f_A if and only if $[x_0, A] = 0$, i.e. if and only if x_0 commutes with A. By the Spectral Theorem for Normal Transformations and the fact that x_0 and A commute, there exists some $g \in U(n + k)$ such that both gx_0g^{-1} and gAg^{-1} are diagonal (see Corollary

32.17 of [37] or Theorem 2.10.2 of [93] and Sections 6.4 and 6.6 of [62]). Subsequent conjugation by permutation matrices can bring gx_0g^{-1} back to x_0 while keeping gAg^{-1} diagonal. Hence, there exists a $g \in U(n+k)$ such that $g \cdot x_0 = x_0$ and $g \cdot A$ is diagonal. By the same reasoning as above, this shows that without loss of generality we may assume that A is diagonal.

By Lemma 8.13, x_0 is a non-degenerate critical point for the function $f_A(x)$ $=< x, A >$ if and only if it is a non-degenerate critical point for the function $g_A(x) = \| x - A \|^2$. So by Lemma 8.16, x_0 is a non-degenerate critical point of f_A if and only if the Jacobian of $E : N \rightarrow u(n+k)$ is non-singular at $(x_0, A - x_0) \in N$.

The computation of the partial derivatives of E was done in Lemma 8.24. To finish the proof we need only pick specific matrices $Y_1, \ldots, Y_{2nk}$ such that $[Y_1, x_0], \ldots, [Y_{2nk}, x_0]$ is a basis for $T_{x_0}(U(n+k) \cdot x_0)$ and check that $[Y_1, A], \ldots, [Y_{2nk}, A], Z_{2nk+1}, \ldots, Z_{(n+k)^2}$ are linearly independent. (Note that $[Y_j, x_0] + [Y_j, A - x_0] = [Y_j, A]$.)

The following choices make linear independence clear. For $j = 1, \ldots, nk$ we choose Y_j by putting a 1 in some entry of the upper right $n \times k$ block, -1 in the corresponding entry of the lower left $n \times k$ block, and zero everywhere else. For $j = nk + 1, \ldots, 2nk$ we put an i in some entry of the upper right $n \times k$ block and also in the corresponding entry of the lower left $n \times k$ block with zeros everywhere else. For $j = 2nk + 1, \ldots, (n+k)^2$ we pick Z_j by either putting a basis element of $u(n)$ in the upper left $n \times n$ block with zeros everywhere else or by putting a basis element of $u(k)$ in the lower right $k \times k$ block with zeros everywhere else.

$\square$

Remark 8.25 The proof of Theorem 8.12 can be modified to work for other values of $x_0 \in u(n+k)$, i.e. for different orbits $U(n+k) \cdot x_0 \subseteq u(n+k)$. The only part of the proof that might need to be modified is the choice of the matrices $Y_1, \ldots, Y_{2nk}$ in the last paragraph.

Moreover, one can show that when $A \in u(n+k)$ has repeated eigenvalues the function $f_A : U(n+k) \cdot x_0 \rightarrow \mathbb{R}$ given by $f_A(x) =< x, A >$ is a Morse-Bott function [80].

8.3   An almost complex structure on the adjoint orbit

Before we pick a specific Morse function and compute the indices of its critical points, we record a few more general facts for future reference.

Lemma 8.26 *Define $J(X) = [X, x]$ for all $X \in T_x(U(n+k) \cdot x_0)$. Then J is an almost complex structure on $U(n+k) \cdot x_0$, i.e. $J^2 = -1$.*

Proof:

By Proposition 8.8, $X \in T_x(U(n+k) \cdot x_0)$ if and only if $X = [Y, x]$ for some $Y \in \mathfrak{u}(n+k)$. Since $g \cdot [Y, x] = [g \cdot Y, g \cdot x]$ for all $g \in U(n+k)$ it suffices to check that $[[[Y, x_0], x_0], x_0] = -[Y, x_0]$ for all $Y \in \mathfrak{u}(n+k)$. The matrix $[Y, x_0]$ is of the form,

$$[Y, x_0] = \begin{pmatrix} 0 & B \\ C & 0 \end{pmatrix}$$

where the upper left block of zeroes is $n \times n$ and the lower right block of zeroes is $k \times k$. Since

$$[[Y, x_0], x_0] = \begin{pmatrix} 0 & -iB \\ iC & 0 \end{pmatrix}$$

we see that $[[[Y, x_0], x_0], x_0] = -[Y, x_0]$.

$\square$

Lemma 8.27 *For all $A \in \mathfrak{u}(n+k)$ and $x \in U(n+k) \cdot x_0$ the projection of A onto $T_x(U(n+k) \cdot x_0)$ is $-[[A, x], x]$.*

Proof:

We first prove the lemma when $x = x_0$. Let $X_1, \ldots, X_{2nk}$ be a basis for $T_{x_0}(U(n+k) \cdot x_0)$. Extend this to a basis of $\mathfrak{u}(n+k)$ using normal vectors $X_{2nk+1}, \ldots, X_{(n+k)^2}$. We have $A = \sum_{j=1}^{(n+k)^2} < A, X_j > X_j$, and the projection of A onto $T_{x_0}(U(n+k) \cdot x_0)$ is $\sum_{j=1}^{2nk} < A, X_j > X_j$. We now compute as follows:

$$
\begin{aligned}
[A, x_0] &= \sum_{j=1}^{(n+k)^2} < A, X_j > [X_j, x_0] \\
&= \sum_{j=1}^{2nk} < A, X_j > [X_j, x_0]
\end{aligned}
$$

because $X_j \in N_{x_0}(U(n+k) \cdot x_0)$ if and only if $[X_j, x_0] = 0$ by Proposition 8.9. Moreover,

$$
\begin{aligned}
[[A, x_0], x_0] &= \sum_{j=1}^{2nk} < A, X_j > [[X_j, x_0], x_0] \\
&= -\sum_{j=1}^{2nk} < A, X_j > X_j.
\end{aligned}
$$

by the preceeding lemma, and the lemma is proved for $x = x_0$.

Now, for any $x \in U(n+k) \cdot x_0$ we have $x = g \cdot x_0$ for some $g \in U(n+k)$. Note that if $X_1, \ldots, X_{2nk}$ is a basis for $T_{x_0}(U(n+k) \cdot x_0)$ and $X_{2nk+1}, \ldots, X_{(n+k)^2}$ is a basis for $N_{x_0}(U(n+k) \cdot x_0)$, then $g \cdot X_1, \ldots, g \cdot X_{2nk}$ is a basis for $T_x(U(n+k) \cdot x_0)$ and $g \cdot X_{2nk+1}, \ldots, g \cdot X_{(n+k)^2}$ is a basis for $N_x(U(n+k) \cdot x_0)$. We have $A = \sum_{j=1}^{(n+k)^2} < A, g \cdot X_j > g \cdot X_j$, and the projection of A onto $T_x(U(n+k) \cdot x)$ is $\sum_{j=1}^{2nk} < A, g \cdot X_j > g \cdot X_j$. As before we have,

$$
\begin{aligned}
[A, x] &= \sum_{j=1}^{(n+k)^2} < A, g \cdot X_j > [g \cdot X_j, x] \\
&= \sum_{j=1}^{2nk} < A, g \cdot X_j > [g \cdot X_j, x],
\end{aligned}
$$

and since $g \cdot [Y, x] = [g \cdot Y, g \cdot x]$ for any Y and x we have,

$$
\begin{aligned}
[[A, x], x] &= \sum_{j=1}^{2nk} < A, g \cdot X_j > [[g \cdot X_j, g \cdot x_0], g \cdot x_0] \\
&= \sum_{j=1}^{2nk} < A, g \cdot X_j > g \cdot [g^{-1} \cdot [g \cdot X_j, g \cdot x_0], x_0] \\
&= \sum_{j=1}^{2nk} < A, g \cdot X_j > g \cdot [[X_j, x_0], x_0] \\
&= -\sum_{j=1}^{2nk} < A, g \cdot X_j > g \cdot X_j.
\end{aligned}
$$

The preceeding proof is a fairly straightforward computation. The following proof is much shorter, but somewhat more subtle. This proof first appeared in [79].

Alternate proof of Lemma 8.27:

Let $A \in \mathfrak{u}(n+k)$ and $x \in U(n+k) \cdot x_0 \subseteq \mathfrak{u}(n+k)$. The projection of A onto $T_x(U(n+k) \cdot x_0)$ is the unique vector $X \in T_x(U(n+k) \cdot x_0)$ such that $A - X \in N_x(U(n+k) \cdot x_0)$. By Proposition 8.9, this is equivalent to $[A - X, x] = 0$, i.e. $[A, x] = [X, x]$. So the lemma is simply the statement that $[A, x] = -[[[A, x], x], x]$, i.e. $[-, x]$ is an almost complex structure on $U(n+k) \cdot x_0$. This was shown in Lemma 8.26.

$\square$

The above lemma can be used to compute the gradient vector field of the function f_A as follows.

Theorem 8.28 *The gradient vector field of f_A is $(\nabla f_A)(x) = -[[A, x], x]$.*

Proof:

The vector $(\nabla f_A)(x)$ is the unique element of $T_x(U(n+k) \cdot x_0)$ which satisfies

$$< (\nabla f_A)(x), X >= D_X f_A$$

for all $X \in T_x(U(n+k) \cdot x_0)$. So by Lemma 8.10,

$$< (\nabla f_A)(x), X >=< A, X >$$

for all $X \in T_x(U(n+k) \cdot x_0)$. That is, $(\nabla f_A)(x)$ is the projection of A onto the tangent space $T_x(U(n+k) \cdot x_0)$. The result now follows from the previous lemma.

$\square$

Remark 8.29 For any $x \in U(n+k) \cdot x_0$, the path $\sigma_x(t) = \exp(t[A, x]) \cdot x$ satisfies $\sigma_x'(0) = -\nabla(f_A)(x)$. However, σ_x is **not** a gradient flow line. To see this we compute $\sigma_x'(t_0)$ as follows. Define $\tilde{\sigma}_x(t) = \sigma_x(t + t_0)$.

$$\begin{aligned}
\sigma_x'(t_0) &= \tilde{\sigma}_x'(0) \\
&= \frac{d}{dt} \exp((t + t_0)[A, x]) \cdot x \Big|_{t=0} \\
&= \frac{d}{dt} \exp(t[A, x]) \cdot (\exp(t_0[A, x]) \cdot x) \Big|_{t=0} \\
&= [[A, x], \exp(t_0[A, x]) \cdot x] \\
&\neq [[A, \sigma_x(t_0)], \sigma_x(t_0)]
\end{aligned}$$

The correct formula for the gradient flow lines of f_A are given in later in this chapter; in Section 8.6 we describe the gradient flow lines in terms of the action of $GL_{n+k}(\mathbb{C})$ on the complex Grassmann manifold $G_{n,n+k}(\mathbb{C})$.

8.4   The critical points and indices of $f_A : U(n+k) \cdot x_0 \to \mathbb{R}$

We will now choose a specific Morse function on the orbit $U(n+k) \cdot x_0$ and compute the critical points and indices of the function. Let

$$A = \begin{pmatrix} i & & & 0 \\ & 2i & & \\ & & \ddots & \\ 0 & & & (n+k)i \end{pmatrix}$$

where $i = \sqrt{-1}$.

Proposition 8.30 *The function $f_A : U(n+k) \cdot x_0 \to \mathbb{R}$ is a Morse function whose critical points are the diagonal matrices in $\mathfrak{u}(n+k)$ which have exactly n entries equal to i and k entries equal to 0 along the diagonal.*

Proof:

By Lemma 8.11, $p \in U(n+k) \cdot x_0$ is a critical point of f_A if and only if p commutes with A. Since A is diagonal with distinct eigenvalues, this implies that p is diagonal. Conjugating with an element of $U(n+k)$ does not change the eigenvalues of a matrix. Hence, p must have exactly n entries equal to i and k entries equal to 0 along its diagonal. By conjugating x_0 with permutation matrices we see that all such diagonal matrices are in the orbit $U(n+k) \cdot x_0$.

$\square$

To compute the indices of the critical points of f_A we will use Lemmas 8.13 and 8.21. To begin this computation we introduce some additional notation. An n-tuple $\sigma = (r_1, \ldots, r_n)$ of integers from 1 to $n+k$ with $r_1 < r_2 < \cdots < r_n$ is called a **Schubert symbol**. For any Schubert symbol σ, we let x_σ be the diagonal matrix in $\mathfrak{u}(n+k)$ with an i in rows $r_1, \ldots, r_n$ and zeros everywhere else. The Schubert symbol σ contains all the information we need to write down the index of the critical point x_σ. To compute the index of x_σ we will need specific matrices $Y_1, \ldots, Y_{2nk}$ such that $[Y_1, x_\sigma], \ldots, [Y_{2nk}, x_\sigma]$ is a basis for $T_{x_\sigma}(U(n+k) \cdot x_0)$. To this end, define the matrix $Y_{r,s}(z)$ to be the matrix in $\mathfrak{u}(n+k)$ which has z in the (r, s) entry, $-\bar{z}$ in the (s, r) entry, and zeros everywhere else. The following proposition is obvious.

Proposition 8.31 *If $D \in \mathfrak{u}(n+k)$ is diagonal with entries $(d_1, \ldots, d_{n+k})$ along the diagonal then,*

$$[Y_{r,s}(z), D] = Y_{r,s}(z(d_s - d_r))$$

for all $z \in \mathbb{C}$.

Theorem 8.32 *The index of x_σ is twice the number of rows above each i which consist entirely of zeros. That is,*

$$\text{index of } x_\sigma = 2 \sum_{j=1}^{n} (r_j - j) = 2 \left(\sum_{j=1}^{n} r_j \right) - n(n+1).$$

Proof:

According to Lemma 8.21, the index of x_σ as a critical point of the function $g_A(x) = \|x - A\|^2$ is given by the number of points B (counted with multiplicity) along the line segment between x_σ and A such that the Jacobian of the function $E : N \to u(n + k)$ given by $E(x, \vec{v}) = x + \vec{v}$ is singular at $(x_\sigma, B - x_\sigma) \in N$. The multiplicity of B is defined as the dimension of the kernel of the Jacobian of E at $(x_\sigma, B - x_\sigma)$. The index of x_σ as a critical point of the function f_A is $2nk$ minus its index as a critical point of g_A by Lemma 8.13.

Consider the matrices $Y_{r,s}(1) \in u(n + k)$ and $Y_{r,s}(i) \in u(n + k)$ where $1 \le r < s \le n + k$ and either $r \in \{r_1, \ldots, r_n\}$ or $s \in \{r_1, \ldots, r_n\}$ but not both. Since $[Y_{r,s}(1), x_\sigma] = \pm Y_{r,s}(i)$ and $[Y_{r,s}(i), x_\sigma] = \pm Y_{r,s}(1)$ it's clear that $\{[Y_{r,s}(1), x_\sigma], [Y_{r,s}(i), x_\sigma]\}$ is a basis for $T_{x_\sigma}(U(n + k) \cdot x_0)$. Pick as a basis of $N_{x_\sigma}(U(n + k) \cdot x_0)$ the obvious matrices $Z_{2nk+1}, \ldots, Z_{(n+k)^2}$ which have only two non-zero entries; those non-zero entries being $1, -1$, or i. Lemma 8.24 tells us the columns of the Jacobian of E at the point

$$(x_\sigma, t(A - x_\sigma)) \in N$$

where $0 < t < 1$ are

$$[Y_{r,s}(1), x_\sigma + t(A - x_\sigma)] \quad r, s \text{ as above}$$
$$[Y_{r,s}(i), x_\sigma + t(A - x_\sigma)] \quad r, s \text{ as above}$$
$$Z_j \quad j = 2nk + 1, \ldots, (n + k)^2.$$

We can work out the commutator terms using Proposition 8.31 as follows.

$$[Y_{r,s}(1), x_\sigma + t(A - x_\sigma)] = \begin{cases} Y_{r,s}(it(s + 1 - r) - i) & \text{if } r \in \{r_1, \ldots, r_n\} \\ Y_{r,s}(i - it(r + 1 - s)) & \text{if } s \in \{r_1, \ldots, r_n\} \end{cases}$$

$$[Y_{r,s}(i), x_\sigma + t(A - x_\sigma)] = \begin{cases} Y_{r,s}(1 - t(s + 1 - r)) & \text{if } r \in \{r_1, \ldots, r_n\} \\ Y_{r,s}(t(r + 1 - s) - 1) & \text{if } s \in \{r_1, \ldots, r_n\} \end{cases}$$

From this description of the Jacobian of E at $(x_\sigma, x_\sigma + t(A - x_\sigma) - x_\sigma)$ it's clear that the Jacobian has non-trivial kernel if and only if one or more of the above commutators is identically zero. The dimension of the kernel at such a point is found by counting the number of commutators which are zero. So to compute the index of x_σ, we count the number of commutators for which there exists a t with $0 < t < 1$ making the commutator zero. Since $r < s$ we see that such a t exists if and only if $r \in \{r_1, \ldots, r_n\}$. For a fixed $r \in \{r_1, \ldots, r_n\}$ the number of allowed s values is equal to the number of rows below row r in x_σ that consist entirely of zeros. Since both $[Y_{r,s}(1), x_\sigma + t(A - x_\sigma)]$ and $[Y_{r,s}(i), x_\sigma + t(A - x_\sigma)]$ are zero for $t = 1/(s + 1 - r)$, the index of x_σ as a critical point of g_A is twice the sum of these numbers.

There are $r_j - j$ rows of zeros above row r_j, and since there are k rows of zeros in the matrix there are $k - (r_j - j)$ rows of zeros below row r_j for all $j = 1, \ldots, n$. Hence, the index of x_σ as a critical point of g_A is

$$2 \sum_{j=1}^{n} (k - (r_j - j)) = 2nk - 2 \sum_{j=1}^{n} (r_j - j),$$

and the index of x_σ as a critical point of f_A is $2 \sum_{j=1}^{n} (r_j - j)$.

$\square$

Corollary 8.33 *There is a bijective correspondence between the critical points of f_A of index λ and the Schubert cells of the complex Grassmann manifold $G_{n,n+k}(\mathbb{C})$ of dimension λ for all $\lambda \in \mathbb{Z}_+$. The correspondence is given by sending x_σ to the Schubert cell $e(\sigma)$.*

Proof:
By Section 6 of [104] we see that a given Schubert symbol $\sigma = (r_1, \ldots, r_n)$ corresponds to a Schubert cell of dimension $2 \sum_{j=1}^{n} (r_j - j)$.

$\square$

Remark 8.34 See Section 8.6 for a description of the Schubert cell $e(\sigma)$.

8.5 A Morse function on the complex Grassmann manifold

We have now established the existence of a Morse function on a certain orbit of the adjoint representation of $U(n+k)$ on its Lie algebra $\mathfrak{u}(n+k)$. We will now relate this function to a Morse function on the complex Grassmann manifold $G_{n,n+k}(\mathbb{C})$.

Let $G_{n,n+k}(\mathbb{C})$ be the complex Grassmann manifold consisting of the n-dimensional complex planes in $\mathbb{C}^{n+k}$, and let $\mathbb{C}^n \subseteq \mathbb{C}^{n+k}$ be the subspace spanned by the first n standard basis vectors. There is a transitive action

$$U(n+k) \times G_{n,n+k}(\mathbb{C}) \to G_{n,n+k}(\mathbb{C})$$

defined by sending $\mathbb{C}^n \in G_{n,n+k}(\mathbb{C})$ to its image under the linear transformation determined by a matrix in $U(n+k)$. It's clear that the stabilizer of $\mathbb{C}^n$ consists of all matrices of the form

$$\begin{pmatrix} U_n & 0 \\ 0 & U_k \end{pmatrix}$$

where $U_n \in U(n)$ and $U_k \in U(k)$, and hence, there is a diffeomorphism

$$\psi_1 : G_{n,n+k}(\mathbb{C}) \to U(n+k)/(U(n) \times U(k))$$

(see for instance Corollary 4.4 of [87] or Theorem 3.62 of [148]). Using this diffeomorphism we can embed $G_{n,n+k}(\mathbb{C})$ into the Lie algebra $\mathfrak{u}(n+k)$ as follows. Define a map $\psi_2 : U(n+k)/(U(n) \times U(k)) \to \mathfrak{u}(n+k)$ by $\psi_2([U]) =$

$U \cdot x_0 = U x_0 U^{-1}$ where $[U]$ denotes the coset represented by $U \in U(n+k)$. The following lemma shows that ψ_2 is a diffeomorphism, and hence,

$$\psi \overset{\text{def}}{=} \psi_2 \circ \psi_1 : G_{n,n+k}(\mathbb{C}) \overset{\psi_1}{\to} U(n+k)/(U(n) \times U(k)) \overset{\psi_2}{\to} U(n+k) \cdot x_0$$

is a diffeomorphism. Using this diffeomorphism we define a Morse function $f_A : G_{n,n+k}(\mathbb{C}) \to \mathbb{R}$ by

$$f_A(x) = < \psi(x), A > = -\text{trace}(\psi(x)A)$$

where $\psi(x)A$ denotes matrix multiplication.

Lemma 8.35 *The map ψ_2 is a well defined diffeomorphism onto the orbit* $U(n+k) \cdot x_0 \subseteq \mathfrak{u}(n+k)$.

Proof:

Whenever a compact Lie group G acts smoothly on a smooth manifold M the quotient space G/G_x is diffeomorphic to the orbit $G \cdot x$ for all $x \in M$ (see for instance Corollary 4.4 of [87] or Theorem 3.62 of [148]). So, all that needs to be checked is that $U(n) \times U(k)$ is the stabilizer of x_0.

Identify $U(n) \times U(k) \subseteq \mathfrak{u}(n+k)$ with matrices of the form $\begin{pmatrix} U_n & 0 \\ 0 & U_k \end{pmatrix}$ where $U_n \in U(n)$ and $U_k \in U(k)$. Then for any $U \in U(n) \times U(k)$ we have,

$$\begin{aligned}
U x_0 U^{-1} &= \begin{pmatrix} U_n & 0 \\ 0 & U_k \end{pmatrix} \begin{pmatrix} iI_n & 0 \\ 0 & 0 \end{pmatrix} \begin{pmatrix} {}^t\bar{U}_n & 0 \\ 0 & {}^t\bar{U}_k \end{pmatrix} \\
&= \begin{pmatrix} iU_n & 0 \\ 0 & 0 \end{pmatrix} \begin{pmatrix} {}^t\bar{U}_n & 0 \\ 0 & {}^t\bar{U}_n \end{pmatrix} \\
&= \begin{pmatrix} iU_n{}^t\bar{U}_n & 0 \\ 0 & 0 \end{pmatrix} \\
&= x_0.
\end{aligned}$$

This shows that $U(n) \times U(k) \subseteq U(n+k)_{x_0}$.

Now assume $U \in U(n+k)$ satisfies $U x_0 U^{-1} = x_0$. Let

$$U = \begin{pmatrix} A_n & B \\ C & D_k \end{pmatrix}$$

where A_n is some $n \times n$ complex matrix and D_k is some $k \times k$ complex matrix. We have,

$$\begin{aligned}
x_0 &= U x_0{}^t\bar{U} \\
&= i \begin{pmatrix} A_n & 0 \\ C & 0 \end{pmatrix} \begin{pmatrix} {}^t\bar{A}_n & {}^t\bar{C} \\ {}^t\bar{B} & {}^t\bar{D}_k \end{pmatrix} \\
&= i \begin{pmatrix} A_n{}^t\bar{A}_n & A_n{}^t\bar{C} \\ C^t\bar{A}_n & C^t\bar{C} \end{pmatrix}.
\end{aligned}$$

In particular, this says that $A_n{}^t\bar{A}_n = I_n$, and hence, $A_n \in U(n)$. So, all the columns and rows of A_n have unit length, and hence, $B = C = 0$. Thus, $U^t\bar{U} = I_{n+k}$ implies that $D_k \in U(k)$, and hence, $U \in U(n) \times U(k)$.

$\square$

The preceeding theorem shows that if $U \in U(n+k)$ gets sent to a point $x \in \mathfrak{u}(n+k)$ by ψ_2, then so does UV where $V \in U(n) \times U(k)$. Note that the computation done in the second part of the proof shows that $\psi([U])$ depends only on the first n columns of U. The following lemma shows that we can use this fact to obtain an explicit formula for $f_A(U \cdot x_0)$.

Lemma 8.36 *If* $U = \begin{pmatrix} A_n & B \\ C & D_k \end{pmatrix} \in U(n+k)$ *where* A_n *is some* $n \times n$ *complex matrix and* D_k *is some* $k \times k$ *complex matrix, then* $f_A(U \cdot x_0)$ *is given by,*

$$-\sum_{j=1}^{n} j (length\ of\ j^{th}\ row\ of\ A_n)^2 - \sum_{j=1}^{k}(j+n)(length\ of\ j^{th}\ row\ of\ C)^2.$$

Proof:
    As in the proof of the preceeding lemma we have,

$$\begin{aligned}
U x_0 U^{-1} &= \begin{pmatrix} A_n & B \\ C & D_k \end{pmatrix} \begin{pmatrix} iI_n & 0 \\ 0 & 0 \end{pmatrix} \begin{pmatrix} {}^t\bar{A}_n & {}^t\bar{C} \\ {}^t\bar{B} & {}^t\bar{D}_k \end{pmatrix} \\
&= i \begin{pmatrix} A_n & 0 \\ C & 0 \end{pmatrix} \begin{pmatrix} {}^t\bar{A}_n & {}^t\bar{C} \\ {}^t\bar{B} & {}^t\bar{D}_k \end{pmatrix} \\
&= i \begin{pmatrix} A_n{}^t\bar{A}_n & A_n{}^t\bar{C} \\ C^t\bar{A}_n & C^t\bar{C} \end{pmatrix}.
\end{aligned}$$

Hence,

$$\begin{aligned}
f_A(U \cdot x_0) &= <U x_0 U^{-1}, A> \\
&= -\text{trace}\left(i \begin{pmatrix} A_n{}^t\bar{A}_n & A_n{}^t\bar{C} \\ C^t\bar{A}_n & C^t\bar{C} \end{pmatrix} (-i) \begin{pmatrix} 1 & & 0 \\ & \ddots & \\ 0 & & n+k \end{pmatrix} \right) \\
&= -\sum_{j=1}^{n} j \sum_{i=1}^{n} \|a_{ji}\|^2 - \sum_{j=1}^{k}(j+n) \sum_{i=1}^{n} \|c_{ji}\|^2
\end{aligned}$$

$\square$

8.6 The gradient flow lines of $f_A : G_{n,n+k}(\mathbb{C}) \to \mathbb{R}$

In this section we describe the gradient flow lines of $f_A : G_{n,n+k}(\mathbb{C}) \to \mathbb{R}$. The description will be given in terms of the $GL_{n+k}(\mathbb{C})$ action on $G_{n,n+k}(\mathbb{C})$, the set of n-dimensional complex planes in $\mathbb{C}^{n+k}$. This action is defined as follows. Given any $G \in GL_{n+k}(\mathbb{C})$ and any plane $P \in G_{n,n+k}(\mathbb{C})$ the product $G \cdot P$ is defined to be $G(P)$, the image of the plane P under the linear transformation determined by G. We begin with the following general lemma.

Lemma 8.37 *Let $\eta : G \times M \to M$ be a smooth action of a compact Lie group G on a smooth manifold M. Let $x \in M$ and let $\eta_x : G \to M$ be induced from η. Then $T_e(G_x) \subseteq \ker d\eta_x(e)$ where $e \in G$ is the identity and G_x is the stabilizer of x. Moreover, $T_e(G_{g \cdot x}) = Ad_g(T_e(G_x))$ for all $g \in G$.*

Proof:
   Let $\gamma(t)$ be a path in G_x with $\gamma(0) = e$. Then $\eta_x(\gamma(t))$ is constant and so $\frac{d}{dt} \eta_x(\gamma(t))|_{t=0} = d\eta_x(e)(\gamma'(0)) = 0$. This shows that $T_e(G_x) \subseteq \ker d\eta_x(e)$. To prove the second statement we need merely recall that $G_{g \cdot x} = gG_x g^{-1}$ and then consider paths of the form $g\sigma(t)g^{-1}$ where $\sigma(t) \in G_x$ and $\sigma(0) = e$.

□

The following theorem describes the gradient flow lines of the function $f_A : G_{n,n+k}(\mathbb{C}) \to \mathbb{R}$ in terms of the $GL_{n+k}(\mathbb{C})$ action on $G_{n,n+k}(\mathbb{C})$. The reader should note that the following proof holds for any matrix $A \in \mathfrak{u}(n+k)$.

Theorem 8.38 *Let $P \in G_{n,n+k}(\mathbb{C})$. The gradient flow of the Morse function $f_A : G_{n,n+k}(\mathbb{C}) \to \mathbb{R}$ through P is given by $\gamma_P(t) = \exp(itA)(P)$.*

Proof:
   Let $\mathbb{C}^n \subseteq \mathbb{C}^{n+k}$ be the vector subspace spanned by the first n standard basis elements of $\mathbb{C}^{n+k}$. Pick any unitary matrix $U \in U(n+k)$ such that $U(\mathbb{C}^n) = P$ and let $x = Ux_0U^{-1} \in \mathfrak{u}(n+k)$. We observed in Remark 8.29 that the path $\sigma_x(t) = \exp(t[A, x]) \cdot x \in \mathfrak{u}(n+k)$ satisfies $\sigma_x'(0) = [[A, x], x]$, i.e. minus the gradient vector of f_A at x. Notice that under the diffeomorphism $U(n+k) \cdot x_0 \approx G_{n,n+k}(\mathbb{C})$ this path goes to the path $\tilde{\sigma}_x(t) = \exp(t[A, x])(P)$. Because of this and the fact that $\gamma_P(t+t_0) = \gamma_{\tilde{P}}(t)$ for some $\tilde{P} \in G_{n,n+k}(\mathbb{C})$ it suffices to show that $\gamma_P'(0) = \tilde{\sigma}_x'(0)$.

   Let $\eta : GL_{n+k}(\mathbb{C}) \times G_{n,n+k}(\mathbb{C}) \to G_{n,n+k}(\mathbb{C})$ be the smooth action of $GL_{n+k}(\mathbb{C})$ on $G_{n,n+k}(\mathbb{C})$ described above, and let $\eta_P : GL_{n+k}(\mathbb{C}) \to G_{n,n+k}(\mathbb{C})$ be induced from η. In this notation, $\gamma_P(t) = \eta_P(\exp(itA))$ and $\tilde{\sigma}_x(t) = \eta_P(\exp(t[A, x]))$. By the chain rule we have $\gamma_P'(0) = d\eta_P(I_{n+k})(iA)$ and $\tilde{\sigma}_x'(0) = d\eta_P(I_{n+k})([A, x])$ where I_{n+k} is the $(n+k) \times (n+k)$ identity matrix. Thus, we will be done once we show $d\eta_P(I_{n+k})(iA - [A, x]) = 0$.

To show that $iA - [A, x] \in \ker d\eta_P(I_{n+k})$ we apply the previous lemma and show that,

$$iA - [A, x] \in T_{I_{n+k}}(GL_{n+k}(\mathbb{C})_P) = U(T_{I_{n+k}}(GL_{n+k}(\mathbb{C})_{\mathbb{C}^n}))U^{-1}.$$

If we use column orientation for $\mathbb{C}^{n+k}$, then the stabilizer of $\mathbb{C}^n \in G_{n,n+k}(\mathbb{C})$ consists of those elements of $GL_{n+k}(\mathbb{C})$ whose lower left $k \times n$ block is zero, and since $GL_{n+k}(\mathbb{C})$ is an open subset of $\mathbb{C}^{(n+k)^2}$ the tangent space at I_{n+k} of the stabilizer of $\mathbb{C}^n$ consists of those matrices in the tangent space whose lower left $k \times n$ block is zero. Therefore all we need to do is to write $iA - [A, x]$ as such a matrix conjugated with U.

$$
\begin{aligned}
iA - [A, x] &= iA - [A, Ux_0U^{-1}] \\
&= iA - U[U^{-1}AU, x_0]U^{-1} \\
&= U(iU^{-1}AU - [U^{-1}AU, x_0])U^{-1}
\end{aligned}
$$

Letting $Y = U^{-1}AU$ one easily sees that $iY - [Y, x_0]$ does have its lower left $k \times n$ block equal to zero.

$\square$

Remark 8.39 It was shown in Remark 8.29 that $\tilde{\sigma}_x'(t)$ is **not** $-\nabla(f)(\tilde{\sigma}_x(t))$ when $t \neq 0$. It is natural to ask why the preceeding proof works for the path $\gamma_P(t) = \exp(itA)(P)$, but it does not work for $\tilde{\sigma}_x(t) = \exp(t[A, x])(P)$, since both of these paths satisfy

$$\gamma_P'(0) = \tilde{\sigma}_x'(0) = -\nabla(f_A)(P).$$

The essential reason is that iA does not depend on the point P, but $[A, x]$ does. So, the path $\gamma_P(t)$ satisfies $\gamma_P(t + t_0) = \gamma_{\tilde{P}}(t)$ for some $\tilde{P} \in G_{n,n+k}(\mathbb{C})$, but the path $\tilde{\sigma}_x(t)$ **does not** satisfy this condition.

Note that the matrix $\exp(itA) \in GL_{n,n+k}(\mathbb{C})$ is **not** a unitary matrix. So, the above description of the gradient flow lines does **not** readily extend to adjoint orbits $U(n + k) \cdot x_0 \subseteq u(n + k)$.

Unstable manifolds and Schubert cells

Our next goal is to describe the unstable manifolds $W^u(x_\sigma)$ of f_A. Given any Schubert symbol $\sigma = (r_1, \ldots, r_n)$ we recall that the associated Schubert cell $e(\sigma) \subseteq G_{n,n+k}(\mathbb{C})$ is defined to be the set of all planes $P \in G_{n,n+k}(\mathbb{C})$ such that $\dim(P \cap \mathbb{C}^{r_j}) = j$ and $\dim(P \cap \mathbb{C}^{r_j - 1}) = j - 1$ for all $j = 1, \ldots, n$ (see for instance Section 1.5 of [67] or Section 6 of [104]). In this definition we view $\mathbb{C}^j \subseteq \mathbb{C}^{n+k}$ as the subspace spanned by the first j standard basis elements.

Next we recall that $G_{n,n+k}(\mathbb{C})$ is topologized as a quotient of the Stiefel manifold $V_{n,n+k}(\mathbb{C}) \subseteq \mathbb{C}^{n(n+k)}$. That is, we have a map $\pi : V_{n,n+k}(\mathbb{C}) \to G_{n,n+k}(\mathbb{C})$ defined by sending an n-tuple of linearly independent vectors in $\mathbb{C}^{n+k}$ to the n-plane in $\mathbb{C}^{n+k}$ which they span. A set $U \subseteq G_{n,n+k}(\mathbb{C})$ is open if and only if $\pi^{-1}(U)$ is open in $V_{n,n+k}(\mathbb{C}) \subset \mathbb{C}^{n(n+k)}$.

For any Schubert symbol σ let P_σ be the plane spanned by the standard basis elements $e_{r_1}, \ldots, e_{r_n}$. Since P_σ corresponds to x_σ under the diffeomorphism $\psi : G_{n,n+k}(\mathbb{C}) \to U(n+k) \cdot x_0$, Theorem 8.32 implies that P_σ is a critical point of $f_A : G_{n,n+k}(\mathbb{C}) \to \mathbb{R}$.

Theorem 8.40 *For any Schubert symbol* $\sigma = (r_1, \ldots, r_n)$,

$$W^u(P_\sigma) = e(\sigma).$$

Proof:

Let $\sigma = (r_1, \ldots, r_n)$ be any Schubert symbol and let $P \in e(\sigma)$. To show that $P \in W^u(P_\sigma)$ we will show that for any open neighborhood $U \subset G_{n,n+k}(\mathbb{C})$ containing P_σ, there exists $T < 0$ such that for all $t < T$ we have $\gamma_P(t) = \exp(itA)(P) \in U$. Since $P \in e(\sigma)$ we can pick a basis $v_1, \ldots, v_n$ of P such that for all $j = 1, \ldots, n$, v_j has a 1 in the r_j^{th} entry and a zero in entries $r_j + 1, \ldots, n + k$. Note that $\exp(itA)$ is simply a diagonal matrix with entries $e^{-t}, e^{-2t}, \ldots, e^{-(n+k)t}$ along the diagonal. Also, note that the vectors

$$e^{r_1 t} \exp(itA)(v_1), \ldots, e^{r_n t} \exp(itA)(v_n).$$

span the plane $\exp(itA)(P)$, and for all $j = 1, \ldots, n$ the mth entry of the vector $e^{r_j t} \exp(itA)(v_j)$ is,

$$
\begin{array}{ll}
ze^{(r_j - m)t} \text{ some } z \in \mathbb{C} & \text{if } m < r_j - 1 \\
1 & \text{if } m = r_j \\
0 & \text{if } m = r_j + 1, \ldots, n + k.
\end{array}
$$

If we consider $\tilde{\gamma}_P(t) = (e^{r_1 t} \exp(itA)(v_1), \ldots, e^{r_n t} \exp(itA)(v_n))$ as a path in $V_{n,n+k}(\mathbb{C})$, then we have $\pi(\tilde{\gamma}_P(t)) = \gamma_P(t)$ for all $t \in \mathbb{R}$. Since we have $(e_{r_1}, \ldots, e_{r_n}) \in \pi^{-1}(U)$ where $\pi^{-1}(U)$ is open in $V_{n,n+k}(\mathbb{C})$ and we have $\lim_{t \to -\infty} \tilde{\gamma}_P(t) = (e_{r_1}, \ldots, e_{r_n})$, we can choose a $T < 0$ such that for all $t < T$ we have $\tilde{\gamma}_P(t) \in \pi^{-1}(U)$. This implies that $\gamma_P(t) \in U$ for all $t < T$.

We have shown that $e(\sigma) \subseteq W^u(P_\sigma)$. Now let $P \in W^u(P_\sigma)$. Since the Schubert cells partition $G_{n,n+k}(\mathbb{C})$ there exists a Schubert symbol $\tilde{\sigma}$ such that $P \in e(\tilde{\sigma})$. But then $P \in W^u(P_{\tilde{\sigma}})$ by the argument in the previous paragraph. Since the unstable manifolds are disjoint this implies that $\tilde{\sigma} = \sigma$, and so $P \in e(\sigma)$.

□

Remark 8.41 The above theorem and the Stable/Unstable Manifold Theorem (Theorem 4.2) can be used to give an alternate proof of Theorem 8.32. To do this, one needs to compute the dimension of the Schubert cell $e(\sigma)$. This computation can be found in Section 6 of [104] or Section 1.5 of [67].

To state the analogous result for the stable manifolds $W^s(P_\sigma)$ of f_A we introduce the following notation. If $e_1, \ldots, e_{n+k}$ is the standard basis for $\mathbb{C}^{n+k}$, then we define the "inverse standard basis" to be, $\tilde{e}_j = e_{n+k-j+1}$ for all $j = 1, \ldots, n + k$. We then have the following "inverse filtration" of $\mathbb{C}^{n+k}$,

$$\tilde{\mathbb{C}}^0 \subset \tilde{\mathbb{C}}^1 \subset \tilde{\mathbb{C}}^2 \subset \cdots \subset \tilde{\mathbb{C}}^{n+k}$$

where $\tilde{\mathbb{C}}^j$ is the subspace spanned by the first j inverse standard basis elements. Let $\sigma = (r_1, \ldots, r_n)$ be a Schubert symbol. We define the "inverse Schubert cell" $\tilde{e}(\sigma)$ by declaring $P \in \tilde{e}(p)$ if and only if for all $j = 1, \ldots, n$ we have,

$$
\begin{array}{lll}
1) & \dim(P \cap \tilde{\mathbb{C}}^{n+k+1-r_j}) & = & n + 1 - j \\
2) & \dim(P \cap \tilde{\mathbb{C}}^{n+k-r_j}) & = & n - j.
\end{array}
$$

The important thing to notice about the inverse Schubert cell $\tilde{e}(\sigma)$ is that $P \in \tilde{e}(\sigma)$ if and only if one can pick a basis $v_1, \ldots, v_n$ for p where v_j has a zero in entries $1, \ldots, r_j - 1$ and a 1 in the r_j^{th} entry for all $j = 1, \ldots, n$. The proof of the following theorem is similar to the previous one.

Theorem 8.42 *For any Schubert symbol* $\sigma = (r_1, \ldots, r_n)$ *we have*

$$W^s(P_\sigma) = \tilde{e}(\sigma).$$

The preceeding two theorems give us the following description of the intersections of the stable and unstable manifolds of f_A.

Corollary 8.43 *If* $\sigma = (r_1, \ldots, r_n)$ *and* $\tilde{\sigma} = (\tilde{r}_1, \ldots, \tilde{r}_n)$ *are Schubert cells, then* $W^u(P_\sigma) \cap W^s(P_{\tilde{\sigma}}) \neq \emptyset$ *if and only if* $r_j \geq \tilde{r}_j$ *for all* $j = 1, \ldots, n$.

Proof:
The plane $P \in W^u(P_\sigma)$ if and only if we can pick a basis $v_1, \ldots, v_n$ for P where v_j has a 1 in the r_j^{th} entry and a zero in entries $r_j + 1, \ldots, n + k$. The plane $P \in W^s(p_{\tilde{\sigma}})$ if and only if we can pick a basis $\tilde{v}_1, \ldots, \tilde{v}_n$ for P where $\tilde{v}_j$ has a zero in entries $1, \ldots, \tilde{r}_j - 1$ and a 1 in the $\tilde{r}_j^{\text{th}}$ entry for all $j = 1, \ldots, n$.

If $r_j \geq \tilde{r}_j$ for all $j = 1, \ldots, n$, then the vectors $w_1, \ldots, w_n$ where w_j has a 1 in entries r_j and $\tilde{r}_j$ and zero in all other entries will be a basis for a plane in $W^u(P_\sigma) \cap W^s(P_{\tilde{\sigma}})$. If $r_j < \tilde{r}_j$ for some $1 \leq j \leq n$, then the following contradiction shows there is no flow from P_σ to $P_{\tilde{\sigma}}$. Assume we have some n-plane $P \in W^u(p_\sigma) \cap W^s(p_{\tilde{\sigma}})$ and let $v_1, \ldots, v_n$ and $\tilde{v}_1, \ldots, \tilde{v}_n$ be as above.

By adding certain multiples of $\tilde{v}_{j+1}, \ldots, \tilde{v}_n$ to $\tilde{v}_j$ we can find a vector $v \in P$ which has a zero in entries $1, 2, \ldots, \tilde{r}_j - 1, \tilde{r}_{j+1}, \tilde{r}_{j+2}, \ldots, \tilde{r}_n$ and a 1 in the $\tilde{r}_j^{\text{th}}$ entry. The vector v cannot be in the span of $v_1 \ldots, v_n$. So the n-plane p would have to contain the $n + 1$ linearly independent vectors $v_1, \ldots, v_n, v$. This contradiction shows that $W^u(P_\sigma) \cap W^s(P_{\tilde{\sigma}}) = \emptyset$.

$\square$

In light of the previous corollary we define a partial ordering on the Schubert cells as follows. For Schubert cells $\sigma = (r_1, \ldots, r_n)$ and $\tilde{\sigma} = (\tilde{r}_1, \ldots, \tilde{r}_n)$ we say that $\sigma \geq \tilde{\sigma}$ if and only if $r_j \geq \tilde{r}_j$ for all $j = 1, \ldots, n$. Notice that with this definition we have $\sigma \geq \tilde{\sigma}$ if and only if $P_\sigma \succeq P_{\tilde{\sigma}}$.

For any two critical points P_σ and $P_{\tilde{\sigma}}$ of $f_A : G_{n,n+k}(\mathbb{C}) \to \mathbb{R}$ we define $W(P_\sigma, P_{\tilde{\sigma}}) = W^u(P_\sigma) \cap W^s(P_{\tilde{\sigma}})$. We will now show that for all critical points P_σ and $P_{\tilde{\sigma}}$ of f_A we have $W^u(P_\sigma) \pitchfork W^s(P_{\tilde{\sigma}})$, i.e. $f_A : G_{n,n+k}(\mathbb{C}) \to \mathbb{R}$ is a Morse-Smale function.

Lemma 8.44 *Let $\pi : E \to B$ be a smooth fiber bundle. Let V, W be submanifolds of B and let $p \in V \cap W$. The manifolds V and W meet transversely at p if and only if there exists some $q \in \pi^{-1}(p)$ with $\pi^{-1}(V) \pitchfork \pi^{-1}(W)$ at q.*

Proof:

First note that since $\pi : E \to B$ is a submersion $\pi^{-1}(V)$ and $\pi^{-1}(W)$ are submanifolds of E by Theorem 5.11. Also, transversality is a local property and so it suffices to prove the lemma for a trivial bundle $E = B \times F$. In this case we have,

$$
\begin{aligned}
\pi^{-1}(V) &= V \times F \subseteq B \times F \\
\pi^{-1}(W) &= W \times F \subseteq B \times F.
\end{aligned}
$$

For any $q = (p, x) \in \pi^{-1}(p)$ we have,

$$
\begin{aligned}
T_q(\pi^{-1}(V)) &= T_p(V) \times T_x(F) \\
T_q(\pi^{-1}(W)) &= T_p(W) \times T_x(F).
\end{aligned}
$$

Clearly,

$$
(T_p(V) \times T_x(F)) \oplus (T_p(W) \times T_x(F)) = T_p(B) \times T_x(F)
$$

if and only if

$$
T_p(V) \oplus T_p(W) = T_p(B).
$$

$\square$

Theorem 8.45 *The function $f_A : G_{n,n+k}(\mathbb{C}) \to \mathbb{R}$ is a Morse-Smale function.*

Proof:

Let $V_{n,n+k}(\mathbb{C}) \subseteq \mathbb{C}^{n(n+k)}$ be the Stiefel manifold. Then $\pi : V_{n,n+k}(\mathbb{C}) \to G_{n,n+k}(\mathbb{C})$ is a smooth fiber bundle. Let $\sigma = (r_1, \ldots, r_n)$ and $\tilde{\sigma} = (\tilde{r}_1, \ldots, \tilde{r}_n)$ be Schubert symbols that satisfy $\sigma \geq \tilde{\sigma}$. For any $P \in W(P_\sigma, P_{\tilde{\sigma}})$ we can pick a basis $v_1, \ldots, v_n$ for P such that v_j has a 1 in the r_j^{th} entry and 0 in entries $r_j + 1, \ldots, n + k$. We can also pick a basis $\tilde{v}_1, \ldots, \tilde{v}_n$ for P which has a zero in entries $1, \ldots, \tilde{r}_j - 1$ and a 1 in the $\tilde{r}_j^{\text{th}}$ entry.

Let $q = (v_1, \ldots, v_n) \in V_{n,n+k}(\mathbb{C})$. We will show

$$\pi^{-1}(W^u(P_\sigma)) \pitchfork \pi^{-1}(W^s(P_{\sigma'}))$$

at q. Recall that $V_{n,n+k}(\mathbb{C}) \subseteq \mathbb{C}^{n(n+k)}$. The tangent space $T_q(\pi^{-1}(W^u(P_\sigma))$ includes all vectors $(v_1, \ldots, v_n) \in \mathbb{C}^{n(n+k)}$ such that v_j has entries $r_j + 1, \ldots, n+k$ equal to zero $(j = 1, \ldots, n)$. The tangent space $T_q(\pi^{-1}(W^s(P_{\sigma'}))$ includes all frames $(\tilde{v}_1, \ldots, \tilde{v}_n) \in \mathbb{C}^{n(n+k)}$ such that $\tilde{v}_j$ has entries $1, \ldots, \tilde{r}_j - 1$ equal to zero $(j = 1, \ldots, n)$. Since $r_j \geq \tilde{r}_j$ for all $j = 1, \ldots, n$ we have $T_q(\pi^{-1}(W^u(P_\sigma)) \oplus T_q(\pi^{-1}(W^s(P_{\tilde{\sigma}})) = \mathbb{C}^{n(n+k)} = T_q(V_{n,n+k}(\mathbb{C}))$.

$\square$

8.7   The homology of $G_{n,n+k}(\mathbb{C})$

The results in the previous sections allow us to easily compute the homology of the complex Grassmann manifold $G_{n,n+k}(\mathbb{C})$ using the Morse-Smale function $f_A : G_{n,n+k}(\mathbb{C}) \to \mathbb{R}$.

Note that Theorem 8.32 shows that the critical points of the Morse function f_A are all of even index, i.e. f_A is a perfect Morse function. This implies that the boundary operators in the chain complex associated to the CW-complex determined by f_A are all zero. So by Theorems 2.15 and 3.28, the homology of $G_{n,n+k}(\mathbb{C})$ can be computed by counting the number of critical points of a given index. This proves the following.

Theorem 8.46 *The homology group $H_j(G_{n,n+k}(\mathbb{C}); \mathbb{Z})$ is isomorphic to the free abelian group generated by the critical points of f_A of index j for all $j \in \mathbb{Z}_+$.*

There are several ways to count the number of critical points of a given index. One way to do this is suggested by Theorem 8.32: For any Schubert symbol $\sigma = (r_1, \ldots, r_n)$, the critical point x_σ is an $(n + k) \times (n + k)$ diagonal matrix with n diagonal entries equal to i and k diagonal entries equal to 0. The index of x_σ is computed by counting the number of rows of zeros that lie above each i and then multiplying by 2. Note that this is the same as counting the number of i's that lie below each row of zeros and then multiplying by 2.

So, to count the number of critical points of index $2j$ we can check how many ways there are to arrange the i's along the diagonal so that the number of rows of zeros that lie above the i's (counted with multiplicity) is j. Alternately, we can check how many ways there are to arrange the zeros along the diagonal so that the number of i's that lie below the zeros (counted with multiplicity) is j.

Example 8.47 Consider $G_{2,5}(\mathbb{C})$, where $n = 2$ and $k = 3$. The homology group $H_6(G_{2,5}(\mathbb{C}); \mathbb{Z})$ is generated by the following critical points.

$$x_{(2,4)} = \begin{pmatrix} 0 & 0 & 0 & 0 & 0 \\ 0 & i & 0 & 0 & 0 \\ 0 & 0 & 0 & 0 & 0 \\ 0 & 0 & 0 & i & 0 \\ 0 & 0 & 0 & 0 & 0 \end{pmatrix} \qquad x_{(1,5)} = \begin{pmatrix} i & 0 & 0 & 0 & 0 \\ 0 & 0 & 0 & 0 & 0 \\ 0 & 0 & 0 & 0 & 0 \\ 0 & 0 & 0 & 0 & 0 \\ 0 & 0 & 0 & 0 & i \end{pmatrix}$$

It is easy to see that these are the only two possible ways to position the i's along the diagonal so that there are 3 rows of zeros above the i's (counted with multiplicity). Alternately, note that the first row of zeros in $x_{(2,4)}$ has two i's below it while the second row of zeros has one i below it. For $x_{(1,5)}$ note that all three rows of zeros have one i below them. Hence, both critical points have index 6.

Another way to express the dimension of $H_j(G_{n,n+k}(\mathbb{C}); \mathbb{Z})$ is in terms of partitions.

Definition 8.48 *A **partition** of $j \in \mathbb{Z}_+$ is an unordered sequence of positive integers with sum j. The number of partitions of j is denoted by $p(j)$.*

The following table gives the value of $p(j)$ for all $j \leq 10$ (see for instance Section 6 of [104]).

j	0	1	2	3	4	5	6	7	8	9	10
$p(j)$	1	1	2	3	5	7	11	15	22	30	42

For example, the integer 5 has seven partitions, namely:

$$1\,1\,1\,1\,1, \quad 1\,1\,1\,2, \quad 1\,1\,3, \quad 1\,4, \quad 1\,2\,2, \quad 2\,3, \quad 5.$$

For every Schubert symbol $(r_1, \ldots, r_n)$ with

$$(r_1 - 1) + (r_2 - 2) + \cdots + (r_n - n) = j$$

we get a partition of j:

$$r_1 - 1 \leq r_2 - 2 \leq r_n - n$$

(if we ignore any leading zeros) consisting of integers less than or equal to k. Conversely, give any partition $i_1 \leq i_2 \leq \cdots \leq i_n$ of j (which we pad with leading zeros to make length n) with integers that are less than or equal to k we have a Schubert symbol:

$$\sigma = (i_1 + 1, i_2 + 2, \ldots, i_n + n).$$

In relation to the critical point x_σ, i_j corresponds to the number of rows of zeros above the jth i along the diagonal for all $i = 1, \ldots, j$. This proves the following.

Theorem 8.49 *For all $j \in \mathbb{Z}_+$, the homology group $H_j(G_{n,n+k}(\mathbb{C}); \mathbb{Z})$ is zero if j is odd and a free abelian group on $\tilde{r}(j/2)$ generators if j is even, where $\tilde{r}(j/2)$ denotes the number of partitions of $j/2$ into at most n integers each of which is less than or equal to k.*

Example 8.50 Consider $G_{2,5}(\mathbb{C})$, where $n = 2$ and $k = 3$. To find the dimension of $H_6(G_{2,5}(\mathbb{C}); \mathbb{Z})$ we consider partitions of 3. There are 3 partitions of 3: 1 1 1, 1 2, and 3. However, only 1 2 and 3 have length less than or equal to 2. Hence, the dimension of $H_6(G_{2,5}(\mathbb{C}); \mathbb{Z})$ is $\tilde{r}(3) = 2$.

A third method of expressing the dimension of $H_j(G_{n,n+k}(\mathbb{C}); \mathbb{Z})$ is closely related to the structure of the cohomology ring of the infinite dimensional complex Grassmann manifold $G_{n,\infty}(\mathbb{C})$, consisting of n-planes in $\mathbb{C}^\infty$.

Let c_l be a variable of degree $2l$ for all $l = 1, \ldots, n$. Consider the graded polynomial algebra $\mathbb{Z}[c_1, \ldots, c_n]$, where the **degree** of a monomial is defined to be

$$\text{degree } c_1^{i_1} c_2^{i_2} \cdots c_n^{i_n} = 2i_1 + 4i_2 \cdots + 2ni_n.$$

By counting the number of i's that lie below each row of zeros in a critical point x_σ we get the following.

Theorem 8.51 *For all $j \in \mathbb{Z}_+$, the dimension of $H_j(G_{n,n+k}(\mathbb{C}); \mathbb{Z})$ is equal to the number of monomials in the graded polynomial algebra $\mathbb{Z}[c_1, \ldots, c_n]$ of degree j.*

Proof:
The bijection is defined as follows: a row of zeros in a critical point x_σ corresponds to c_l where l is the number of i's that lie below the row of zeros. The matrix x_σ is mapped to the product of the c_l's determined by the rows of zeros that have at least one i below them. It's clear that this map is a bijection between the critical points of index j and the monomials in $\mathbb{Z}[c_1, \ldots, c_n]$ of degree j for all $j \in \mathbb{Z}_+$.

□

Example 8.52 Consider $G_{2,5}(\mathbb{C})$ once again. To find the dimension of the homology group $H_6(G_{2,5}(\mathbb{C}); \mathbb{Z})$ we can count the number of monomials in the polynomial algebra $\mathbb{Z}[c_1, c_2]$ of degree 6. The monomials of degree 6 are:

$$c_1 c_2 \quad \text{and} \quad c_1^3.$$

The critical point $x_{(2,4)}$ corresponds to the monomial $c_1 c_2$, and the critical point $x_{(1,5)}$ corresponds to the monomial c_1^3.

We end this section by stating the following theorem. For a proof see for instance Theorem 20.3.2 of [81] or Sections 7 and 14 of [104].

Theorem 8.53 *The cohomology ring* $H^*(G_{n,\infty}(\mathbb{C}); \mathbb{Z})$ *is isomorphic to the graded polynomial algebra* $\mathbb{Z}[c_1, \ldots, c_n]$, *where* c_j *denotes the* j^{th} *Chern class of the universal n-plane bundle over* $G_{n,\infty}(\mathbb{C})$.

8.8   Further generalizations and applications

Most of the results in this chapter can be viewed as special cases of a more general theory applying Morse homology to Lie groups, symmetric spaces, and symplectic geometry. In this section we briefly outline some of these applications. A detailed discussion of these results would require significant background in the theory of Lie groups, symmetric spaces, and symplectic geometry and is hence outside the scope of this book.

Lie groups and symmetric spaces

Some of the first applications of Morse homology theory to the topology of Lie groups and symmetric spaces can be found in the work of Bott and Samelson from the 1950's. In fact, in his lectures delivered at Bonn in 1958 [27] Bott used the same results we discussed in Sections 8.1 and 8.2 to prove the following: Let $U(n)$ denote the unitary group in n complex variables, and consider $U(n_1) \times \cdots \times U(n_k)$ as a subgroup of $U(n)$, where $n_1 + \cdots + n_k = n$. Then the quotient space

$$W(n_1, \ldots, n_k) = U(n)/(U(n_1) \times \cdots \times U(n_k))$$

is the complex flag manifold of type $n_1, \ldots, n_k$, and we have the following:

Theorem 8.54 $H_j(W(n_1, \ldots, n_k); \mathbb{Z}) = 0$ *for j odd, and $W(n_1, \ldots, n_k)$ has no torsion.*

Note that when $k = 2$ the complex flag manifolds $W(n_1, \ldots, n_k)$ are Grassmann manifolds.

Now let G be a compact Lie group, $T' \subset G$ a torus in G, and $C(T')$ the centralizer of T' in G. In [22] and [23] Bott proved the following.

Theorem 8.55 *If G is a connected, simply connected, compact Lie group, then the loop space $\Omega(G)$ and the space $G/C(T')$ have the following properties:*

a. *They are free of torsion;*

b. *Their odd Betti numbers vanish;*

c. *Their Betti numbers can be read off from the diagram of G.*

Bott proved this theorem by exhibiting a Morse function whose critical points are all of even index and applying the lacunary principle of Morse (Theorem 3.39). In a sequel to [22] Bott and Samelson used similar methods to show that analogous results hold for symmetric spaces [28] [29], and in [24] Bott used Morse theory to prove his famous periodicity theorems.

Bott's results in [22] and [23] were also generalized by Frankel [57], using the Morse-Bott Inequalities (Theorem 3.53), to apply to Kähler manifolds. Let M be a compact connected Kähler manifold, and assume that Φ is a connected 1-parameter group of isometries of M. This second assumption is equivalent to the assumption that there is a toral group Φ' that acts complex analytically on M. The fixed point set F of Φ on M is the set of points in M that are left fixed by all the elements in Φ, and this set F coincides with the fixed point set of Φ' (there is an averaging process to make Φ' act by isometries). In [57] Frankel proved the following.

Lemma 8.56 *If F is non-empty, then the fixed point set F coincides with the non-degenerate critical set of a real C^∞ function on M, i.e. the components of F are the critical submanifolds of a Morse-Bott function. Moreover, the critical submanifolds are all of even index.*

Using the preceeding lemma and the Morse-Bott Inequalities (Theorem 3.53) Frankel proved the following.

Theorem 8.57 *If F is non-empty and $F = \coprod_j F_j$ are the connected components of F, then the Betti numbers satisfy*

$$b_k(M; \Lambda) = \sum_j b_{k-\lambda_j}(F_j; \Lambda)$$

for all k, where λ_j denotes the index of the critical submanifold F_j and $\Lambda = \mathbb{Q}$ or $\Lambda = \mathbb{Z}_p$ where p is prime.

As a corollary to this theorem Frankel proved the following.

Corollary 8.58 *If the fixed points are isolated, then M has no torsion and its odd dimensional Betti numbers vanish.*

Frankel also proved the same corollary for $G/C(T')$ (as in Theorem 8.55) when G is a compact semisimple Lie group.

Most of the preceeding results were proved using different Morse (Morse-Bott) functions than the functions f_A considered earlier in this chapter. In fact, there are many different Morse (Morse-Bott) functions that have been used to study the topology of Lie groups and symmetric spaces over the past 50 years.

In [74] Hangan showed that Bott's perfect Morse functions (considered in Example 3.7) when composed with the Plüker embedding yield Morse functions on the both the real and complex Grassmann manifolds, and in [58] Frankel proved that the trace function on the classical matrix groups $SO(n)$, $U(n)$, and $Sp(n)$ is a Morse-Bott function. Moreover, Frankel showed that the critical submanifolds of the trace function on these groups are Grassmann manifolds, and he proved that the Morse-Bott Inequalities (Theorem 3.53) are in fact equalities, i.e. $R(t) = 0$, or in other words, the trace function on these groups is perfect. Results analogous to Frankel's were proved by Ramanujam for certain symmetric spaces using the trace function in [120]. Ramanujam also showed that the length function g_A described in Lemma 8.13 is perfect for certain homogeneous spaces (including the complex flag manifolds) [121] and certain matrix groups [119].

Symplectic geometry

A symplectic manifold is a pair (M, ω) where M is a smooth manifold and ω is a non-degenerate closed 2-form.

Coadjoint orbits of $U(n)$ as symplectic manifolds

The adjoint and coadjoint orbits of a compact Lie group G have a natural symplectic structure ω called the **Kostant-Kirillov-Souriau symplectic structure** [88] [134]. To describe this symplectic structure for coadjoint orbits of $G = U(n)$, first note that the dual of the Lie algebra $\mathfrak{u}(n)$ can be identified with the set of $n \times n$ Hermitian matrices, and every Hermitian matrix is conjugate via a unitary matrix to a diagonal matrix. Hence, the coadjoint orbits $\mathcal{H}_\lambda$ of $U(n)$ are characterized by n-tuples $\lambda = (\lambda_1, \ldots, \lambda_n) \in \mathbb{R}^n$ corresponding to the eigenvalues of a Hermitian matrix in the orbit. Define

$$\mathcal{H}_\lambda = \{\text{Hermitian matrices with spectrum } \lambda\}.$$

If we let $\mu_1 < \cdots < \mu_k$ be the distinct eigenvalues in the n-tuple λ and n_j the multiplicity of μ_j, then there is a diffeomorphism

$$\mathcal{H}_\lambda \approx U(n)/(U(n_1) \times \cdots \times U(n_k)).$$

Thus, the coadjoint orbits of $U(n)$ are the complex flag manifolds. The case where $k = 2$ corresponds to the complex Grassmann manifolds.

Note: Similar statements hold for the adjoint orbits of $U(n)$, but for the following discussion of the moment map it's more convenient to work with coadjoint orbits.

Recall from Proposition 8.8 that for any $h \in \mathcal{H}_\lambda$ the tangent space is given by

$$T_h\mathcal{H}_\lambda = \{[X, h] \mid X \text{ is a Hermitian matrix}\}.$$

The Kostant-Kirillov-Souriau symplectic structure on $T_h\mathcal{H}_\lambda$ is given by

$$\omega_h([X, h], [Y, h]) = \text{trace}([X, Y]ih)$$

where X and Y are Hermitian matrices and $i = \sqrt{-1}$. For a proof that the preceeding formula defines a non-degenerate closed 2-form on $\mathcal{H}_\lambda$ see Section 3.2 of [12].

There is also an almost complex structure J on the flag manifold $\mathcal{H}_\lambda$ that is an isometry compatible with the symplectic structure, i.e.

$$\omega_h(J[X, h], J[Y, h]) = \omega_h([X, h], [Y, h])$$

for all $[X, h], [Y, h] \in T_h\mathcal{H}_h$ and

$$g_h([X, h], [Y, h]) \overset{\text{def}}{=} \omega_h([X, h], J[Y, h])$$

is a Riemannian metric on $\mathcal{H}_\lambda$. The almost complex structure J is defined using the Lie algebra structure, but the formulas are somewhat more complicated than those given in Section 8.3. See Section 3.3 of [12] for more details.

Adjoint orbits of general compact Lie groups

According to Section 4 of [7], many of the results in this chapter generalize to arbitrary compact Lie groups as follows.

Let G be a compact Lie group and consider the adjoint action

$$G \times \mathfrak{g} \to \mathfrak{g}.$$

Every orbit of the adjoint action is of the form $G/C(T_0)$ where $C(T_0)$ is the centralizer of some subtorus $T_0 \subseteq T$ of a maximal torus T. Fixing a positive Weyl chamber $\mathcal{W}$ in the Lie algebra of T fixes a parabolic subgroup P of the complexification G_c of G with $C(T_0) \subseteq P$. We can then identify the adjoint orbit $G/C(T_0)$ with G_c/P, and we see that every adjoint orbit of G is a complex manifold that is homogeneous under the action of G_c. This complex structure together with the Kostant-Kirillov-Souriau symplectic structure gives a Kähler metric $< , >$ on the adjoint orbit.

Fixing a point λ in the interior of the positive Weyl chamber $\mathcal{W}$ determines a function $\phi(x) =< x, \lambda >$ analogous to the function considered in Section 8.2. Moreover, there is a formula for the gradient flow of ϕ analogous to the one given in Theorem 8.38, and the unstable manifolds of ϕ are the Bruhat cells of the adjoint orbit (compare with Theorem 8.40). See Section 4 of [7] for more details.

The moment map and coadjoint orbits

A **symplectic vector field** on a symplectic manifold (M, ω) is a vector field X that satisfies one of the following three equivalent conditions:

1. Its local 1-parameter group φ_t preserves the symplectic form.

2. The Lie derivative in the direction X is zero, i.e. $L_X \omega = 0$.

3. The 1-form $\imath(X)\omega$ is closed.

A vector field X on M is said to be a **Hamiltonian vector field** if and only if $\imath(X)\omega$ is exact, i.e. $\imath(X)\omega = dH$ for some function $H : M \to \mathbb{R}$ (called a **Hamiltonian**). The space $C^\infty(M)$ of smooth functions on M (endowed with the Poisson bracket) is a Lie algebra, and so is the set $\text{Ham}(M, \omega)$ of all Hamiltonian vector fields. Moreover, the map $H \mapsto X_H$ from $C^\infty(M)$ to $\text{Ham}(M, \omega)$, assigning a function H to its Hamiltonian vector field X_H (defined by the equation $\imath(X_H)\omega = dH$), is a surjective Lie algebra homomorphism with kernel $\mathbb{R}$ provided that M is connected.

A smooth action of a connected Lie group G on a symplectic manifold (M, ω) is called **weakly Hamiltonian** if and only if the fundamental vector field $X^\#$ associated to any element $X \in \mathfrak{g}$ is a Hamiltonian vector field. The fundamental vector field $X^\#$ associated to $X \in \mathfrak{g}$ is defined by

$$X^\#(x) = \frac{d}{dt}(\exp(tX) \cdot x)\Big|_{t=0}$$

where $\exp(tX) \cdot x$ denotes the action of G on M. Thus, a weakly Hamiltonian action determines a Lie algebra homomorphism

$$\mathfrak{g} \to \text{Ham}(M, \omega).$$

If there is a Lie algebra homomorphism

$$u : \mathfrak{g} \to C^\infty(M)$$

sending $X \in \mathfrak{g}$ to a Hamiltonian for $X^\#$, i.e. $\imath(X^\#)\omega = du(X)$, then the action of G on M is called a **Hamiltonian action**, and the map

$$\mu : M \to \mathfrak{g}^*$$

defined by $\mu(x)(X) = u(X)(x)$ for $x \in M$ and $X \in \mathfrak{g}$ is called "the" **moment map** (or the **momentum mapping**). If M is compact, then one can show that a weakly Hamiltonian action is in fact a Hamiltonian action (see for instance Section 26 of [73]). Moreover, the moment map is always equivariant with respect to the action of G on M and the coadjoint action of G on $\mathfrak{g}^*$ (see for instance Section 26 of [73] or Section 15 of [88]).

If G is a compact connected Lie group and $M \subset \mathfrak{g}^*$ is a coadjoint orbit of G, then the Kostant-Kirillov-Souriau symplectic structure ω on M is defined as follows. The tangent space at a point $\xi \in M$ is given by

$$T_\xi M = \{\mathrm{ad}(X)^*\xi | \, X \in \mathfrak{g}\}$$

where $\mathrm{ad}(X) : \mathfrak{g} \to \mathfrak{g}$ is the linear map given by $\mathrm{ad}(X)(Y) = [X, Y]$ (compare with Proposition 8.8), and for $\xi \in M$ and $\mathrm{ad}(X)^*\xi$, $\mathrm{ad}(Y)^*\xi \in T_\xi M$ we have

$$\omega_\xi(\mathrm{ad}(X)^*\xi, \mathrm{ad}(Y)^*\xi) = \xi([X, Y]).$$

The action of G on the coadjoint orbit M is Hamiltonian (with respect to the Kostant-Kirillov-Souriau symplectic structure), and the inclusion $M \subset \mathfrak{g}^*$ is a moment map $\mu : M \to \mathfrak{g}^*$. See Section 15 of [88], or Section 5.2 of [99] for more details.

Now suppose that we have a Hamiltonian action

$$\mathbb{T}^n \times M \to M$$

of an n-dimensional torus $\mathbb{T}^n$ on a symplectic manifold (M, ω) of dimension m with moment map $\mu : M \to \mathfrak{t}^*$ (where $\mathfrak{t}$ denotes the Lie algebra of the torus $\mathbb{T}^n$). One can show that for every $X \in \mathfrak{t}$, the component of the moment map along X

$$\mu^X : M \to \mathbb{R}$$

defined by $\mu^X(x) = \mu(x)(X)$ for all $x \in M$ is a Morse-Bott function whose critical submanifolds are all of even index. Moreover, the critical set of μ^X is the fixed point set of the closure of the subgroup of $\mathbb{T}^n$ generated by X. See Homework 21 of [134] for more details.

The restriction of the coadjoint action of $U(n + k)$ on a coadjoint orbit in $\mathfrak{u}(n + k)^*$ to a maximal torus $\mathbb{T}^{n+k} \subset U(n + k)$ is a Hamiltonian action with moment map

$$\mu : M \to \mathfrak{t}^*$$

given by projecting onto $\mathfrak{t}^*$. Using an Ad-invariant inner product on the Lie algebra $\mathfrak{u}(n+k)$ the adjoint orbits of $U(n+k)$ can be identified with coadjoint orbits. Under this identification, one can show that the Morse-Bott functions μ^X determined by the Hamiltonian action of $\mathbb{T}^{n+k}$ on a coadjoint orbit of

$U(n+k)$ correspond to the Morse functions $f_A : U(n+k) \cdot x_0 \to \mathbb{R}$ considered in Section 8.2 [154].

As a final example of how Morse theory can be applied to symplectic geometry, we note that the Morse theory of the functions μ^X (the components of the moment map) played a crucial role in the proof of the famous Atiyah-Guillemin-Sternberg Convexity Theorem:

Theorem 8.59 (Atiyah-Guillemin-Sternberg Convexity Theorem)   *Let (M, ω) be a compact connected symplectic manifold and assume that there is a Hamiltonian torus action*

$$\mathbb{T}^n \times M \to M$$

on M with moment map $\mu : M \to \mathfrak{t}^* \approx \mathbb{R}^n$. Then

1. *The set of fixed points of the action is a finite union of connected symplectic submanifolds $C_1, \ldots, C_k$.*

2. *The moment map is constant on each connected component $C_1, \ldots, C_k$.*

3. *The image of μ is the convex hull of the images of the components $C_1, \ldots, C_k$*

$$\mu(M) = \left\{ \sum_{j=1}^{k} \lambda_j f(C_j) \,\middle|\, \lambda_j \geq 0, \sum_{j=1}^{k} \lambda_j = 1 \right\}.$$

For more details see [7], Chapter 3 of [11], [73], Section 5.4 of [99], or Section 27 of [134]. See Problem 26 of Chapter 3 for another version of this theorem.

Problems

1. Show that the Lie algebra of the unitary group $U(n)$ is the space of skew-Hermitian matrices $\mathfrak{u}(n)$. Conclude that the dimension of $U(n)$ is n^2. Note: $U(n)$ is the space of $n \times n$ complex matrices that satisfy ${}^t\bar{A}A = I_{n \times n}$, and $\mathfrak{u}(n)$ is the space of $n \times n$ complex matrices that satisfy ${}^t\bar{A} + A = 0$.

2. Show that the Lie algebra of the special unitary group $SU(n)$ is the Lie algebra of skew-Hermitian matrices with trace zero $\mathfrak{su}(n)$. What is the dimension of $SU(n)$? Note: $SU(n)$ is the subgroup of $U(n)$ consisting of those matrices which have determinant one.

3. Show that the Lie algebra of the special orthogonal group $SO(n)$ is the Lie algebra of skew-symmetric matrices $\mathfrak{so}(n)$. Conclude that the dimension of $SO(n)$ is $\frac{1}{2}n(n-1)$. Note: $SO(n)$ is the space of $n \times n$ real matrices of determinant one that satisfy ${}^tAA = I_{n \times n}$, and $\mathfrak{so}(n)$ is the space of $n \times n$ real matrices that satisfy ${}^tA + A = 0$.

4. Show that the Lie algebra of the symplectic group $Sp(n)$ is the Lie algebra $\mathfrak{sp}(n)$. Conclude that the dimension of $Sp(n)$ is $2n^2 + n$. Note: If J is the almost complex structure given by

$$J = \begin{pmatrix} 0 & I_n \\ -I_n & 0 \end{pmatrix},$$

then

$$Sp(n) = \left\{ A \in U(2n) \mid {}^t A J A = J \right\}$$

and

$$\mathfrak{sp}(n) = \left\{ A \in u(2n) \mid {}^t A J + J A = 0 \right\}.$$

5. Verify that $u(n)$, $\mathfrak{su}(n)$, $\mathfrak{so}(n)$, and $\mathfrak{sp}(n)$ are all Lie algebras when the bracket is defined to be the usual commutator for matrices, i.e. $[A, B] = AB - BA$.

6. Show that $\mathbb{R}^3$ is a Lie algebra where the Lie bracket is defined to be the cross product, i.e.

$$[\vec{x}, \vec{y}] \stackrel{\text{def}}{=} \vec{x} \times \vec{y}$$

for all $\vec{x}, \vec{y} \in \mathbb{R}^3$.

7. Show that $\mathfrak{so}(3)$ is isomorphic to the Lie algebra $\mathbb{R}^3$ where the Lie bracket is defined to be the cross product.

8. Show that the Lie algebra $\mathfrak{so}(3)$ is isomorphic to $\mathfrak{su}(2)$.

9. Show that $Sp(1)$ is the same as $SU(2)$.

10. Show that the Killing form on $\mathfrak{su}(2)$ is negative definite.

11. Show that the Killing form on $\mathfrak{su}(2)$ is 4 times the trace form.

12. Show that the Killing form on $u(2)$ is **not** negative definite.

13. Show that both the Killing form and the trace form are associative bilinear forms.

14. Let $V = M_{n \times n}(\mathbb{C})$ be the n^2-dimensional complex vector space consisting of all $n \times n$ matrices with entries in $\mathbb{C}$. Show that

$$< , >: V \times V \to \mathbb{C}$$

defined by $< A, B > = \operatorname{tr}({}^t \bar{B} A)$ for all $A, B \in V$ is an inner product on V.

15. Show that for any two skew-Hermitian matrices $A, B \in u(n)$ the trace of the product AB is a real number.

16. Show that the Lie algebra $\mathfrak{g}$ of an abelian Lie group G has a trivial Lie bracket, i.e. if G is abelian, then $[X, Y] = 0$ for all $X, Y \in \mathfrak{g}$.

17. Prove the following theorem: If M is a smooth manifold embedded in some Euclidean space $\mathbb{R}^r$, then for almost all $A \in \mathbb{R}^r$ the function $g_A : M \to \mathbb{R}$ given by $g_A(x) = \|x - A\|^2$ is a Morse function.

18. Suppose that A is a diagonal matrix with distinct eigenvalues. Show that any matrix B that commutes with A must be diagonal.

19. Suppose that $p, A \in \mathfrak{u}(n + k)$ where $[p, A] = 0$ and A has distinct eigenvalues. Show that there exists a unitary matrix $g \in U(n + k)$ such that both gpg^{-1} and gAg^{-1} are diagonal. Is the conclusion still true if A has repeated eigenvalues?

20. Show that
$$U(n + k)/(U(n) \times U(k))$$
is diffeomorphic to
$$SU(n + k)/(SU(n + k) \cap (U(n) \times U(k))).$$
What is the relationship between the trace form and the Killing form on $\mathfrak{su}(n + k)$ for $n + k \geq 2$?

21. Prove Theorem 8.54.

22. Show that the Schubert cells are the cells of a CW-complex whose underlying space is $G_{n,n+k}(\mathbb{C})$.

23. Show that for all $j \leq 2nk$ the j-skeleton of $G_{n,n+k}(\mathbb{C})$ is the j-skeleton of $G_{n,n+k'}(\mathbb{C})$ for all $k \leq k'$.

24. The universal n-plane bundle γ_n over the Grassmann manifold $G_{n,n+k}(\mathbb{C})$ is a bundle whose total space consists of pairs (P, v) where $P \in G_{n,n+k}(\mathbb{C})$ and $v \in P$. Show that γ_n is a locally trivial vector bundle on $G_{n,n+k}(\mathbb{C})$.

25. Show that for any n-plane bundle ξ over a compact base space B there exists a map $f : B \to G_{n,n+k}(\mathbb{C})$ such that $\xi \approx f^*(\gamma_n)$ provided that k is sufficiently large.

26. Show that the following polynomial is the Poincaré polynomial of the complex Grassmann manifold $G_{n,n+k}(\mathbb{C})$.
$$P_{n,n+k}(t) = \frac{\prod_{j=1}^{n+k} \left(1 - t^{2j}\right)}{\prod_{j=1}^{n} \left(1 - t^{2j}\right) \prod_{j=1}^{k} \left(1 - t^{2j}\right)}$$

27. Give an example of a perfect Morse function $f : M \to \mathbb{R}$ such that $H_k(M; \mathbb{Z}) \neq 0$ for all $0 \leq k \leq \dim M$.

Chapter 9

An Overview of Floer Homology Theories

9.1 Introduction to Floer homology theories

Floer homology theories are attempts to build in infinite dimensions the equivalent of the Morse-Smale-Witten chain complex $(C_*(f), \partial_*)$. The finite dimensional manifold M is replaced by an infinite dimensional manifold $\mathcal{M}$ and the Morse-Smale function $f : M \to \mathbb{R}$ is replaced by some "functional" on $\mathcal{M}$.

There are several Floer homology theories.

Symplectic Floer homology

The symplectic Floer homology was invented by Andreas Floer [55] [56] as a tool to solve the Arnold conjecture concerning a lower bound on the number of fixed points, fix(φ), of a Hamiltonian diffeomorphism $\varphi : M \to M$ of a compact symplectic manifold (M, ω). The conjecture asserts that if all the fixed points are non-degenerate, then

$$\#\text{fix}(\varphi) \geq \sum_{j \geq 0} \dim H_j(M; \mathbb{Q}).$$

Since φ is isotopic to the identity, the Lefschetz Fixed Point Theorem provides the lower bound

$$\#\text{fix}(\varphi) \geq \sum_{j \geq 0} (-1)^j \dim H_j(M; \mathbb{Q})$$

(see Theorem 5.56), but this bound is not as good as the one conjectured by Arnold.

The preceeding formula compares with the Poincaré-Hopf Index Theorem which asserts that if V is a C^∞ vector field on a compact manifold M whose zeroes are all non-degenerate, then the total number of zeros is bounded from below by the alternating sum

$$\sum_{j \geq 0} (-1)^j b_j(M)$$

of the Betti numbers $b_j(M) = \dim H_j(M, \mathbb{Q})$. However, Morse theory gives $\sum_{j \geq 0} b_j(M)$ as a lower bound for the number of critical points of a Morse function on M (see Section 3.4 on the Morse inequalities). This motivated Floer to try to use Morse theory to solve the Arnold conjecture.

Recall that the Morse-Smale-Witten chain complex $(C_*(f), \partial_*)$ of a Morse-Smale function $f : M \to \mathbb{R}$ on a finite dimensional compact smooth Riemannian manifold (M, g) has for k-chains $C_k(f)$ the free abelian group generated over $\mathbb{Z}$ by the critical points of index k. The Morse Homology Theorem (Theorem 7.4) states that the Morse-Smale-Witten chain complex $(C_*(f), \partial_*)$ computes the singular homology $H_*(M; \mathbb{Z})$. If $\mathrm{Cr}_k(f)$ is the set of critical points of f of index k, then clearly:

$$\#\mathrm{Cr}_k(f) = \mathrm{rank}\, C_k(f) \geq \dim H_k((C_*(f), \partial_*)) = \dim H_k(M; \mathbb{Z}).$$

Hence, the total number of critical points of f is bounded from below by the sum of the Betti numbers .

The fixed points of a Hamiltonian diffeomorphism of a symplectic manifold M are in 1-1 correspondence with critical points of the action functional a_H on the space $\mathcal{L}M$ of smooth contractible loops in M. The symplectic Floer homology, outlined in Section 9.2, is an attempt to construct a "Morse-Smale-Witten chain complex" $(C_*(a_H), \partial_*)$ and show that this complex computes the singular homology of M. The Arnold conjecture will then follow.

Floer homology for Lagrangian intersections

Since the graph Γ_φ of a symplectic diffeomorphism $\varphi : M \to M$ is a Lagrangian submanifold of $M \times M$ and fixed points of φ are the intersection points of the Lagrangian submanifolds Γ_φ and Γ_{id} (assuming that Γ_φ and Γ_{id} meet transversally), the existence of a symplectic Floer homology suggests the existence of a Floer homology theory for Lagrangian intersections.

If L and L' are two Lagrangian submanifolds of a symplectic manifold (M, ω), we consider the infinite dimensional manifold of smooth paths

$$\mathcal{M}(L, L') = \{\gamma : [0, 1] \to M \mid \gamma(0) \in L \text{ and } \gamma(1) \in L'\}.$$

For $\gamma, \gamma' \in \mathcal{M}(L, L')$, we assume that there exists a smooth map $\Gamma : [0, 1] \times [0, 1] \to M$ with $\Gamma(s, t) = \gamma_s(t)$, $\gamma_s \in \mathcal{M}(L, L')$, $\gamma_0 = \gamma$ and $\gamma_1 = \gamma'$. We fix $\gamma_0 \in \mathcal{M}(L, L')$ and define $\sigma_{\gamma_0} : \mathcal{M}(L, L') \to \mathbb{R}$ by

$$\sigma_{\gamma_0}(\gamma) = \int_{[0,1] \times [0,1]} \Gamma^* \omega.$$

This is the function which is considered in the Floer homology for Lagrangian intersections. One checks that the set of critical points coincides with the intersection points of L and L'.

Instanton Floer homology

The last Floer homology considered here is the instanton Floer homology. Here the infinite dimensional manifold $\mathcal{M}$ is the gauge equivalence classes of $SU(2)$ (or $SO(3)$) connections of $SU(2)$ (or $SO(3)$) bundles over some 3-dimensional manifold N. The functional on $\mathcal{M}$ is the Chern-Simons functional

$$cs(\theta) = \frac{1}{4\pi^2} \int_N \mathrm{tr}(\frac{1}{2}\theta \wedge d\theta + \frac{1}{3} \theta \wedge \theta \wedge \theta)$$

for all $\theta \in \mathcal{M}$. Here an element of $\mathcal{M}$ is identified with 1-forms on N with values in the Lie algebra $su(2)$ (or $(so(3))$ of $SU(2)$ (or $(SO(3))$).

In Section 9.2 we discuss in some detail the symplectic Floer homology, leaving out technical details and proofs. In Section 9.3 we give a brief supplement on the Floer homology of Lagrangian intersections, and in Section 9.4 we develop the instanton Floer homology in some detail and show a connection with the symplectic Floer homology. The following is an outline of the general approach that is common to all the Floer homologies.

We saw in Proposition 6.2 that if $f : M \to \mathbb{R}$ is a Morse-Smale function on a finite dimensional compact smooth Riemannian manifold (M, g) and p, q are critical points of f, then

$$W(q,p) = \{x \in M | \lim_{t \to -\infty} \gamma_x(t) = q \text{ and } \lim_{t \to \infty} \gamma_x(t) = p\}$$

is a smooth manifold, where γ_x satisfies $\gamma_x'(t) = -(\nabla f)(\gamma_x(t))$ and $\gamma_x(0) = x$. If $W(q,p)$ is non-empty, then the dimension of $W(q,p)$ is

$$\dim W(q,p) = \lambda_q - \lambda_p$$

where λ_q and λ_p denote the Morse indices of the critical points q and p respectively.

The main difficulty with Floer homologies is to make sense of this statement in infinite dimensional situations. One shows that $W(q,p)$ is the inverse image of a regular value of some Fredholm operator $\mathcal{F}$ between Banach manifolds, and hence, by The Inverse Function Theorem, $W(q,p)$ is a smooth finite dimensional manifold of dimension equal to the index of $\mathcal{F}$. This allows one to make sense of the **difference** of indices, i.e. the relative index of two critical points. The index of an individual critical point is usually defined modulo some number, and it is usually not equal to the dimension of the unstable manifold (which is usually of infinite dimension).

Another important fact (needed to construct the boundary operator ∂) is that $W(q, p)$ is compact when $\lambda_q - \lambda_p = 1$. One proves this by studying certain compactifications of the space $W(q, p)$ and then noting that when $\lambda_q - \lambda_p = 1$ the compactification is equal to $W(q, p)$. This compactification problem is handled in the symplectic case using the theory of pseudo-holomorphic curves (due to Gromov [68]) and in the instanton case by using the Uhlenbeck compactification principle [147].

One also needs to study the boundary of the 2-dimensional part of the moduli space in order to show that $\partial^2 = 0$. There are many more subtleties and complications... Please read on.

9.2   Symplectic Floer homology

The main references for the material in this section are [55], [56], [63], [128], and [129].

Preliminaries

A symplectic form on a smooth manifold M is a 2-form ω that satisfies

1. $d\omega = 0$ (i.e. ω is closed)

2. ω is non-degenerate.

Condition (2) means that the bundle map $\tilde{\omega} : T_* M \to T^* M$ assigning a vector field V the 1-form $\tilde{\omega}(V)$, also denoted $i(V)\omega$, such that $\tilde{\omega}(V)(W) = \omega(V, W)$ for all vector fields W, is an isomorphism. The couple (M, ω) is called a symplectic manifold.

If $f : M \to \mathbb{R}$ is a smooth function on a symplectic manifold (M, ω), then we can define a vector field X_f, called the Hamiltonian vector field of f, by the equation

$$X_f = \tilde{\omega}^{-1}(df).$$

This also means that X_f is defined by $i(X_f)\omega = df$. The Hamiltonian vector field X_f satisfies

$$L_{X_f}\omega = d(i(X_f)\omega) + i(X_f)d\omega = d(df) = 0$$

where L_{X_f} is the Lie derivative in the direction X_f. Therefore, the 1-parameter group of diffeomorphisms φ_t generated by X_f satisfies $\varphi_t^*(\omega) = \omega$, i.e. φ_t is a symplectic diffeomorphism. Clearly, the fixed points of $\varphi = \varphi_1$ are the critical points of f. Hence, if f is a Morse function, then

$$\#\text{fix}(\varphi) \geq \sum_{i \geq 0} \dim H_i(M; \mathbb{Q}).$$

Now let $H_t : M \to \mathbb{R}$ be a smooth family of smooth functions, and assume that H_t is 1-periodic, i.e. $H_{t+1} = H_t$. Denote by X_t the family of vector fields defined by

$$i(X_t)\omega = dH_t$$

and consider the system of differential equations

$$\dot{x}(t) = X_t(x(t)). \tag{9.1}$$

Let $\psi_t : M \to M$ be the family of symplectic diffeomorphisms obtained by solving this system of differential equations. We have

$$\frac{d\psi_t}{dt}(x) = X_t(\psi_t(x)), \qquad \psi_0(x) = x.$$

The fixed points of $\varphi = \psi_1$ are in 1-1 correspondence with the periodic solutions of (9.1) with period 1. We denote this space as follows:

$$\mathcal{P}(H) = \{x : \mathbb{R}/\mathbb{Z} \to M \mid \dot{x}(t) = X_t(x(t))\}.$$

The **non-degeneracy** condition for $x \in \mathcal{P}(H)$ is

$$\det(d\psi_1(x(0)) - \mathrm{id}) \neq 0,$$

i.e. $d\psi_1(x(0))$ does not have $+1$ as an eigenvalue. We can now state the following.

Conjecture 9.1 (Arnold Conjecture [5]) Let (M, ω) be a compact symplectic manifold and let $H_t : M \to \mathbb{R}$ be a periodic Hamiltonian. Suppose that all the periodic solutions of (9.1) are non-degenerate, then

$$\#\mathcal{P}(H) \geq \sum_{i \geq 0} \dim H_i(M; \mathbb{Q}).$$

Today this conjecture seems completely proved. We will outline the solution for monotone symplectic manifolds, i.e. symplectic manifolds (M, ω) such that

$$< c_1, A > = \tau < \omega, A >$$

for some positive constant τ, for all $A \in H_2^s(M)$, where $H_2^s(M)$ is the image of the Hurewicz homomorphism $\pi_2(M) \to H_2(M; \mathbb{Z})$ and $c_1 \in H^2(M; \mathbb{Z})$ is the first Chern class of (M, ω).

Recall that on any symplectic manifold (M, ω) there are infinitely many almost complex structures J such that $\omega(JX, JY) = \omega(X, Y)$ and $g(X, Y) = \omega(X, JY)$ is a Riemannian metric on M. The space $\mathcal{J}(M, \omega)$ of all such almost complex structures is contractible [10] [99], and an almost complex structure $J \in \mathcal{J}(M, \omega)$ is said to be **compatible** with ω. The Chern classes of (M, ω) are defined to be the Chern classes of the complex bundle $(T_* M, J)$ for some $J \in \mathcal{J}(M, \omega)$. The Chern classes are independent of the choice of J since $\mathcal{J}(M, \omega)$ is contractible. We will assume that (M, ω) is monotone and normalize ω so that $< \omega, A > \in \mathbb{Z}$ for all $A \in H_2^s(M; \mathbb{Z})$. We define the **minimal Chern number** N of (M, ω) to be ∞ if $< \omega, A > = 0$ for all $A \in H_2^s(M; \mathbb{Z})$ and

$$N = \inf\{< \omega, A > \mid A \in H_2^s(M; \mathbb{Z}) \text{ and } < \omega, A > \text{ is positive}\}$$

otherwise.

Let $\mathcal{L}M$ be the space of smooth contractible loops in M, i.e. if $x \in \mathcal{L}M$ is a smooth map $x : \mathbb{R}/\mathbb{Z} \to M$, then there exists a smooth map $u : D \to M$ where $D = \{z \in \mathbb{C} \mid \|z\| \leq 1\}$ such that $u(e^{2\pi i t}) = x(t)$. The well-known **action functional** from the calculus of variations is a map

$$a_H : \mathcal{L}M \to \mathbb{R}/\mathbb{Z}$$

given by

$$a_H(x) = -\int_D u^* \omega - \int_0^1 H_t(x(t))\, dt$$

for some $u : D \to M$ with $u(e^{2\pi i t}) = x(t)$. An easy calculation shows that

$$d_x a_H = 0 \quad \text{if and only if} \quad \dot{x}(t) = X_t(x(t))$$

i.e. the critical points of a_H are precisely the elements of $\mathcal{P}(H)$.

Gradient flows and connecting orbits

Let $J = (J_t)$ be a 1-periodic family of almost complex structures compatible with ω, i.e. $J_t = J_{t+1}$, and let $g_t(X, Y) = \omega(X, J_t Y)$ be the associated family of Riemannian metrics. The tangent space $T_x \mathcal{L}M$ consists of vector fields along the path $x(t)$, and hence, for $\xi, \eta \in T_x \mathcal{L}M$ we can define a metric

$$< \xi, \eta >= \int_0^1 g_t(\xi_{x(t)}, \eta_{x(t)})\, dt.$$

The gradient of a_H with respect to this metric is

$$\text{grad } a_H(x)(t) = J_t(x(t))\dot{x}(t) - \nabla H_t(x(t))$$

where ∇H_t is the gradient of H_t with respect to the metric g_t. The gradient flows are smooth curves $u : \mathbb{R} \to \mathcal{L}M$ satisfying the partial differential equation

$$\overline{\partial}_{H,J}(u) = \frac{\partial u}{\partial s} + J_t(u)\frac{\partial u}{\partial t} - \nabla H_t(u) = 0 \qquad (9.2)$$

where $u(s, t) \in M$ satisfies the periodicity condition $u(s, t) = u(s, t + 1)$. Observe that if $u(s, t) = x(t)$ is independent of s, then (9.2) reduces to (9.1), and if $H_t \equiv 0$ and $J_t = J$ for all t, then (9.2) reduces to

$$\frac{\partial u}{\partial s} + J(u)\frac{\partial u}{\partial t} = 0.$$

A map $u : \mathbb{R} \times S^1 \to M$ satisfying this equation is called a J-holomorphic curve [10] [98].

In general, if (S, i) is a Riemann surface S with its complex structure i, then a smooth map $u : S \to M$ into a symplectic manifold (M, ω) with a compatible almost complex structure J is said to be a J-holomorphic curve if and only if it satisfies the Cauchy-Riemann equation:

$$J_{u(x)} \circ d_x u = (d_x u) \circ i_x$$

for all $x \in S$.

In finite dimensional Morse theory we saw that every gradient flow $\gamma_x(t)$ of a Morse function f begins and ends at a critical point (Proposition 3.19), i.e. $\lim_{t \to \pm\infty} \gamma_x(t)$ exist, and they are both critical points of f. In symplectic Floer homology this is not always true. In fact, this happens if and only if the solution $u(s, t)$ to (9.2) has finite energy $E(u)$, where

$$E(u) = \frac{1}{2}\int_0^1 \int_{-\infty}^{\infty} \left(\left|\frac{\partial u}{\partial s}\right|^2 + \left|\frac{\partial u}{\partial t} - X_t(u)\right|^2 \right) \, ds \, dt.$$

More precisely, there exists $x^{\pm} \in \mathcal{P}(H)$ such that

$$\lim_{s \to \pm\infty} u(s, t) = x^{\pm}(t)$$

and $\lim_{s \to \pm\infty} \partial_s u(s, t) = 0$ where both limits are uniform in t if and only if $E(u) < \infty$.

Let $\mathcal{M}(x^-, x^+, H, J) = \mathcal{M}(x^-, x^+)$ be the space of all finite energy solutions $u(s, t)$ of (9.2) satisfying $\lim_{s \to \pm\infty} u(s, t) = x^{\pm}(t)$. This is the analogue of the manifold $W(q, p)$ from Chapter 6. The first important result of the theory is the following.

Theorem 9.2 *There exists a subset $\mathcal{H}_{reg}(J) \subseteq C^{\infty}(M \times \mathbb{R}/\mathbb{Z})$ of Baire second category of smooth periodic Hamiltonians such that*

1. *each $x \in \mathcal{P}(H)$ is non-degenerate for $H \in \mathcal{H}_{reg}(J)$.*

2. *each connected component of $\mathcal{M}(x^-, x^+, H, J)$, where $x^{\pm} \in \mathcal{P}(H)$ and $H \in \mathcal{H}_{reg}(J)$, is a smooth finite dimensional manifold, and the dimension of the component $\mathcal{M}_U(x^-, x^+, H, J)$ of $\mathcal{M}(x^-, x^+, H, J)$ containing U is*

$$\dim \mathcal{M}_U(x^-, x^+, H, J) = \mu_{CZ}(x^-, u^-) - \mu_{CZ}(x^+, u^- \# U)$$

where $\mu_{CZ}(z)$ is the Conley-Zehnder index of a critical point $z \in \mathcal{P}(H)$.

Moreover, there exists a function $\eta_H : \mathcal{P}(H) \to \mathbb{R}$ such that

$$dim \, \mathcal{M}_U(x^-, x^+, H, J) = \eta_H(x^-) - \eta_H(x^+) + 2\tau E(U)$$

where $\tau > 0$ is the number such that $< c_1, A >= \tau < \omega, A >$.

In the preceeding theorem $u^- : D \to M$ is the extension of x^- to D and $u = u^- \# U : D \to M$ is the extension of x^+ pictured below.

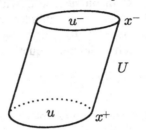

Remark 9.3 The first step in symplectic Floer homology theory is to show the existence of $\mathcal{H}_{reg}(J)$. Then one proves that for all $x^{\pm} \in \mathcal{P}(H)$ and $H \in \mathcal{H}_{reg}(J)$ the linearization D_U of the operator $\bar{\partial}_{J,H}$ at a solution u of (9.2) is a linear Fredholm operator. This implies that $\mathcal{M}_U(x^-, x^+, H, J)$ is a finite dimensional smooth manifold of dimension equal to the index of D_U. We have,

$$ind \, D_U = \mu_{CZ}(x^-, u^-) - \mu_{CZ}(x^+, u^- \# U).$$

The Conley-Zehnder index

Let $Sp(n)$ be the set of all $2n \times 2n$ symplectic matrices,

$$Sp(n) = \{A | \, ^tAJ_0A = J_0\}$$

where J_0 is the standard complex structure on $\mathbb{R}^{2n}$. The Polar Decomposition Theorem asserts that each $A \in Sp(n)$ can be written in a unique way as

$$A = B \, \exp(J_0 C)$$

where $B \in U(n)$ is a unitary matrix and C is a symmetric matrix. It follows that $U(n)$ is a deformation retract of $Sp(n)$. Hence, $\pi_1(Sp(n)) \approx \pi_1(U(n)) \approx \mathbb{Z}$.

The determinant map

$$det : U(n) \to S^1$$

admits a continuous extension

$$\rho : Sp(n) \to S^1$$

such that

(i) $\rho(ABA^{-1}) = \rho(B)$

(ii) if $A_i \in Sp(V_i, \omega_i)$, $i = 1, 2$, and $A(\xi_1, \xi_2) = (A_1(\xi_1), A_2(\xi_2)) \in Sp(V_1 \oplus V_2, \omega_1 \oplus \omega_2)$, then $\rho(A) = \rho_1(A_1)\rho_2(A_2)$

(iii) $\rho(A) = \pm 1$ if A is a symplectic matrix with no eigenvalues on the unit circle.

In fact, the extension of the determinant map with these three properties is unique [130].

Let $Sp^*(n) = \{A \in Sp(n)| \det(\mathrm{id} - A) \neq 0\}$. This set has two connected components: $Sp^+(n) = \{A \in Sp(n)| \det(\mathrm{id} - A) > 0\}$ and $Sp^-(n) = \{A \in Sp(n)| \det(\mathrm{id} - A) < 0\}$, and each connected component is simply connected. A path $\psi : [0, 1] \to Sp(n)$ is said to be an **admissible** path if and only if $\psi(0) = \mathrm{id}$ and $\psi(1) \in Sp^*(n)$. Consider $W^+ = -\mathrm{id} \in Sp^+(n)$ and $W^- = \mathrm{diag}(2, -1, \cdots, -1, 1/2, -1, \ldots, -1) \in Sp^-(n)$. An admissible path $\psi : [0, 1] \to Sp(n)$ admits an extension, unique up to homotopy $\hat{\psi} : [0, 2] \to Sp(n)$ with $\hat{\psi}(s) \in Sp^*(n)$, $s > 1$ defined as follows. If $\psi(1) \in Sp^+(n)$, choose any path φ from $\psi(1)$ to W^+ in $Sp^+(n)$. Any other path from $\psi(1)$ to W^+ will be homotopic to φ since $Sp^+(n)$ is simply connected. Then define $\hat{\psi} : [0, 2] \to Sp(n)$ as the composition $\psi * \varphi$ of the paths ψ and φ. In the case $\psi(1) \in Sp^-(n)$, choose φ to be a path in $Sp^-(n)$ from $\psi(1)$ to W^- and do the same construction as above. Since $\rho(W^\pm) = \pm 1$, it follows that $\rho^2 \circ \psi : [0, 2] \to S^1$ is a loop. We define the Conley-Zehnder index of ψ to be the degree

$$\mu_{CZ}(\psi) = \deg(\rho^2 \circ \psi).$$

Let $x \in \mathcal{P}(H)$ be a non-degenerate trajectory of the dynamical system

$$\dot{x}(t) = X_t(x(t)).$$

Then $x(t) = \psi_t(x(0))$, where ψ_t is the flow of X_t. The differential

$$d_{x(0)}\psi_t : T_{x(0)}M \to T_{x(t)}M$$

gives a family of linear symplectic maps. To get matrices, we need a trivialization of $x^*(T_*M)$. Choose an extension of $x(t)$ to a smooth map $u : D \to M$ where $D = \{z \in \mathbb{C}| \|z\| \leq 1\}$ and $x(t) = u(e^{2\pi it})$, and consider the trivial bundle $u^*(T_*M)$ over D. This choice specifies a matrix $\psi_x(t) \in Sp(n)$ representing the symplectic map $d_{x(0)}\psi_t$. This way we get an admissible path $\psi_x : [0, 1] \to Sp(n)$. We define the **Conley-Zehnder index** of the periodic solutions $x \in \mathcal{P}(H)$, together with the extension $u : D \to M$ to be the number

$$\mu_H(x, u) = n - \mu_{CZ}(\psi_x).$$

Remark 9.4 Without specifying the extension $u : D \to M$ of $x \in \mathcal{P}(H)$ the Conley-Zehnder index $\mu_{CZ}(x)$ can still be defined modulo $2N$, where N is the minimal Chern number.

The definition of symplectic Floer homology

In order to define the boundary operator we need the following result.

Theorem 9.5 *Let* $x^{\pm} \in \mathcal{P}(H)$, $H \in \mathcal{H}_{reg}(J)$, *and suppose that* (M, ω) *is monotone. The real numbers act on the 1-dimensional connected components* $\mathcal{M}_u^1(x^-, x^+, H, J)$ *of* $\mathcal{M}(x^-, x^+, H, J)$ *by reparameterization, and the quotient space* $\hat{\mathcal{M}}^1(x^-, x^+, H, J) = \mathcal{M}^1(x^-, x^+, H, J)/\mathbb{R}$ *is a compact 0-dimensional manifold. Thus, the set of connecting orbits is finite.*

The idea behind the proof is the following "bubbling" theorem.

Theorem 9.6 *Let* $u_j \in \mathcal{M}(x^-, x^+, H, J)$ *be a sequence such that*

$$\sup_j E(u_j) < \infty.$$

Then there exist finitely many points $z_l \in \mathbb{R} \times S^1$, $l = 1, \ldots, k$ *and a solution of equation (9.2) such that a subsequence of* u_j *converges to* u, *uniformly in all derivatives, on a compact subset of* $\mathbb{R} \times S^1 - \{z_1, \ldots z_k\}$. *Moreover,* $E(u)$ *is finite (hence* u *is a connecting orbit).*

If (M, ω) is monotone and $\{u_j\}$ is a sequence in some component of the space $\mathcal{M}^1(x^-, x^+, H, J)$, then one proves that $k = 0$, i.e. $\{u_j\}$ converges on $\mathbb{R} \times S^1$. Hence, $\hat{\mathcal{M}}^1(x^-, x^+, H, J)$ is compact. Now one has to prove that each connected component of $\hat{\mathcal{M}}^1(x^-, x^+, H, J)$ can be oriented and all the orientations are compatible with the "gluing" operation on orbits. With this in mind it is possible to assign an integer $\varepsilon([u]) = \pm 1$ to each $[u] \in \hat{\mathcal{M}}^1(x^-, x^+, H, J)$.

Let $CF_k(M)$ be the free abelian group generated over a principal ideal domain $\mathbb{F}$ by the critical points $x \in \mathcal{P}(H)$ with $\mu_H(x) = k (\mathrm{mod}\, 2N)$. Define a homomorphism

$$\partial : CF_k(M) \to CF_{k-1}(M)$$

by

$$\partial(y) = \sum_{\substack{x \in \mathcal{P}(H) \\ \mu_H(x) = k-1 (\mathrm{mod}\, 2N)}} \sum_{[u] \in \hat{\mathcal{M}}^1(y, x, H, J)} \varepsilon([u]) x.$$

Theorem 9.7 (Floer [55]) *If* (M, ω) *is monotone and* $H \in \mathcal{H}_{reg}(J)$, *then*

$$\partial \circ \partial = 0.$$

The key point is to understand the ends of the 2-dimensional moduli space $\mathcal{M}^2(z, x) = \mathcal{M}^2(z, x, H, J)$. One shows that the closure of $\hat{\mathcal{M}}^2(z, x) = \mathcal{M}^2(z, x)/\mathbb{R}$ is a compact 1-dimensional manifold whose boundary is given by

$$\partial \hat{\mathcal{M}}(z, x) = \bigcup_{y \in \mathcal{P}(H)} \hat{\mathcal{M}}^1(z, y) \times \hat{\mathcal{M}}^1(y, x).$$

This is a consequence of Floer's gluing theorem [55]. This implies that

$$\sum_{\substack{y \in \mathcal{P}(H) \\ \mu_H(y) = k (\mathrm{mod}\ 2N)}} \sum_{v \in \hat{\mathcal{M}}^1(z,y)} \sum_{u \in \hat{\mathcal{M}}^1(y,x)} \varepsilon(v)\varepsilon(u) = 0$$

for all pairs $x, y \in \mathcal{P}(H)$ with $\mu_H(z) = k + 1 (\mathrm{mod}\ 2N)$ and $\mu_H(x) = k - 1 (\mathrm{mod}\ 2N)$.

We can now define the Floer homology groups of the regular pair (H, J)

$$HF_*(M, \omega, H, J; \mathbb{F}) = \ker \partial_* / \mathrm{im}\ \partial_*.$$

One of the main results of the theory is the following.

Theorem 9.8 (Continuation Theorem, Floer [55]) *Let (M, ω) be a monotone symplectic manifold. For any $(H, J), (H', J') \in \mathcal{H}_{reg}(J)$ there exists a natural isomorphism*

$$HF_*(M, \omega, H, J; \mathbb{F}) \approx HF_*(M, \omega, H', J'; \mathbb{F}).$$

Since the Floer homology is independent of the choice of the regular pair (H, J), we may choose H and J to be time independent: $H_t = H$ and $J_t = J$. Then an element $x \in \mathcal{P}(H)$ is just a critical point of H. Moreover, if H is a time independent Morse function with sufficiently small second derivatives, then one can show that the 1-dimensional moduli space $\mathcal{M}^1(x, y, H, J)$ of connecting orbits in Floer homology between $x, y \in \mathcal{P}(H)$ is exactly the space of gradient flow lines between the critical points x and y. Therefore, computing the homology of the Floer chain complex reduces to computing the homology of the Morse chain complex. By Theorem 7.4, the latter computes the singular homology. This observation leads to the main result of Floer's symplectic homology theory.

Theorem 9.9 (Floer [55])

$$HF_*(M, \omega, H, J; \mathbb{F}) \approx \bigoplus_{j = k (\mathit{mod}\ 2N)} H_j(M; \mathbb{F})$$

where $H_j(M; \mathbb{F})$ is the singular homology.

9.3   Floer homology for Lagrangian intersections

The main references for the material in this section are [50], [52], [53], and [54].

A Lagrangian submanifold L of a symplectic manifold (M, ω) is a submanifold of dimension equal to half of the dimension of M such that $i^*\omega = 0$, where $i : L \to M$ is the inclusion. The following example shows a link between this section and the previous section.

Example 9.10 If $\varphi : M \to M$ is a symplectic diffeomorphism, then its graph $\Gamma_\varphi = \{(x, y) \in M \times M \mid y = \varphi(x)\}$ is a Lagrangian submanifold of $(M \times M, \omega \ominus \omega)$ where $\omega \ominus \omega$ denotes $\pi_1^*\omega - \pi_2^*\omega$, and $\pi_i : M \times M \to M$ are the projections onto each factor for $i = 1, 2$. The graph of the identity is the diagonal $\triangle$, and $\Gamma_\varphi \cap \triangle$ is the set of fixed points of φ.

Given two Lagrangian submanifolds $L, L' \subset M$ we consider the space

$$\mathcal{M}(L, L') = \{\gamma \in C^\infty([0, 1], M) \mid \gamma(0) \in L \text{ and } \gamma(1) \in L'\}.$$

We will assume that given $\gamma, \gamma' \in \mathcal{M}(L, L')$, there exists a smooth map $\Gamma :$ $[0, 1] \times [0, 1] \to M$ with $\Gamma(s, t) = \gamma_s(t)$ and $\gamma_s \in \mathcal{M}(L, L')$, $\gamma_0 = \gamma$ and $\gamma_1 = \gamma'$. Now fix some $\gamma_0 \in \mathcal{M}(L, L')$ and define

$$\sigma_{\gamma_0} : \mathcal{M}(L, L') \to \mathbb{R}$$

by

$$\sigma_{\gamma_0}(\gamma) = \int_{[0,1] \times [0,1]} \Gamma^*\omega$$

where $\Gamma(s, t) = \gamma_s(t)$, $\gamma_0 = \gamma_0$ and $\gamma_1 = \gamma$.

One can show that σ_{γ_0} is well-defined and is independent of γ_0 up to a constant. Hence, its differential $d\sigma_{\gamma_0}$ is well-defined (independent of γ_0).

$$(d\sigma_{\gamma_0})(x)(\xi) = \int_0^1 \omega(\dot{x}(t), \xi_t) \, dt$$

$x \in \mathcal{M}(L, L')$ is a critical point of σ_{γ_0} if and only if x is a constant map. We have $x(0) = x(1) \in L \cap L'$, i.e. the set of critical points of σ_{γ_0} coincides with the intersection points $L \cap L'$, and x is a non-degenerate critical point if and only if L intersects L' transversally at x.

Clearly, for $x \in \mathcal{M}(L, L')$

$$T_x\mathcal{M}(L, L') = x^*(T_*M).$$

We define a metric on $T_*\mathcal{M}(L, L')$ as we did in the previous section:

$$< V, W > = \int_0^1 \omega(V_{x(t)}, JW_{x(t)}) \, dt$$

where J is an almost complex structure compatible with ω.

The gradient of σ_{γ_0} is

$$(\text{grad } \sigma_{\gamma_0})(x_0) = J\dot{x}(t).$$

As usual, we consider gradient flows and connecting orbits. For $p, q \in L \cap L'$ we define

$$\mathcal{M}(p, q, L, L') = \left\{ x : [0, 1] \to \mathcal{M}(L, L') \mid \frac{\partial x}{\partial s} = -J \frac{\partial x}{\partial t} \right\}$$
$$= \{ h : \mathbb{R} \times [0, 1] \to M \mid dh \circ i = J \circ dh \}$$

where i is the complex structure on $\mathbb{R} \times [0, 1]$. In [50] and [52] Floer proved that under several conditions (for instance, $\pi_1(L) = \pi_1(L') = \pi_2(M) = 0$ and $L \pitchfork L'$) there is a grading map $\mu : L \cap L' \to \mathbb{Z}$ and each $\mathcal{M}(p, q)$ is a smooth manifold of dimension

$$\dim \mathcal{M}(p, q) = \mu(p) - \mu(q).$$

Moreover, he shows that $\mathcal{M}(p, q)$ can be compactified. Therefore, as before he can construct a homology theory $HF_*(L, L')$, and the sum of the Betti numbers for this homology theory gives a lower bound on the number of points in $L \cap L'$.

Applying the preceeding setup to the Example 9.10 we get a Floer homology theory, we denote $HF_*(\varphi, M)$, for every symplectic diffeomorphism φ of a symplectic manifold (M, ω) under the condition that $\pi_1(M) = 0$ and $(M \times M, \omega \ominus \omega)$ is monotone. The Euler characteristic of $HF_*(\varphi, M)$ is equal to the Lefschetz number of φ.

9.4 Instanton Floer homology

The main references for the material in this section are [47], [48], [49], and [51].

Let $\mathcal{M}(N)$ be the gauge equivalence classes of $SU(2)$ (or $SO(3)$) connections on an $SU(2)$ (or $SO(3)$) principal bundle over an oriented Riemannian manifold (N, g) of dimension 3. Recall that any $SU(2)$ bundle is equivalent to the trivial bundle $N \times SU(2)$ [110].

The space of connections is denoted by $\mathcal{A}$ and the gauge group by $\mathcal{G}$, so $\mathcal{M}(N) = \mathcal{A}/\mathcal{G}$. A connection A can be identified with a 1-form θ on N with values in the Lie algebra $su(2)$ (or $so(3)$) of $SU(2)$ (or $SO(3)$). The function

$$\theta \mapsto \frac{1}{4\pi^2} \int_N \text{tr}(\frac{1}{2}\theta \wedge d\theta + \frac{1}{3}\theta \wedge \theta \wedge \theta)$$

induces a well-defined map

$$cs : \mathcal{M}(N) = \mathcal{A}/\mathcal{G} \to S^1$$

called the **Chern-Simons functional**. Note: For convenience, we will drop the $1/4\pi^2$ in the following formulas.

An easy calculation shows that the critical points of cs are flat connections. Now let $\mathcal{R} \subseteq \mathcal{M}(N)$ denote the space of gauge equivalence classes of flat connections on an $SU(2)$ (or $SO(3)$) bundle $\pi : P \to N$. A metric on $\mathcal{M}(N)$ is defined as follows. Let $a, b \in T_A \mathcal{M}(N)$, a and b are 1-forms on N with values in the Lie algebra $su(2)$ (or $SO(3)$). We set,

$$< a, b > = \int_N \mathrm{tr}(a \wedge *b)$$

where $*$ is the Hodge star operator of the oriented manifold N.

One can easily see that the gradient of the cs functional is

$$\mathrm{grad}\, cs(A) = *F_A$$

where F_A is the curvature of the connection A. The equation for the gradient flow is

$$\frac{dA(t)}{dt} + *F_A = 0.$$

Consider now the 4-dimensional manifold $N \times \mathbb{R}$ and the bundle $P \times \mathbb{R} \to N \times \mathbb{R}$ which is a trivial extension of $P \to N$. A connection A on $P \to N$ extends to a connection $A(t)$ on $P \times \mathbb{R}$ by letting the second (t-component) be zero. The covariant derivative in the $\mathbb{R}$-direction is just the partial derivative $\frac{\partial}{\partial t}$. We give $N \times \mathbb{R}$ the product metric of the metric g on N and the flat metric on $\mathbb{R}$, and consider the corresponding Hodge star operator on $N \times \mathbb{R}$. Then the gradient flow equation above takes the form

$$F + *F = 0 \qquad\qquad (9.3)$$

where F is the curvature of $A(t)$ on $P \times \mathbb{R}$. This equation is called the "instanton" equation. (In the literature, instanton/anti-instanton refers to self/anti-self dual connections of $SU(2)$ (or $SO(3)$) connections over 4-manifolds.)

The energy of an instanton F (i.e. a solution to equation (9.3)) is defined to be

$$E(F) = \int_{N \times [0,1]} F \wedge *F.$$

One shows that for a connection $A(x, t)$ on $P \times \mathbb{R}$ the limits $\lim_{t \to \pm\infty} A(x, t)$ exist and are flat connections $A^{\pm}$ if and only if $E(F) < \infty$, where $E(F)$ is the energy of the curvature $F = F_{A(x,t)}$. Hence, we can consider for $A^{\pm}$ flat connections the space $\mathcal{M}_g(N \times \mathbb{R})$ of all gauge equivalence classes of connections A on $P \times \mathbb{R}$ such that

$$(1) \quad \int_{N \times [0,1]} F \wedge *F = E(F) < \infty$$

$$(2) \quad F + *F = 0.$$

Then we see

$$\mathcal{M}_g(N \times \mathbb{R}) = \bigcup_{(A^-, A^+) \in \mathcal{R}} \mathcal{M}(A^-, A^+)$$

where $\mathcal{M}(A^-, A^+)$ is the space of finite energy connections $A(x, t)$ interpolating A^- and A^+.

The following problems are as before.

(i) Prove that the moduli space $\mathcal{M}_g(A^-, A^+)$ is a finite dimensional manifold and find a formula for the dimension.

(ii) Show that $\mathcal{M}_g(A^-, A^+)$ can be oriented.

(iii) Show that the 1-dimensional part of the moduli space is compact.

For (i), one shows that there is a grading

$$\mu : \mathcal{R} \to \mathbb{Z}/n_p\mathbb{Z}$$

where n_p is an integer depending on the bundle $P \to N$.

1. If $P \to N$ is the trivial $SU(2)$ bundle, then $n_p = 8$.

2. If $P \to N$ is a trivial $SO(3)$ bundle over a homology sphere, then $n_p = 8$.

3. If $P \to N$ is a non-trivial $SO(3)$ bundle with $w_2(P) \neq 0$, then $n_p = 4$.

In each case,

$$\dim \mathcal{M}_g(A^-, A^+) = \mu(A^-) - \mu(A^+).$$

To solve problem (iii) one uses the Uhlenbeck compactification principle (the equivalent of "bubbling") [147]. Now all the ingredients are there to define a chain complex which has length n_p. The homology of this chain complex is called the instanton homology $FI_*(N)$. This homology is independent of the choice of the Riemannian metric g.

The Casson invariant

We define the "Euler characteristic" $\mathcal{X}_F(N)$ of $FI_*(N)$ to be

$$\mathcal{X}_F(N) = \sum_{i=0}^{n_p} (-1)^i \dim FI_i(N).$$

When P is the trivial bundle $P = N \times SU(2)$ over a homology 3-sphere N, Taubes [144] proved that $\mathcal{X}_F(N)$ is a topological invariant equal to twice the Casson invariant $\lambda(N)$ of N. A similar result was proved by Fintushel and Stern for trivial $SO(3)$ bundles in [49].

Recall that the Casson invariant $\lambda(N)$ is a count, with suitable signs, of the number of $SU(2)$-equivalence classes of representations of the fundamental group $\pi_1(N)$ into $SU(2)$ [3]. Denote by $\text{Hom}(\pi_1(N), SU(2))$ the space of all representations of $\pi_1(N)$ into $SU(2)$. The holonomy of a flat connection is an element of $SU(2)$ parameterized by a loop in N, i.e. it defines an element of $\text{Hom}(\pi_1(N), SU(2))$. Thus, we get a natural map from the space of gauge equivalence classes of flat connections $\mathcal{R}$ to $\text{Hom}(\pi_1(N), SU(2))$. The group $SU(2)$ acts on $\text{Hom}(\pi_1(N), SU(2))$ by adjoint representations. The map above factors through a bijection:

$$\mathcal{R} \xrightarrow{\approx} \text{Hom}(\pi_1(N), SU(2))/SU(2).$$

Therefore, the Casson invariant is a count with suitable signs of flat connections, i.e. of critical points of the Chern-Simons functional.

The scenario of Taubes' theorem is reminiscent of the Poincaré-Hopf Index Theorem (on a compact finite dimensional smooth manifold M) which implies that the algebraic sum of the zeros of the gradient vector field of a Morse function is equal to the Euler characteristic of M. Here the situation is much more complex. Taubes proves first that he can perturb the Chern-Simons functional so to have only non-degenerate critical points, and then he uses the "spectral flow" to make sense of the "signs" of the zeros.

9.5   A symplectic flavor of the instanton homology

The main reference for the material in this section is [41].

Let $\pi : P \to \Sigma$ be a non-trivial $SO(3)$ principal bundle over a compact oriented Riemann surface Σ of genus k. Up to isomorphism, there is just one such bundle. As before, denote by $\mathcal{A}$ the space of all $SO(3)$-connections of P, by $\mathcal{G}$ the gauge group, and by $\mathcal{F} \subseteq \mathcal{A}$ the set of flat connections. We let

$$\mathcal{M}(\Sigma) = \mathcal{F}/\mathcal{G}_0$$

be the quotient of $\mathcal{F}$ by the identity component $\mathcal{G}_0 \subseteq \mathcal{G}$.

Atiyah and Bott [9] have extensively studied the topology of the moduli space $\mathcal{M}(\Sigma)$:

1. $\mathcal{M}(\Sigma)$ is a compact manifold of dimension $6k - 6$ provided that $k \geq 2$.

2. $\mathcal{M}(\Sigma)$ is a Kähler manifold with Kähler form

$$\omega(A)[\xi, \eta] = \int_\Sigma \text{tr}(\xi \wedge \eta)$$

where $A \in \mathcal{M}(\Sigma)$ and $\xi, \eta \in T_A\mathcal{M}(\Sigma)$.

3. If $k \geq 2$, then the space $\mathcal{M}(\Sigma)$ is connected, simply connected, and we have $\pi_2(\mathcal{M}(\Sigma)) = \mathbb{Z}$.

Let $f : P \to P$ be an automorphism of the bundle $\pi : P \to\to \Sigma$ and let $h : \Sigma \to \Sigma$ be the induced diffeomorphism. The automorphism $f : P \to P$ induced a diffeomorphism $\varphi_f : \mathcal{M}(\Sigma) \to \mathcal{M}(\Sigma)$ by $\varphi_f([A]) = [f^*A]$ for $A \in \mathcal{F}$. This diffeomorphism is a symplectic diffeomorphism, and it is isotopic to the identity. Since $\mathcal{M}(\Sigma)$ is simply connected, φ_f is a Hamiltonian diffeomorphism, i.e. there exists a smooth family of functions $H_t : \mathcal{M}(\Sigma) \to \mathbb{R}$ such that $\varphi_f = \Psi_1$ where $\Psi_t : \mathcal{M}(\Sigma) \to \mathcal{M}(\Sigma)$ is defined by the differential equation

$$\frac{d\Psi}{dt}(a) = X_{H_t}(\Psi_t(A)), \qquad \Psi_0(a) = a$$

and $i(X_{H_t})\omega = dH_t$. For more information on Hamiltonian diffeomorphisms see [15].

We assume that $H = H_t$ is 1-periodic and that all the fixed points of φ_f are non-degenerate. We denote by $H_*^{\text{sym}}(\mathcal{M}(\Sigma); \varphi_f)$ the Floer homology $H_*(\mathcal{M}(\Sigma), H, J)$.

Let $\hat{\pi} : Q \to M$ be the $SO(3)$ principal bundle over the mapping torus $M = \Sigma \times [0, 1]/(x, 0) \sim (h(x), 1)$ of $h : \Sigma \to \Sigma$, with total space Q also the mapping torus of $f : P \to P$. One checks that flat connections of the bundle $\hat{\pi} : Q \to M$ correspond in a natural way with fixed points of φ_f.

We thus see that the chains for the instanton homology of the bundle $\hat{\pi}$ over the 3-manifold M coincide with the chains of the symplectic Floer homology $H_*^{\text{sym}}(\mathcal{M}(\Sigma); \varphi_f)$.

Theorem 9.11 (Dostoglou-Salamon [41]) *There is a natural isomorphism*

$$H_*^{sym}(\mathcal{M}(\Sigma); \varphi_f) \approx FI_*(M).$$

Problems

1. Suppose H is a Morse function with sufficiently small second derivatives. For each critical point x of index λ_x, consider the trivial loop $x(t) = x$ for all t with trivial extension $u : D^2 \to M$, i.e. $u(y) = x$ for all $y \in D^2$. Show that the Maslov index of the pair (x, u) is equal to $m - \lambda_x$.

2. A diffeomorphism $\varphi : M \to M$ of a symplectic manifold (M, ω) is said to be a **Hamiltonian diffeomorphism** if and only if $\varphi = \psi_1$, where $\psi_t : M \to M$ is defined by $\frac{d\psi_t}{dt}(x) = X_t(\psi_t(x))$; $\psi_0(x) = x$ and there is a smooth family of functions H_t such that $i(X_t)\omega = dH_t$. Show that one can always choose H to be 1-periodic in t, i.e. $H \in C^\infty(M \times S^1, \mathbb{R})$.

3. Show that if $\varphi : M \to M$ is a Hamiltonian diffeomorphism of a compact symplectic manifold (M, ω) that is C^1-close to the identity, then

$$\#\text{fix}(\varphi) \geq \text{cat}(M).$$

4. Let $u : \mathbb{R} \times S^1 \to M$ be a smooth solution to equation (9.2), and let $\varphi_t = \varphi_{t+1}$ be a Hamiltonian loop with Hamiltonian $K_t = K_{t+1}$. Prove that the function $\tilde{u}(s, t) = \varphi_t^{-1}(u(s, t))$ satisfies

$$\frac{\partial \tilde{u}}{\partial s} + \tilde{J}_t(\tilde{u})\frac{\partial \tilde{u}}{\partial t} + \nabla \tilde{H}_t(\tilde{u}) = 0$$

where $\tilde{J}_t = \varphi_t^* J_t$ (i.e. $\tilde{J}_t(x) = (d_x\varphi_x)^{-1} \circ J_{\varphi_t(x)} \circ d_x\varphi_t$) and $\tilde{H}_t = (H_t - K_t) \circ \varphi_t$, and the gradient is computed with respect to the metric induced by $\tilde{J}_t$. Deduce that the Hamiltonian loop φ_t induces an isomorphism of Floer homologies.

$$HF_*(M, \omega, H_t, J_t) \to HF_*(M, \omega, (H_t - K_t) \circ \varphi_t, \varphi_t^* J_t)$$

5. Write down explicitly the functional σ_{γ_0} and its gradient flow equation for the Lagrangian Floer homology of Example 9.10, i.e. $(M, \omega) = (N \times N, \omega \ominus \omega)$, where (N, ω) is a symplectic manifold and the Lagrangian submanifolds are $L = \Delta$, the diagonal in $N \times N$, and L' is the graph of a symplectic diffeomorphism $\varphi : N \to N$.

Hints and References for Selected Problems

Chapter 2: The CW-Homology Theorem

1. This is a diagram chase. See Section IV.5 of [30] or Section 4.5 of [138].

2. This is another diagram chase. See Section IV.5 of [30] or Section 4.5 of [138].

3. See Section IV.18 of [30] or Section 4.6 of [138].

4. Take A as an open neighborhood of the northern hemisphere and B as an open neighborhood of the southern hemisphere such that $A \cap B \simeq S^{n-1}$.

5. $(X, A) \simeq (X, X)$.

6. Use local coordinates around the point $[0 : \cdots : 0 : 1] \in \mathbb{R}P^n$ to define a homeomorphism $h : \mathbb{R}P^n \to \mathbb{R}P^{n-1} \cup_f D^n$. See Chapter 8 for a description of the coordinate charts on $\mathbb{R}P^n$. See Proposition 19.10 of [66] if you get stuck.

7. This is a standard example that can be found in almost any book on algebraic topology. For instance, this is Example IV.10.5 of [30]. $H_0(T^2) \approx \mathbb{Z}$, $H_1(T^2) \approx \mathbb{Z} \oplus \mathbb{Z}$, $H_2(T^2) = \mathbb{Z}$, and $H_k(T^2) = 0$ for all $k > 2$.

8. This is a standard example that can be found in almost any book on algebraic topology. For instance, this is Example IV.10.6 of [30]. $H_0(K^2) \approx \mathbb{Z}$, $H_1(K^2) = \mathbb{Z} \oplus \mathbb{Z}_2$, and $H_k(K^2) = 0$ for all $k > 1$.

9. This is a standard example that can be found in almost any book on algebraic topology. For instance, this is Example IV.10.4 of [30]. $H_0(\mathbb{R}P^2) = 0$, $H_1(\mathbb{R}P^2) \approx \mathbb{Z}_2$, and $H_k(\mathbb{R}P^2) = 0$ for all $k > 1$.

10. This computation can be found in Section IV.14 of [30]. When n is even, $H_0(\mathbb{R}P^n) \approx \mathbb{Z}$, $H_k(\mathbb{R}P^n) \approx \mathbb{Z}_2$ for k odd and $0 < k < n$, and all the

other homology groups are zero. When n is odd the answer is the same except that $H_n(\mathbb{R}P^n) \approx \mathbb{Z}$.

11. This is easy since all the boundary maps in the CW-chain complex are zero.

12. See Section II.1 of [150].

13. See Section VII.4 of [30] or Section 1.6 of [138].

14. See Section VII.10 of [30] or Section 7.4 of [138].

15. See Section VII.11 of [30] or Section 7.5 of [138].

16. See Section VII.10 of [30] or Section 7.6 of [138].

17. See Section 6.10 of [38].

18. See Section 6.10 of [38].

19. Add a 1-cell connecting the two points of S^0 for the first part, and add a 2-cell whose boundary agrees with S^1 for the second part. This problem comes from [40].

20. Add an $(n-1)$-cell whose boundary agrees with S^{n-1}. For the more general statement use Theorem 2.30. This problem comes from [40].

21. If X and Y are finite CW-complexes, then there is a CW-structure on the product $X \times Y$ given by the product of cells in X with cells in Y. In fact, this is true if either X or Y is locally finite. (See for instance Section IV.12 of [30].) Show that $S^2 \times S^4 \approx S^2 \vee S^4 \cup_f D^6$ for some attaching map $f : S^5 \to S^2 \vee S^4$. This problem comes from [40].

22. Note that $S^2 \times S^4 - \{pt.\} \simeq S^2 \vee S^4$ by the previous problem, and $\mathbb{C}P^3 - \{pt.\} \simeq \mathbb{C}P^2$. Moreover, the attaching map for the 2-cell in $\mathbb{C}P^2$ is non-trivial. This problem comes from [40].

23. If there exists a retraction $r : D^n \to S^{n-1}$, then $r \circ i = id_{S^{n-1}}$ where $i : S^{n-1} \to D^n$ is the inclusion. Apply the homology functor and use the results from Example 2.2 to get a contradiction. See Corollary IV.6.7 of [30] for more details.

24. See Section XV.6 of [43].

25. See Section XV.6 of [43]. For one direction compose with the retraction. For the other direction take $Z = X$ and f the inclusion.

26. The deformation retraction projects to the quotient space.

27. Use Lemma 2.23.

28. Consider the mapping cylinder M_f.

29. See Section I.5 of [150].

30. This is Theorem 1 of [141].

Chapter 3: Basic Morse Theory

1. In local coordinates the differential of f is given by $\left(\frac{\partial f}{\partial x_1}, \ldots, \frac{\partial f}{\partial x_m} \right)$.

2. Compare this with Example 3.7.

3. The zero matrix is the only critical point. The Hessian is a 4×4 matrix, and it has index 2. When $n > 2$, det is not a Morse function.

5. This is Lemma 2.4 from [100].

6. Proof: Since this is a local property, we may assume that f is defined on a convex neighborhood of 0 in $\mathbb{R}^m$, $p = 0$, $f(0) = 0$, $df(0) = 0$, and $A = \left(\frac{\partial^2 f}{\partial x_i \partial x_j}(0) \right)$ is a diagonal matrix with entries ± 1 on the diagonal. By Remark 3.13 $f(x) = {}^t x S_x x$ where S_x is a symmetric matrix depending smoothly on x such that

$$S_0 = \left(\frac{\partial^2 f}{\partial x_i \partial x_j}(0) \right) = A.$$

Thus, there is a neighborhood U_0 of 0 such that if $x \in U_0$, then $S_x \in U$ where U is the neighborhood of A in Proposition 3.16. If P is the map in Proposition 3.16, then $P(S_x) = Q_x$ satisfies ${}^t Q_x S_x Q_x = A$ and $Q_0 = I_{m \times m}$. Let $\phi : U \to \mathbb{R}^m$ be defined by $\phi(x) = Q_x^{-1} x$. Clearly, $\phi(0) = Q_0^{-1}(0) = 0$, and for any $v \in \mathbb{R}^m$ we have

$$
\begin{aligned}
(d\phi_0)(v) &= \lim_{t \to 0} \frac{\phi(tv) - \phi(0)}{t} \\
&= \lim_{t \to 0} \frac{\phi(tv)}{v} \\
&= \lim_{t \to 0} \frac{Q_{(tv)}^{-1}(tv)}{t} \\
&= \lim_{t \to 0} \frac{t Q_{(tv)}^{-1}(v)}{t} \\
&= \lim_{t \to 0} Q_{(tv)}^{-1}(v) \\
&= Q_0^{-1}(v) \\
&= v.
\end{aligned}
$$

Hence, $d\phi_0$ is the identity map, and the Inverse Function Theorem implies that ϕ restricted to a smaller neighborhood is a coordinate system near 0. Let $y = \phi(x) = Q_x^{-1}(x)$. Then,

$$
\begin{aligned}
\sum_{i=1}^{m} a_i y_i^2 &= {}^t y A y \\
&= {}^t(Q_x^{-1}x) A(Q_x^{-1}x) \\
&= {}^t x {}^t Q_x^{-1} [{}^t Q_x S_x Q_x] Q_x^{-1} x \\
&= {}^t x S_x x \\
&= f(x).
\end{aligned}
$$

7. Use the weak Morse inequalities.

8. Note that the function descends to a Morse function on $\mathbb{R}P^n$ and use the weak Morse inequalities with $\mathbb{Z}_2$ coefficients. The homology groups of $\mathbb{R}P^n$ with $\mathbb{Z}_2$ coefficients are $H_k(\mathbb{R}P^n; \mathbb{Z}_2) \approx \mathbb{Z}_2$ for $k = 0, \ldots, n$. See for instance Example 19.28 of [66]. Note that in this problem we get a stronger result using $\mathbb{Z}_2$ coefficients than we would get using coefficient in $\mathbb{Z}$. (See the hint for Problem 11 of Chapter 2.)

9. Start with an arbitrary Morse function and perturb the Morse function using bump functions defined on small neighborhoods of the critical points.

10. See Section IV.3 of [91]. Alternately, see Theorem 2.5 of [102].

11. Viewing γ as a diffeomorphism onto its image and restricting f to the image of γ we have $h^{-1} = \gamma^{-1} \circ f^{-1}$.

12. This computation is similar to one in the proof of Theorem 3.20. Let $s = h(t)$. Use the chain rule to show that

$$
\frac{d}{ds} h^{-1}(s) = \frac{1}{\frac{d}{dt}h(t)},
$$

and use the computation from the proof of Proposition 3.18 to show that

$$
\frac{d}{dt} h(t) = -\|(\nabla f)(\gamma(t))\|^2.
$$

13. This is a local problem. So, choose an isometric chart and do the computation in $\mathbb{R}^n$.

14. This is an easy computation using the local form given by Lemma 3.11. Note however, as in Remark 3.24, that isometric Morse charts usually do

not exist since the gradient flow depends on the choice of the Riemannian metric g but a Morse chart does not. We will see in Chapter 4 that $W^s(p) \approx \mathbb{R}^{m-\lambda_p}$ and $W^u(p) \approx \mathbb{R}^{\lambda_p}$ for any Riemannian metric g, but the proof takes up most of that chapter.

15. This problem comes from Section 6.1 of [77]. There is a hint there involving 2-jets.

16. See [101] Theorem 1.

17. See [45].

18. See Section 4 of [102].

19. Use the results in Section 3.4.

20. This is a trivial consequence of Theorem 3.53. Alternately, try using the perturbation technique described in Section 3.5.

21. Use the previous problem and Example 3.50.

22. This is proved in [58]. The critical submanifolds are Grassmann manifolds.

23. See Section III.2.3 of [11].

24. Use the previous problem. See [8] or Section III.3.1 of [11].

25. See Section III.2.2 of [11] or [57].

26. Following Atiyah [7], consider the following equivalent statements:
    (A_n): $f^{-1}(t)$ is either empty or connected for all $t \in \mathbb{R}^n$.
    (B_n): $f(M)$ is convex.
    Prove that (A_1) holds as a consequence of the results in the previous two problems, and then prove (A_n) by induction. See Section III.4.2 of [11] for more details. See Theorem 8.59 of this book for another version of this theorem.

27. See Section 3.1 of [2].

28. See Section 3.1 of [2] or Section 11.2 of [99].

29. See Section 3.1 of [2] or Section 11.2 of [99].

Chapter 4: The Stable/Unstable Manifold Theorem

1. The graph of $f : M \to N$ is $\{(x,y) \in M \times N | f(x) = y\}$.

2. This follows from the Mean Value Theorem. See for instance Section 5.4 of [94] or Section 2.2 of [117].

3. The solution is $\Phi(t) = e^{At}$ where

$$e^{At} = \sum_{k=0}^{\infty} \frac{A^k t^k}{k!}.$$

   For more details see Sections 1.3 and 1.4 of [117].

4. Find the eigenvalues and eigenvectors of the matrices.

5. See Remark 4.14. Such a basis of generalized eigenvalues can be found for any $m \times m$ matrix A with entries in $\mathbb{R}$. One way to prove this is to consider the Jordan canonical form of the matrix A. See for instance Section 1.8 of [117] or Appendix III of [78]. The solution to the system of differential equations $\frac{d}{dt}\vec{x} = A\vec{x}$ is $\vec{x}(t) = e^{At}\vec{x}_0$. See Section 1.9 of [117] or Section 4.I of [83] for more details.

   The **Center Manifold Theorem** asserts that under certain conditions a similar decomposition exists for non-linear systems of differential equations in terms of the stable, center, and unstable manifolds. For more details concerning the Center Manifold Theorem see for instance Section 2.7 of [6], [69], [126], or Appendix III to Chapter 5 of [133].

6. The matrix valued function $\Phi(t) = e^{At}$ solves the homogeneous linear system of differential equations $\frac{d}{dt}\vec{x} = A\vec{x}$. The computation is fairly straightforward. For more details see for instance Section 1.10 of [117].

7. This computation is similar to the one in the proof of Lemma 4.5. Note that for every $t \in \mathbb{R}$ we have $\vec{x}(t, \vec{y}) \in \mathbb{R}^m$ whereas $\Phi(t, \vec{y})$ is an $m \times m$ matrix. For more applications of this technique see Section 2.3 of [117].

8. This is also known as the **Shrinking Lemma** or the **Contraction Mapping Principle**. Proofs can be found in Appendix C of [83], Section 6.1 of [94], or [124].

9. For more properties of the norm of a linear transformation see for instance Section 1.3 of [117].

10. Consider for instance the linear diffeomorphism $T : \mathbb{R}^2 \to \mathbb{R}^2$ determined by the matrix

$$A = \begin{pmatrix} 0 & -1 \\ 2 & 0 \end{pmatrix}.$$

The origin is a hyperbolic fixed point for T because the eigenvalues of $dT_0 = T$ both have length $\sqrt{2} > 1$. However, T is not diagonalizable, and it is neither expanding nor contracting with respect to the standard norm on $\mathbb{R}^2$ since $\|T\| = 2$ and $\|T^{-1}\| = 1$! This does not contradict Remark 4.14, because Remark 4.14 only implies that there exists some norm $|\ \ |'$ on $\mathbb{R}^2$ (inducing the same topology as the standard norm on $\mathbb{R}^2$) such that T is expanding with respect to the norm $|\ \ |'$. The reader should find such a norm for T and verify that T^{-1} satisfies the conclusion of the Contraction Mapping Theorem.

11. A similar result also holds for complex Banach spaces. See the Appendix on Spectral Theory to Chapter 4 of [83].

12. See Chapter 2 of [139].

13. Use the following theorem: Let $f : X \to Y$ be a bijective continuous function. If X is compact and Y is Hausdorff, then f is a homeomorphism. See for instance Theorem 3.5.6 of [108].

14. There is only one choice for the attaching map.

15. The 0-skeleton consists of the critical points p and q, and the 1-skeleton also includes the points on the flow lines from r to q and from q to p. For the 1-cells there is only one possibility for the attaching maps. To attach the 2-cell consider T^2 as a quotient of the unit square in $\mathbb{R}^2$. Fix some closed disk around s, and define the attaching map by projecting radially onto the boundary of the unit square.

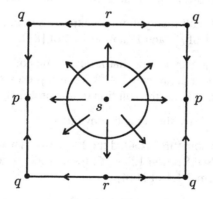

Do **not** try to use the gradient flow as the attaching map for the 2-cell! The endpoint map coming from the gradient flow maps to the critical points, and the attaching map for the 2-cell must map surjectively onto the 1-skeleton.

16. For a diffeomorphism we cannot have any complex eigenvalues of length 1. So, we need $a \neq \pm 1$ and $b \neq \pm 1$ for case (1). For case (2) we need $a \neq \pm 1$, and for case (3) we must have $a^2 + b^2 \neq 1$. For a vector field we cannot have any complex eigenvalues that are purely imaginary. So, we need $a \neq 0$ and $b \neq 0$ for case (1). For case (2) and (3) we need $a \neq 0$. In Chapter 4 of [83] Irwin identifies the linear equivalence classes of linear diffeomorphism and linear vector fields on $\mathbb{R}^m$ for $m = 1, 2$ and 3 using the Jordan normal form.

17. The inverse is given by $\varphi^{-1}(x, y) = y + x - x^2, x)$.

18. The first statement follows from the Inverse Function Theorem. To show that $0 \in \mathbb{R}^m$ is a hyperbolic fixed point, show that $d\varphi_0$ doesn't have any eigenvalues of length 1.

19. A critical point $\vec{p}$ of a vector field $\xi : \mathbb{R}^m \to \mathbb{R}^m$ is hyperbolic if and only if the $m \times m$ matrix $\frac{\partial}{\partial \vec{x}} \xi \big|_{\vec{p}}$ does not have any purely imaginary complex numbers as eigenvalues. See Example 4.27.

20. Note that $d\varphi_p^{-1} = (d\varphi_p)^{-1}$.

21. A **flow** on a differentiable manifold M is a C^1 function $\varphi : \mathbb{R} \times M \to M$ that satisfies

    1. $\varphi_0(x) = x$ for all $x \in M$
    2. $\varphi_t \circ \varphi_s = \varphi_{s+t}$ for all $s, t \in \mathbb{R}$

    where $\varphi_t(x) = \varphi(t, x)$.

22. See the previous problem for the definition of a flow.

23. This is more difficult than you might expect. See p. 22 of [143]. Also, see Proposition 1.5.1 of [6] and Example 1.25 of [83].

24. Topological conjugacy means that h takes orbits of φ onto orbits of ψ. If φ and ψ depend on a parameter $t \in \mathbb{R}$, then h preserves the parameter. See Example 1.5.1 of [6] for the solution to this problem.

25. Similar matrices have the same eigenvalues.

26. This theorem is sometimes called the Flow Box Theorem or the Rectification Theorem. See Section 1.5 of [6] for a proof. Also, see Theorem 5.8 of [83] and Theorem 2.1.1 of [114].

27. The Hartman-Grobman Theorem says that every diffeomorphism is topologically equivalent, near a hyperbolic fixed point, to a linear diffeomorphism. See Section 2.8 of [117] for a proof of the theorem in $\mathbb{R}^m$. See Section 5.II of [83] for a proof of the theorem on a Banach manifold.

28. A **regular point** is a point where grad $f \neq 0$. This result should be compared to Example 4.27: For a Hamiltonian system of differential equations the Hamiltonian remains constant along the flow lines of the system. For a gradient system of differential equations the flow lines are orthogonal to the level sets of the function f. See Section 2.12 of [117] for more information on Hamiltonian and gradient systems of differential equations.

29. Consider the result in the previous problem.

$$\frac{d\vec{x}}{dt} = -\frac{\partial H}{\partial \vec{x}}$$
$$\frac{d\vec{y}}{dt} = -\frac{\partial H}{\partial \vec{y}}$$

Chapter 5: Basic Differential Topology

1. Pick local coordinates and consider the determinants of the minors.

2. See Section 1.7 of [71].

3. This can be generalized to the category of smooth manifolds with boundary and neat submanifolds. See section II.2 of [91] for more details.

4. A map is open if and only if it maps open sets to open sets. See Remark 5.8.

5. Use the previous problem.

6. Use the previous problem.

7. Consider the unit sphere $S^2 \subset \mathbb{R}^3$ and stereographic projection from the north pole $h : S^2 - \{(0,0,1)\} \to \mathbb{R}^2 \times \{0\} \subset \mathbb{R}^3$. Identify $\mathbb{R}^2 \times \{0\}$ with the plane of complex numbers, and show that any polynomial map $P : \mathbb{R}^2 \times \{0\} \to \mathbb{R}^2 \times \{0\}$ determines a smooth map $f : S^2 \to S^2$ whose critical points correspond to the zeroes of the derivative $P'(z)$. Conclude that $f : S^2 \to S^2$ has only a finite number of critical points and use Remark 5.8 to conclude that f is surjective. See Section 1 of [103] if you get stuck.

8. See Section II.2 of [91].

9. See Theorem IV.1.6 of [91].

10. In local coordinates this reduces to a computation involving vectors in $\mathbb{R}^n \times \mathbb{R}^n$. See Section 3.3 of [71] for more details.

11. Use the previous problem and Theorem 5.11.

12. Consider the normal bundles and use the second part of Theorem 5.11.

13. This is false. It is possible to draw counterexamples in $\mathbb{R}^2$.

14. This is also false. It is possible to draw counterexamples in $\mathbb{R}^2$. Note that this problem says that the converse to Corollary 5.12 is false.

15. See Section 2 of [101] or Section II.2 of [91].

16. See Section 3.7 of [36] or Section IV.1 of [91].

17. See [97] for an excellent introduction to differential topology on Banach manifolds with corners.

18. Assume such a map exists and consider the preimage of a regular value. See [76] or Section 2 of [101] for more details. This result can also be found in Section 3.9 of [36].

19. Assume that f doesn't have any fixed points and use this assumption to construct a smooth map $g : D^n \to S^{n-1}$ that restricts to the identity on S^{n-1}. This contradicts the result in the previous problem. See Section 2 of [101] for more details. This result can also be found in Section 3.9 of [36].

20. Any continuous map $f : D^n \to D^n$ can be pointwise approximated by a smooth map from D^n to D^n. See Section 2 of [101] for more details.

21. See Remark 5.8. Start with the assumption that y is also a regular value for a smooth homotopy $F : M \times [0,1] \to N$ between f and g, and use the fact that a compact 1-dimensional manifold with boundary has an even number of boundary points. If y is not a regular value of F, then use the fact that $\#f^{-1}(y)$ is locally constant and find a value y' near y that is a regular value for f, g, and F. See Section 4 of [101] or Section 2.4 of [71] for more details.

22. Use the previous problem and the following Homogeneity Lemma: Let y and z be interior points of a finite dimensional smooth manifold with boundary N. Then there exists a diffeomorphism $h : N \to N$ that is smoothly homotopic to the identity such that $h(y) = z$. See Section 4 of [101] or Section 2.4 of [71] for more details. The Homogeneity Lemma can also be found in Section 3.8 of [36].

23. Use the previous two problems.

24. This is Lemma 2.3 of [104].

25. Use the previous problem. This is Lemma 3.1 of [104].

26. Use Problem 24. This is Theorem 2.2 of [104].

27. The dimension of the kernel is constant by Theorem 5.5.

28. See Section III.5 of [91]. This result is due to Ehresmann [46].

29. See Proposition 4.2 of [148].

30. See Section 3.3 of [71].

31. Use the previous problem.

32. Note that $\deg(f; y) = I(f, \{y\})$, and use the Inverse Function Theorem and Theorem 5.44. See Section 5 of [103] or Section 3.3 of [71] for more details.

33. Assume that a non-zero tangent vector field exists, and use the vector field and a tubular neighborhood of M embedded in some $\mathbb{R}^n$ to define a map $f : M \to M$ that doesn't have any fixed points. Then apply Theorem 5.53 to get a contradiction. See Section IV.23 of [30] for more details.

34. Use the Inverse Function Theorem.

35. See Section 3.4 of [71] or Section VI.12 of [30].

36. First, show that the index of a critical point of a vector field is well defined, i.e. it does not depend on the choice of the coordinate chart. See Section 3.5 of [71] or Section 6 of [103] for more details.

Chapter 6: Morse-Smale Functions

3. The 0-skeleton is the critical point p of index 0, and the 1-skeleton also includes the points in $W^u(r)$ and $W^u(q)$. For the 1-cells there is only one choice for the attaching maps. To attach the 2-cell consider T^2 as a quotient of the unit square in $\mathbb{R}^2$. Fix some closed disk around s, and define the attaching map by projecting radially onto the 1-skeleton.

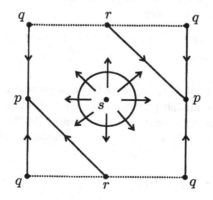

Note that, unlike Problem 15 from Chapter 4, there are no flow lines from r to q when the torus is tilted. In above diagram the solid lines represent the two circles that make up the 1-skeleton.

4. See Theorem 7.3 of [83].

5. Use the Tubular Flow Theorem. See Proposition 3.1.1 of [114].

6. This is not easy. See [115], [116], and Chapter 4 of [114]. Note that the assumption that M has dimension 2 is essential for proving that structural stability implies the Morse-Smale condition.

7. This is a difficult result due to Palis and Smale [113].

8. The Hartman-Grobman Theorem can be proved using the Local Stable Manifold Theorem (Theorem 4.11 and The λ-Lemma (Theorem 6.17). See Section 2.7 of [114] and [112] for more details.

9. According to Remark 2.29, what we need is an open neighborhood $U \subseteq F_{k+1}$ of F_k such that F_k is a strong deformation retract of U. The Morse-Smale transversality condition implies that a gradient flow line starting at a point in F_{k+1} will end in F_k, but the endpoint map of the gradient flow always maps to the critical points. So, the gradient flow **cannot** be used to define a homotopy $R : U \times I \to U$ that fixes F_k. If we add the assumption that the metric is compatible with the Morse charts for f, then according to Remark 3 of Laudenbach's Appendix in [20] the unstable manifolds of f form a CW-structure on M. In this case the index filtration coincides with the skeleta of the CW-structure, and the inclusion maps in the index filtration are cofibrations by Theorem 2.25. See Problem 35 for an interesting counterexample.

10. The assumption that the metric is compatible with the Morse charts for f implies that locally we have

$$\nabla f = -x_1 \frac{\partial}{\partial x_1} - \cdots - x_k \frac{\partial}{\partial x_k} + x_{k+1} \frac{\partial}{\partial x_{k+1}} + \cdots + x_m \frac{\partial}{\partial x_m}.$$

The stable manifold of $\vec{0}$ consists of those point where the first k coordinates are zero, and the unstable manifold consists of those points where the last $m - k$ coordinates are zero. Thus, assuming that the Riemannian metric is compatible with the Morse charts for f drastically simplifies the proof of Theorem 4.2.

11. See Corollary IX.1.4 of [91].

12. Note that $N_0 \times \mathbb{R}$ is a neat submanifold of $M \times \mathbb{R}$ which carries a framing induced by the projection $\pi_1 : M \times \mathbb{R} \to M$. Only transitivity is non-trivial. See Lemma IX.2.2 of [91].

13. Use Theorem 5.19 to show that there exists disjoint representatives for the framed cobordism classes when $2k < m$ (see Proposition IX.2.3 of [91]). The zero element in the group is represented by the empty manifold. The main thing to check is the existence of an inverse cobordism. See Theorem IX.3.1 of [91] or Chapter II of [140].

14. See Lemma IX.5.2 of [91].

15. First prove the result when $y = y'$, i.e. prove that the framed cobordism class is the same for any positive basis of $T_y S^{m-n}$. Next show that if y is sufficiently close to y', then $g^{-1}(y)$ is framed cobordant to $g^{-1}(y')$. Then prove that if g and $\tilde{g}$ are smoothly homotopic and y is a regular value for both g and $\tilde{g}$, then $g^{-1}(y)$ is framed cobordant to $\tilde{g}^{-1}(y)$. See Theorem A in Section 7 of [103].

16. See Theorem IX.5.5 of [91], Theorem B in Section 7 of [101], or Chapter II of [140].

17. See Corollary IX.5.8 of [91] or Section 7 of [103].

18. The result is not true without the assumption that the tangent bundle of M is stably trivial. See Proposition 3.4 of [59].

19. It's clear that g and h induce a homeomorphism

$$(X \times I) \amalg Y \xrightarrow{\approx} (X' \times I) \amalg Y'.$$

Use the fact that $hf = f'g$ to show that this homeomorphism induces a homeomorphism on the quotient space.

20. This result was communicated to us by Mark Johnson. By Lemma 3.29 the homotopy type of $X \cup_{f_\partial \circ \sigma} D^n$ depends only on the homotopy class of σ. So, we can pick a representative $\sigma : S^{n-1} \to S^{n-1}$ for $-1 \in \pi_{n-1}(S^{n-1})$ that is a homeomorphism. Now apply the result from the previous problem to the following diagram.

$$
\begin{array}{ccccc}
S^{n-1} & \xrightarrow{f_\partial \circ \sigma} Y & \longrightarrow C_{f_\partial \circ \sigma} & \xleftarrow{\approx} & Y \cup_{f_\partial \circ \sigma} D^n \\
\approx \downarrow \sigma & \approx \| & \downarrow & & \\
S^{n-1} & \xrightarrow{f_\partial} Y & \longrightarrow C_{f_\partial} & \xleftarrow{\approx} & Y \cup_{f_\partial} D^n
\end{array}
$$

21. If $y \notin D = \{x \in \mathbb{R}^n | |x| \leq R\}$, then $\Psi(y) = y$ by assumption. So assume $y \in D$, and let $D' = \{x \in \mathbb{R}^n | |x| \leq R'\}$ where $R' > 0$ is large enough to ensure that

$$D \cup \Psi(D) \cup \{x - \Psi(x) + y | x \in D\} \subset D'.$$

Define $g : D' \to \mathbb{R}^n$ by setting $g(x) = x - \Psi(x) + y$. If $x \in D$, then $g(x) \in D'$ by the way R' was chosen. If $x \notin D$, then $g(x) = y \in D \subset D'$. Thus, g maps D' to D'. By the Brouwer Fixed Point Theorem (see Problem 19 of Chapter 5) there is a point $x \in D'$ such that $g(x) = x$. Hence, $x - \Psi(x) + y = x$, i.e. $\Psi(x) = y$. This shows that $\Psi : \mathbb{R}^n \to \mathbb{R}^n$ is surjective.

22. By the previous problem Ψ is surjective. Once we show that Ψ is injective we will know that Ψ^{-1} exists. The Inverse Function Theorem then implies that Ψ^{-1} is smooth.

To show that Ψ is injective, let $D = \{x \in \mathbb{R}^n | |x| \leq R\}$, and note that $D \cup \Psi(D)$ is compact. First we check that card $\Psi^{-1}(y) = 1$ for $y \notin D \cup \Psi(D)$: We know that $\Psi(y) = y$. Suppose that $\Psi(x) = y$ for some $x \neq y$. Then we must have $x \in D$. But then $\Psi(x) \in \Psi(D)$ while $y \notin \Psi(D)$, a contradiction.

Next we check that if card $\Psi^{-1}(y) > 1$, then i) $y \in D \cup \Psi(D)$ and ii) $\Psi^{-1}(y) \subset D \cup \Psi(D)$. The first statement was proved above. For the second: If $\Psi(x) = y$ and $x \in D$, then $x \in D \cup \Psi(D)$. If $\Psi(x) = y$ and $x \notin D$, then $x = \Psi(x) = y$, so $x \in D \cup \Psi(D)$.

Note that we have reduced the problem to showing that $\Psi : D \cup \Psi(D) \to D \cup \Psi(D)$ is injective. Using the Inverse Function Theorem, the compactness of $D \cup \Psi(D)$, and the fact that $\Psi(x) = x$ for $x \notin D$ we see that there is some constant $\rho > 0$ (independent of x) so that $\Psi(x) = \Psi(x')$ with $x \neq x'$ implies that $|x - x'| \geq \rho$. Thus, using the Inverse Function Theorem again and the compactness of $D \cup \Psi(D)$ we see that the set $\{y \in \mathbb{R}^n | \text{card } \Psi^{-1}(y) > 1\}$ is open.

Finally, suppose that card $\Psi^{-1}(y_n) > 1$ and $\lim_{n \to \infty} y_n = y$. Then there exists $\{x_n, x'_n\} \subset D \cup \Psi(D)$ with $|x_n - x'_n| \geq \rho$ such that $\Psi(x_n) = \Psi(x'_n) = y_n$ for all n. Since $D \cup \Psi(D)$ is compact we may assume that there exists x, x' with $\lim_{n \to \infty} x_n = x$ and $\lim_{n \to \infty} x'_n = x'$. Then $|x - x'| \geq \rho$ and $\Psi(x) = \Psi(x') = y$. Thus, $\{y \in \mathbb{R}^n | \text{card } \Psi^{-1}(y) > 1\}$ is also closed in $\mathbb{R}^n$, which is connected. Thus, $\{y \in \mathbb{R}^n | \text{card } \Psi^{-1}(y) > 1\}$ is either $\emptyset$ or $\mathbb{R}^n$. The second option is impossible, and hence, Ψ is injective.

28. First show that $d\Psi_{r,\xi}(x)$ is the $n \times n$ identity matrix for $|x| \geq \sqrt{2}r$. Then use the following fact: If $L : \mathbb{R}^n \to \mathbb{R}^n$ is a linear transformation that

satisfies

$$\|L\| = \sup\left\{\frac{|Lv|}{|v|}\ \bigg|\ v \in \mathbb{R}^n - \{0\}\right\} < 1,$$

then the linear transformation $1 + L : \mathbb{R}^n \to \mathbb{R}^n$ is invertible. While you're at it, prove this fact.

33. **Part a:** Every value of f is regular, so $(\nabla_g f)(x, t) \neq 0$ for all $(x, t) \in N \times [0, 1]$. If $\xi \in \ker df_{(x,t)}$, then $g((\nabla_g f)(x, t), \xi) = df_{(x,t)}(\xi) = 0$. However, $T_x N \times t = \ker df_{(x,t)}$. Hence, $(\nabla_g f)(x, t) \perp_g (T_x N \times t)$. Thus, $(\nabla_g f)(x, t) = a\frac{\partial}{\partial t}$ for some $a \in \mathbb{R} - \{0\}$. Then we have

$$|(\nabla_g f)(x, t)|_g^2 = df_{(x,t)}((\nabla_g f)(x, t)) = df_{(x,t)}\left(a\frac{\partial}{\partial t}\right) = a.$$

Thus,

$$\frac{(\nabla_g f)(x, t)}{|(\nabla_g f)(x, t)|_g^2} = \frac{\partial}{\partial t}.$$

Part b: Again, since every value of f is regular we have $df_{(x,t)} \neq 0$ for all $(x, t) \in N \times [0, 1]$. Hence, $(\nabla_{\tilde{g}} f)(x, t) \neq 0$ for all $(x, t) \in N \times [0, 1]$. Again, $(\nabla_{\tilde{g}} f)(x, t) \perp_{\tilde{g}} (T_x N \times t)$ so that $(\nabla_{\tilde{g}} f)(x, t) = a\left(\frac{d\psi_t(x)}{dt}, \frac{\partial}{\partial t}\right)$ for some $a \in \mathbb{R} - \{0\}$ by condition ii). As above,

$$|(\nabla_{\tilde{g}} f)(x, t)|_{\tilde{g}}^2 = df_{(x,t)} a\left(\frac{d\psi_t(x)}{dt}, \frac{\partial}{\partial t}\right) = a\frac{\partial t}{\partial t} = a.$$

so that

$$\frac{(\nabla_{\tilde{g}} f)(x, t)}{|(\nabla_{\tilde{g}} f)(x, t)|_{\tilde{g}}^2} = \left(\frac{d\psi_t(x)}{dt}, \frac{\partial}{\partial t}\right) = \frac{d}{ds}\bigg|_{s=0}(\psi_{t+s}(x), t + s).$$

Then uniqueness of solutions for ordinary differential equations implies that $\varphi_s(\psi_t(x), t) = (\psi_{t+s}(x), t + s)$.

34. Let $\Psi : f^{-1}(c) \times [0, 1] \to f^{-1}(c) \times [0, 1]$ be a smooth isotopy in $f^{-1}(c)$ from the identity of $f^{-1}(c)$ to $\psi : f^{-1}(c) \to f^{-1}(c)$. Replacing Ψ, if necessary, with $\Psi(x, t) = (\psi_{F_{\frac{1}{2}, \frac{3}{4}}(t)}(x), t)$ we may assume that $\psi_t(x) = x$ for all $x \in f^{-1}(c)$ and for t near 0, and that $\psi_t(x) = \psi(x)$ for all $x \in f^{-1}(c)$ and t near 1.

Let $\Phi : f^{-1}(c) \times [0, 1] \to f^{-1}([c, c + \varepsilon])$ be the diffeomorphism given by $\Phi(x, t) = \varphi_t(x)$, where φ_t is the flow determined by the vector field $\varepsilon\nabla_g f / |\nabla_g f|_g^2$. It is easy to check that the pullback Riemannian metric $\Phi^* g$ makes $T_x f^{-1}(c) \times t$ and $\frac{\partial}{\partial t}$ orthogonal,

$$d\Phi_{(x,t)}\left(\frac{\partial}{\partial t}\right) = \varepsilon\frac{\nabla_g f}{|\nabla_g f|_g^2}(\Phi(x, y)),$$

and the composite $h = f \circ \Phi$ is given by $(x, t) \mapsto c + t\varepsilon$ (see the proof of Theorem 3.20). Moreover, since

$$d\Phi_{(x,t)}(\nabla_{\Phi^*g} h) = (\nabla_g f)(\Phi(x, t))$$

and

$$|\nabla_{\Phi^*g} h|^2_{\Phi^*g} = |\nabla_g f|^2_g,$$

we have

$$d\Phi_{(x,t)} \left(\frac{\nabla_{\Phi^*g} h}{|\nabla_{\Phi^*g} h|^2_{\Phi^*g}} \right) = \frac{\nabla_g f}{|\nabla_g f|^2_g}(\Phi(x, t)).$$

Thus,

$$\varepsilon \frac{\nabla_{\Phi^*g} h}{|\nabla_{\Phi^*g} h|^2_{\Phi^*g}} = \frac{\partial}{\partial t}$$

since $d\Phi_{(x,t)}$ is an isomorphism.

We now turn the isotopy upside down by writing $\tilde{\Psi}(x, t) = \Psi(x, 1 - t)$; we do so because we want ψ to appear at the level $f^{-1}(c) \times 0$, which corresponds to $f^{-1}(c) \subset M$, rather than the level set $f^{-1}(c) \times 1$, which corresponds to $f^{-1}(c + \varepsilon) \subset M$. Using this isotopy we alter the metric Φ^*g to a new metric g' as in the previous problem satisfying conditions i), ii), and iii) of that problem with respect to the metric Φ^*g and the isotopy $\tilde{\Psi}$.

Since the isotopy $\tilde{\Psi}$ is constant near $t = 0$ and near $t = 1$, the metric g' agrees with Φ^*g near $t = 0$ and near $t = 1$. Since $\Phi : f^{-1}(c) \times [0, 1] \to f^{-1}([c, c+\varepsilon])$ is a diffeomorphism, we can push g' forward to a metric Φ_*g' on $f^{-1}([c, c + \varepsilon])$; and since g' agrees with Φ^*g near $t = 0$ and near $t = 1$, the new metric Φ_*g' agrees with g near $f^{-1}(c)$ and near $f^{-1}(c + \varepsilon)$. Thus, Φ_*g' extends smoothly by g to all of M to produce the required metric $\tilde{g}$.

Finally, the orbits of $\nabla_{\tilde{g}} f$ are those of $\nabla_g f$ outside of $f^{-1}([c, c + \varepsilon])$ since the metrics agree on that set. Within $f^{-1}([c, c + \varepsilon])$, the orbits of $\nabla_{\tilde{g}} f$ are the images of the orbits of $\nabla_{g'} h$ since $\Phi_*g' = \tilde{g}$ on $f^{-1}([c, c + \varepsilon])$, and hence, $\Phi^*\tilde{g} = g'$. However, the orbits of $\nabla_{g'} h$, by the previous problem, are the orbits of the isotopy $\tilde{\Psi}$, carrying the identity map at $f^{-1}(c) \times 1$ to the diffeomorphism ψ at $f^{-1}(c) \times 0$. Thus, the orbits of $\nabla_{\tilde{g}} f$ within $f^{-1}([c, c + \varepsilon])$ continue the orbits of $\nabla_g f$ outside so that an orbit of $\nabla_g f$ which intersects $f^{-1}(c)$ at a point $x \in f^{-1}(c)$ becomes an orbit of $\nabla_{\tilde{g}} f$ which intersects $f^{-1}(c)$ at the point $\psi(x) \in f^{-1}(c)$. Thus,

$$W^u_{\tilde{g}}(q) \cap f^{-1}(c) = \psi(W^u_g(q) \cap f^{-1}(c)).$$

This completes the solution to the problem.

35. Compare this result with Problem 9 of this chapter. Note that the Riemannian metric in part e is different from the Riemannian metric in part f. The function does **not** satisfy the Morse-Smale condition with respect to the Riemannian metric produced in part f.

Chapter 7: The Morse Homology Theorem

1. See Problem 1 of Chapter 6.

2. See Problem 2 of Chapter 6.

3. Changing orientations changes the signs of the coefficients $n(q, p)$.

4. Changing the orientation of $T_q^u S^2$ also changes the orientation of $T_q^s S^2$ because of the orientation convention $T_q S^2 = T_q^s S^2 \oplus T_q^u S^2$.

5. This is an easy application of Theorem 7.15.

6. Use the index pairs described in Example 7.17 and index pairs of the form (M^b, M^a) where a and b are regular values of the Morse function. See [128] Theorem 2.1 for more details.

7. See [128] p. 119 or [122] for the details of this alteration. See also [56].

8. This follows Theorem 2.30, Theorem 7.15, and the fact that the connecting homomorphism is a natural transformation.

9. This was proved in Section 7.5. The definition of a connected simple system is taken from Definition 2.6 of [127]

10. Start by showing that N_0 is homotopic to a discrete set of points.

11. Use the Connecting Manifold Theorem (Theorem 6.40) to show that the diagram given in Section 7.1 commutes. This approach to proving the Morse Homology Theorem first appeared in [34].

12. The proof is essentially the same as for integer coefficients. See [128].

13. The orientation double cover has $\mathbb{Z}_2$ as its group of deck transformations.

14. See Lemma 3.1 of [127].

15. Note that flowing decreases the index. See Proposition 3.4 of [127].

16. Use the results in Section 6.3.

17. Statement a) is true and statement b) is false.

18. See Theorem 4.3 of [127].

19. See Corollary 4.4 of [127].

20. See Section 5 of [127].

21. See Section 2.2 of [13].

22. Note that in Section 6 of [102] the Morse-Smale function is assumed to be self-indexing. The results need to be modified to apply to general Morse-Smale functions.

Chapter 8: Morse Theory On Grassmann Manifolds

1. The lie algebra is the tangent space at the identity. Let $\gamma(t)$ be a path of $n \times n$ matrices with entries in $\mathbb{C}$ that satisfies ${}^t\overline{\gamma(t)}\gamma(t) = I_{n \times n}$ for all t and $\gamma(0) = I_{n \times n}$. Take the derivative using the product rule to show that $T_{I_{n \times n}} U(n) \subseteq \mathfrak{u}(n)$. For the other inclusion use the exponential map. See Example I.2.16 of [31] or Section 3.37 of [148] for more details. The unitary group $U(n)$ corresponds to the set of automorphisms that preserve the standard Hermitian product on $\mathbb{C}^n$. See Section I.1.8 of [31] for more details.

2. This is similar to the previous problem, but one also has to show that the condition $\det(A) = 1$ on $SU(n)$ corresponds to the trace of a matrix in $T_{I_{n \times n}} SU(n)$ being zero. Use the fact that

$$\det e^A = e^{\text{trace } A}.$$

See Example I.2.18 of [31] or Sections 3.35 through 3.37 of [148] for more details.

3. See Example I.2.15 of [31] or Section 3.37 of [148]. The Lie algebra of $SO(n)$ and the Lie algebra of the orthogonal group $O(n)$ are the same since $SO(n)$ is the identity component of $O(n)$. The orthogonal group $O(n)$ corresponds to the set of automorphisms that preserve the standard inner product on $\mathbb{R}^n$. See Section I.1.8 of [31] for more details.

4. See Example I.2.19 of [31]. The symplectic group $Sp(n)$ corresponds to the set of automorphisms that preserve the standard symplectic inner product on $\mathbb{H}^n$ where $\mathbb{H}$ denotes the quaternions. See Sections I.1.9 through I.1.12 of [31] for more details.

5. Show that each set is closed under commutation. See Theorem 24.2.1 of [42] for more details.

6. The Jacobi identity can be proved using elementary identities from vector analysis.

7. The Lie algebra of the rotation group $SO(3)$ is the Lie algebra $so(3)$ of all 3×3 real skew-symmetric matrices where the Lie bracket is just the usual commutator for matrices. A basis for $so(3)$ is given by the following matrices.

$$X_1 = \begin{pmatrix} 0 & 0 & 0 \\ 0 & 0 & -1 \\ 0 & 1 & 0 \end{pmatrix}, \quad X_2 = \begin{pmatrix} 0 & 0 & 1 \\ 0 & 0 & 0 \\ -1 & 0 & 0 \end{pmatrix}, \quad X_3 = \begin{pmatrix} 0 & -1 & 0 \\ 1 & 0 & 0 \\ 0 & 0 & 0 \end{pmatrix}$$

Note that these matrices satisfy $[X_1, X_2] = X_3$, $[X_2, X_3] = X_1$, and $[X_3, X_1] = X_2$, just like $\hat{i} \times \hat{j} = \hat{k}$, $\hat{j} \times \hat{k} = \hat{i}$, and $\hat{k} \times \hat{i} = \hat{j}$. See Example 24.2.3 of [42] for more details.

8. A basis for $su(2)$ is given by the following matrices:

$$s_1 = \begin{pmatrix} i & 0 \\ 0 & -i \end{pmatrix}, \quad s_2 = \begin{pmatrix} 0 & 1 \\ -1 & 0 \end{pmatrix}, \quad s_3 = \begin{pmatrix} 0 & i \\ i & 0 \end{pmatrix}.$$

These matrices satisfy $[s_1, s_2] = 2s_3$, $[s_2, s_3] = 2s_1$, and $[s_3, s_1] = 2s_2$. See Theorem 24.2.4 of [42] for more details.

9. See Section I.1.9 of [31].

10. Use the basis for $su(2)$ given above. The Killing form $B : su(2) \times su(2) \to \mathbb{R}$ is given by $B(A, B) = \text{trace}([A, [B, -]])$. Show that s_1, s_2, and s_3 are orthogonal with respect to the Killing form B, and show that $B(s_1, s_1) = B(s_2, s_2) = B(s_3, s_3) = -8$.

11. Using the solution to the previous problem it suffices to check that

$$\text{trace}(s_1 s_1) = \text{trace}(s_2, s_2) = \text{trace}(s_3 s_3) = -2$$

and

$$\text{trace}(s_1 s_2) = \text{trace}(s_1 s_3) = \text{trace}(s_2 s_3) = 0.$$

12. Consider the following basis for $u(2)$:

$$s_1 = \begin{pmatrix} i & 0 \\ 0 & 0 \end{pmatrix}, \quad s_2 = \begin{pmatrix} 0 & 1 \\ -1 & 0 \end{pmatrix}, \quad s_3 = \begin{pmatrix} 0 & i \\ i & 0 \end{pmatrix}, \quad s_4 = \begin{pmatrix} 0 & 0 \\ 0 & i \end{pmatrix}.$$

Show that this basis is orthogonal with respect to the Killing form B, $B(s_1, s_1) = -2$, and $B(s_2, s_2) = -8$. Hence,

$$B(-4s_1 + s_2, -4s_1 + s_2) = 0.$$

The Killing form B of a compact connected Lie group G is negative definite if and only if G is semisimple (see Example 8.6). Of the Lie groups considered here: $U(n)$, $SU(n)$, $SO(n)$, and $Sp(n)$, $SU(n)$ is semisimple for $n \geq 2$, $SO(n)$ is semisimple for $n \geq 3$, and $Sp(n)$ is semisimple for $n \geq 1$. See Section I.1.2 of [90] for more details.

13. See Sections III.4 and III.5 of [84].

14. This inner product is known as the **Frobenius inner product**. See for instance Example 6.1.5 of [62].

15. Find a basis for $\mathfrak{u}(n)$ that is orthogonal with respect to the trace form. Remember that $\mathfrak{u}(n)$ is a real vector space even though it is constructed from complex numbers.

16. See Sections I.2.10 through I.2.14 of [31].

17. Use Lemma 8.16 and Sard's Theorem. For more details see Theorem 6.6 of [100].

18. By assumption, $AB = BA$. The ith row of the matrix AB is the ith row of B times a_{ii}. The jth column of the matrix BA is the jth column of B times a_{jj}. Hence,
$$a_{ii}b_{ij} = a_{jj}b_{ij}$$
for all i and j. Since $a_{ii} \neq a_{jj}$ if $i \neq j$ this implies that $b_{ij} = 0$ if $i \neq j$.

19. Suppose that $A \in \mathfrak{u}(n+k)$ has distinct eigenvalues. Since A is skew-Hermitian, the Spectral Theorem for Normal Operators (see for instance Corollary 32.17 of [37] or Theorem 2.10.2 of [93]) implies that there exists a unitary matrix $g \in U(n+k)$ such that $g \cdot A = gAg^{-1}$ is diagonal. Since $[g \cdot p, g \cdot A] = g \cdot [p, A] = 0$, the matrix $g \cdot p$ commutes with the matrix $g \cdot A$, where $g \cdot A$ is diagonal with distinct eigenvalues. This implies that $g \cdot p$ is diagonal (see the previous problem).

20. For the first part see Example 3.9.6 of [87].

22. See Section 1.5 of [67] or Section 6 of [104]. Note that this is an example where the unstable manifolds are the cells in a CW-structure.

23. Any critical point x_σ of index $j \leq 2nk$ in $G_{n,n+k'}(\mathbb{C})$ lies in the submanifold $G_{n,n+k}(\mathbb{C})$ since increasing k just adds extra rows of zeros to the bottom of the matrix.

24. See Section 3.1 of [81] or Section 5 of [104].

25. See Section 3.5 of [81] or See Section 5 of [104].

26. This can be proved directly using combinatorics. Alternately, it can be proved inductively using the Morse-Bott Inequalities (Theorem 3.53). To

do this, consider the Morse-Bott function f_A where A is matrix whose $(1, 1)$ entry is i and all the other entries are 0. The Morse-Bott function f_A has two critical submanifolds. One is diffeomorphic to $G_{n-1,n+k-1}(\mathbb{C})$, and the other is diffeomorphic to $G_{n,n+k-1}(\mathbb{C})$. The Morse-Bott Inequalities for the function f_A show that the Poincaré polynomial satisfies the following inductive formula.

$$P_{n,n+k}(t) = P_{n,n+k-1}(t) + t^{2k}P_{n-1,n+k-1}(t)$$

For more details see Section 3.3 of [70].

27. The height function on the torus T^2 is a perfect Morse function. If the critical points of a Morse function are all of even index, then the Morse function is perfect. However, the converse is false.

Chapter 9: An Overview of Floer Homology Theories

3. Show that there exists a closed 1-form $\mu(\varphi)$ whose zeroes are in 1-1 correspondence with the set $\text{Fix}(\varphi)$ of fixed points of φ, and that $\mu(\varphi)$ is exact if and only if φ is Hamiltonian [14].

4. See [132].

Bibliography

[1] Ralph Abraham and Joel Robbin, *Transversal mappings and flows*, W. A. Benjamin, Inc., New York-Amsterdam, 1967. MR 39 #2181

[2] B. Aebischer, M. Borer, M. Kälin, Ch. Leuenberger, and H. M. Reimann, *Symplectic geometry*, Progress in Mathematics, vol. 124, Birkhäuser Verlag, Basel, 1994, ISBN 3-7643-5064-4. MR 96a:58082

[3] Selman Akbulut and John D. McCarthy, *Casson's invariant for oriented homology 3-spheres*, Mathematical Notes, vol. 36, Princeton University Press, Princeton, NJ, 1990, ISBN 0-691-08563-3. MR 90k:57017

[4] A. Andronov and L. Pontrjagin, *Systémes grossiers*, Dokl. Akad. Nauk SSR **14** (1937), 247{251.

[5] Vladimir Arnol'd, *Sur une propriété topologique des applications globalement canoniques de la mécanique classique*, C. R. Acad. Sci. Paris **261** (1965), 3719{3722. MR 33 #1861

[6] D. K. Arrowsmith and C. M. Place, *An introduction to dynamical systems*, Cambridge University Press, Cambridge, 1990, ISBN 0-521-30362-1; 0-521-31650-2. MR 91g:58068

[7] M. F. Atiyah, *Convexity and commuting Hamiltonians*, Bull. London Math. Soc. **14** (1982), 1{15. MR 83e:53037

[8] _____, *Angular momentum, convex polyhedra and algebraic geometry*, Proc. Edinburgh Math. Soc. (2) **26** (1983), 121{133. MR 85a:58027

[9] M. F. Atiyah and R. Bott, *The Yang-Mills equations over Riemann surfaces*, Philos. Trans. Roy. Soc. London Ser. A **308** (1983), 523{615. MR 85k:14006

[10] M. Audin and J. Lafontaine Editors, *Holomorphic curves in symplectic geometry*, Progress in Mathematics, vol. 117, Birkhäuser Verlag, Basel, 1994, ISBN 3-7643-2997-1. MR 95i:58005

[11] Michèle Audin, *The topology of torus actions on symplectic manifolds*, Progress in Mathematics, vol. 93, Birkhäuser Verlag, Basel, 1991, ISBN 3-7643-2602-6. MR 92m:57046

[12] _____, *Symplectic and almost complex manifolds*, Hcsgolomorphic urves in ymplectic eometry, Progr. Math., vol. 117, Birkhäuser, Basel, 1994, pp. 41{74. MR 1 274 926

[13] D. M. Austin and P. J. Braam, *Morse-Bott theory and equivariant cohomology*, Tfmvhe loer emorial olume, Birkhäuser, Basel, 1995, pp. 123{183. MR 96i:57037

[14] Augustin Banyaga, *On fixed points of symplectic maps*, Invent. Math. **56** (1980), 215{229. MR 81a:58023

[15] _____, *The structure of classical diffeomorphism groups*, Mathematics and its Applications, vol. 400, Kluwer Academic Publishers Group, Dordrecht, 1997, ISBN 0-7923-4475-8. MR 98h:22024

[16] Augustin Banyaga and David E. Hurtubise, *A proof of the Morse-Bott Lemma*, Expositiones Mathematicae (2004).

[17] S. Berceanu and A. Gheorghe, *On the construction of perfect Morse functions on compact manifolds of coherent states*, J. Math. Phys. **28** (1987), 2899{2907. MR 88m:58026

[18] Rolf Berndt, *An introduction to symplectic geometry*, Graduate Studies in Mathematics, vol. 26, American Mathematical Society, Providence, RI, 2001, ISBN 0-8218-2056-7. MR 2001f:53158

[19] P. Biran, *Lagrangian barriers and symplectic embeddings*, Geom. Funct. Anal. **11** (2001), 407{464. MR 2002g:53153

[20] Jean-Michel Bismut and Weiping Zhang, *An extension of a theorem by Cheeger and Müller*, Astérisque (1992), 235. MR 93j:58138

[21] Raoul Bott, *Nondegenerate critical manifolds*, Ann. of Math. (2) **60** (1954), 248{261. MR 16,276f

[22] ———, *On torsion in Lie groups*, Proc. Nat. Acad. Sci. U. S. A. **40** (1954), 586{588. MR 16,12a

[23] ———, *An application of the Morse theory to the topology of Lie-groups*, Bull. Soc. Math. France **84** (1956), 251{281. MR 19,291a

[24] ———, *The stable homotopy of the classical groups*, Ann. of Math. (2) **70** (1959), 313{337. MR 22 #987

[25] ———, *Lectures on Morse theory, old and new*, Bull. Amer. Math. Soc. (N.S.) **7** (1982), 331{358. MR 84m:58026a

[26] ———, *Morse theory indomitable*, Inst. Hautes Études Sci. Publ. Math. (1988), 99{114 (1989). MR 90f:58027

[27] ———, *Raoul Bott: collected papers. Vol. 1*, Contemporary Mathematicians, Birkhäuser Boston Inc., Boston, MA, 1994, ISBN 0-8176-3613-7. MR 95i:01027

[28] Raoul Bott and Hans Samelson, *Applications of the theory of Morse to symmetric spaces*, Amer. J. Math. **80** (1958), 964{1029. MR 21 #4430

[29] ———, *Correction to "Applications of the theory of Morse to symmetric spaces"*, Amer. J. Math. **83** (1961), 207{208. MR 30 #589

[30] Glen E. Bredon, *Topology and geometry*, Springer-Verlag, New York, 1993, ISBN 0-387-97926-3. MR 94d:55001

[31] Theodor Bröcker and Tammo tom Dieck, *Representations of compact Lie groups*, Graduate Texts in Mathematics, vol. 98, Springer-Verlag, New York, 1995, ISBN 0-387-13678-9. MR 97i:22005

[32] Morton Brown, *Locally flat imbeddings of topological manifolds*, Ann. of Math. (2) **75** (1962), 331{341. MR 24 #A3637

[33] J. W. Cannon and G. R. Conner, *The combinatorial structure of the Hawaiian earring group*, Topology Appl. **106** (2000), 225{271. MR 2001g:20020

[34] Ralph L. Cohen, *Topics in Morse theory: Lecture notes*, Stanford University, 1991.

[35] Charles Conley, *Isolated invariant sets and the Morse index*, CBMS Regional Conference Series in Mathematics, vol. 38, American Mathematical Society, Providence, R.I., 1978, ISBN 0-8218-1688-8. MR 80c:58009

[36] Lawrence Conlon, *Differentiable manifolds: a first course*, Birkhäuser Advanced Texts: Basler Lehrbücher. [Birkhäuser Advanced Texts: Basel Textbooks], Birkhäuser Boston Inc., Boston, MA, 1993, ISBN 0-8176-3626-9. MR 94d:58001

[37] C.W. Curtis, *Linear algebra: An introductory approach*, Springer-Verlag, New York, 1984.

[38] James F. Davis and Paul Kirk, *Lecture notes in algebraic topology*, Graduate Studies in Mathematics, vol. 35, American Mathematical Society, Providence, RI, 2001, ISBN 0-8218-2160-1. MR 2002f:55001

[39] Albrecht Dold, *Lectures on algebraic topology*, Classics in Mathematics, Springer-Verlag, Berlin, 1995, ISBN 3-540-58660-1. MR 96c:55001

[40] Wojciech Dorabiala, *Personal communication*, Penn State Altoona (2003).

[41] Stamatis Dostoglou and Dietmar Salamon, *Instanton homology and symplectic fixed points*, Sgymplectic eometry, London Math. Soc. Lecture Note Ser., vol. 192, Cambridge Univ. Press, Cambridge, 1993, pp. 57{93. MR 96a:58065

[42] B. A. Dubrovin, A. T. Fomenko, and S. P. Novikov, *Modern geometry—methods and applications. Part I*, Graduate Texts in Mathematics, vol. 93, Springer-Verlag, New York, 1992, ISBN 0-387-97663-9. MR 92h:53001

[43] James Dugundji, *Topology*, Allyn and Bacon Inc., Boston, Mass., 1978, ISBN 0-205-00271-4. MR 57 #17581

[44] Nelson Dunford and Jacob T. Schwartz, *Linear operators. Part I*, Wiley Classics Library, John Wiley & Sons Inc., New York, 1988, ISBN 0-471-60848-3. MR 90g:47001a

[45] James Eells Jr. and Nicolaas H. Kuiper, *Manifolds which are like projective planes*, Inst. Hautes Études Sci. Publ. Math. (1962), 5{46. MR 26 #3075

[46] C. Ehresmann, *Les connexions infinitésimales dans un espace fibré différentiable*, Ctbolloque de opologie, ruxelles, 1950, pp. 29{55.

[47] Ronald Fintushel and Ronald J. Stern, *Instanton homology of Seifert fibred homology three spheres*, Proc. London Math. Soc. (3) **61** (1990), 109{137. MR 91k:57029

[48] ———, *Invariants for homology 3-spheres*, Gldmdeometry of ow-imensional anifolds, 1 (urham, 1989), London Math. Soc. Lecture Note Ser., vol. 150, Cambridge Univ. Press, Cambridge, 1990, pp. 125{148. MR 93e:57025

[49] ———, *Integer graded instanton homology groups for homology three-spheres*, Topology **31** (1992), 589{604. MR 93f:57018

[50] Andreas Floer, *Viterbo's index and the Morse index for the symplectic action*, Pshsrticeriodic olutions of amiltonian ystems and elated opics (l iocco, 1986), NATO Adv. Sci. Inst. Ser. C Math. Phys. Sci., vol. 209, Reidel, Dordrecht, 1987, pp. 147{152. MR 89j:58030

[51] ———, *An instanton-invariant for 3-manifolds*, Comm. Math. Phys. **118** (1988), 215{240. MR 89k:57028

[52] ———, *Morse theory for Lagrangian intersections*, J. Differential Geom. **28** (1988), 513{547. MR 90f:58058

[53] ———, *A relative Morse index for the symplectic action*, Comm. Pure Appl. Math. **41** (1988), 393{407. MR 89f:58055

[54] ———, *The unregularized gradient flow of the symplectic action*, Comm. Pure Appl. Math. **41** (1988), 775{813. MR 89g:58065

[55] ———, *Symplectic fixed points and holomorphic spheres*, Comm. Math. Phys. **120** (1989), 575{611. MR 90e:58047

[56] ———, *Witten's complex and infinite-dimensional Morse theory*, J. Differential Geom. **30** (1989), 207{221. MR 90d:58029

[57] Theodore Frankel, *Fixed points and torsion on Kähler manifolds*, Ann. of Math. (2) **70** (1959), 1{8. MR 24 #A1730

[58] ———, *Critical submanifolds of the classical groups and Stiefel manifolds*, Dctshmmifferential and ombinatorial opology (a ymposium in onor of arston orse), Princeton Univ. Press, Princeton, N.J., 1965, pp. 37{53. MR 33 #4952

[59] John M. Franks, *Morse-Smale flows and homotopy theory*, Topology **18** (1979), 199{215. MR 80k:58063

[60] Robert Franzosa, *Index filtrations and the homology index braid for partially ordered Morse decompositions*, Trans. Amer. Math. Soc. **298** (1986), 193{213. MR 88a:58121

[61] Robert D. Franzosa, *The connection matrix theory for Morse decompositions*, Trans. Amer. Math. Soc. **311** (1989), 561{592. MR 90a:58149

[62] Stephen H. Friedberg, Arnold J. Insel, and Lawrence E. Spence, *Linear algebra*, Prentice Hall Inc., Upper Saddle River, NJ, 1997, ISBN 0-13-233859-9. MR 1 434 064

[63] Kenji Fukaya and Kaoru Ono, *Arnold conjecture and Gromov-Witten invariant for general symplectic manifolds*, Tathe rnoldfest (oronto, on, 1997), Fields Inst. Commun., vol. 24, Amer. Math. Soc., Providence, RI, 1999, pp. 173{190. MR 2000m:53121

[64] William Fulton and Joe Harris, *Representation theory*, Graduate Texts in Mathematics, vol. 129, Springer-Verlag, New York, 1991, ISBN 0-387-97527-6; 0-387-97495-4. MR 93a:20069

[65] M. Goresky and R. MacPherson, *Stratified morse theory*, Springer-Verlag, New York, 1988.

[66] Marvin J. Greenberg and John R. Harper, *Algebraic topology*, Benjamin/Cummings Publishing Co. Inc. Advanced Book Program, Reading, Mass., 1981, ISBN 0-8053-3558-7; 0-8053-3557-9. MR 83b:55001

[67] Phillip Griffiths and Joseph Harris, *Principles of algebraic geometry*, Wiley-Interscience [John Wiley & Sons], New York, 1978, ISBN 0-471-32792-1. MR 80b:14001

[68] M. Gromov, *Pseudoholomorphic curves in symplectic manifolds*, Invent. Math. **82** (1985), 307{347. MR 87j:53053

[69] John Guckenheimer and Philip Holmes, *Nonlinear oscillations, dynamical systems, and bifurcations of vector fields*, Applied Mathematical Sciences, vol. 42, Springer-Verlag, New York, 1990, ISBN 0-387-90819-6. MR 93e:58046

[70] Martin A. Guest, *Morse theory in the 1990s*, Igtnvitations to eometry and opology, Oxf. Grad. Texts Math., vol. 7, Oxford Univ. Press, Oxford, 2002, pp. 146{207. MR 1 967 749

[71] Victor Guillemin and Alan Pollack, *Differential topology*, Prentice-Hall Inc., Englewood Cliffs, N.J., 1974. MR 50 #1276

[72] Victor Guillemin and Shlomo Sternberg, *Geometric asymptotics*, American Mathematical Society, Providence, R.I., 1977. MR 58 #24404

[73] _____, *Symplectic techniques in physics*, Cambridge University Press, Cambridge, 1984, ISBN 0-521-24866-3. MR 86f:58054

[74] Theodor Hangan, *A Morse function on Grassmann manifolds*, J. Differential Geometry **2** (1968), 363{367. MR 39 #6357

[75] Sigurdur Helgason, *Differential geometry, Lie groups, and symmetric spaces*, Pure and Applied Mathematics, vol. 80, Academic Press Inc. [Harcourt Brace Jovanovich Publishers], New York, 1978, ISBN 0-12-338460-5. MR 80k:53081

[76] Morris W. Hirsch, *A proof of the nonretractibility of a cell onto its boundary* , Proc. Amer. Math. Soc. **14** (1963), 364{365. MR 26 #3033

[77] _____, *Differential topology*, Springer-Verlag, New York, 1994, ISBN 0-387-90148-5. MR 96c:57001

[78] Morris W. Hirsch and Stephen Smale, *Differential equations, dynamical systems, and linear algebra* , Academic Press [A subsidiary of Harcourt Brace Jovanovich, Publishers], New York-London, 1974. MR 58 #6484

[79] David E. Hurtubise, *The Floer homotopy type of height functions on complex Grassmann manifolds* , Trans. Amer. Math. Soc. **349** (1997), 2493{2505. MR 97h:58013

[80] _____, *The flow category of the action functional on $\mathcal{L}G_{n,n+k}(\mathbb{C})$*, Illinois J. Math. **44** (2000), 33{50. MR 2001i:57047

[81] Dale Husemoller, *Fibre bundles*, Graduate Texts in Mathematics, vol. 20, Springer-Verlag, New York, 1994, ISBN 0-387-94087-1. MR 94k:55001

[82] M. C. Irwin, *On the stable manifold theorem*, Bull. London Math. Soc. **2** (1970), 196{198. MR 42 #6873

[83] _____, *Smooth dynamical systems*, Pure and Applied Mathematics, vol. 94, Academic Press Inc. [Harcourt Brace Jovanovich Publishers], New York, 1980, ISBN 0-12-374450-4. MR 82c:58018

[84] Nathan Jacobson, *Lie algebras*, Dover Publications Inc., New York, 1979, ISBN 0-486-63832-4. MR 80k:17001

[85] I. M. James, *General topology and homotopy theory*, Springer-Verlag, New York, 1984, ISBN 0-387-90970-2. MR 86d:55001

[86] Mei-Yue Jiang, *Morse homology and degenerate Morse inequalities*, Topol. Methods Nonlinear Anal. **13** (1999), 147{161. MR 2000f:57037

[87] Katsuo Kawakubo, *The theory of transformation groups*, The Clarendon Press Oxford University Press, New York, 1991, ISBN 0-19-853212-1. MR 93g:57044

[88] A. A. Kirillov, *Elements of the theory of representations*, Springer-Verlag, Berlin, 1976. MR 54 #447

[89] Frances Clare Kirwan, *Cohomology of quotients in symplectic and algebraic geometry*, Mathematical Notes, vol. 31, Princeton University Press, Princeton, NJ, 1984, ISBN 0-691-08370-3. MR 86i:58050

[90] Anthony W. Knapp, *Representation theory of semisimple groups*, Princeton Mathematical Series, vol. 36, Princeton University Press, Princeton, NJ, 1986, ISBN 0-691-08401-7. MR 87j:22022

[91] Antoni A. Kosinski, *Differential manifolds*, Academic Press Inc., Boston, MA, 1993, ISBN 0-12-421850-4. MR 95b:57001

[92] Ivan Kupka, *Contribution à la théorie des champs génériques*, Contributions to Differential Equations **2** (1963), 457{484. MR 29 #2818a

[93] Peter Lancaster, *Theory of matrices*, Academic Press, New York, 1969. MR 39 #6885

[94] Serge Lang, *Real analysis*, Addison-Wesley Publishing Company Advanced Book Program, Reading, MA, 1983, ISBN 0-201-14179-5. MR 87b:00001

[95] Janko Latschev, *Gradient flows of Morse-Bott functions*, Math. Ann. **318** (2000), 731{759. MR 2001m:58026

[96] Rafael de la Llave and C. Eugene Wayne, *On Irwin's proof of the pseudostable manifold theorem*, Math. Z. **219** (1995), 301{321. MR 96g:58102

[97] Juan Margalef Roig and Enrique Outerelo Dom"nguez, *Differential topology*, North-Holland Mathematics Studies, vol. 173, North-Holland Publishing Co., Amsterdam, 1992, ISBN 0-444-88434-3. MR 93g:58005

[98] Dusa McDuff and Dietmar Salamon, *J-holomorphic curves and quantum cohomology*, University Lecture Series, vol. 6, American Mathematical Society, Providence, RI, 1994, ISBN 0-8218-0332-8. MR 95g:58026

[99] ———, *Introduction to symplectic topology*, Oxford Mathematical Monographs, The Clarendon Press Oxford University Press, New York, 1998, ISBN 0-19-850451-9. MR 2000g:53098

[100] J. Milnor, *Morse theory*, Princeton University Press, Princeton, N.J., 1963. MR 29 #634

[101] ———, *Differential topology*, Lmmviectures on odern athematics, ol. i, Wiley, New York, 1964, pp. 165{183. MR 31 #2731

[102] John Milnor, *Lectures on the h-cobordism theorem*, Princeton University Press, Princeton, N.J., 1965. MR 32 #8352

[103] John W. Milnor, *Topology from the differentiable viewpoint*, Based on notes by David W. Weaver, The University Press of Virginia, Charlottesville, Va., 1965. MR 37 #2239

[104] John W. Milnor and James D. Stasheff, *Characteristic classes*, Princeton University Press, Princeton, N. J., 1974. MR 55 #13428

[105] Marston Morse, *Relations between the critical points of a real function of n independent variables*, Trans. Amer. Math. Soc. **27** (1925), 345{396. MR 1 501 318

[106] ———, *The calculus of variations in the large*, American Mathematical Society Colloquium Publications, vol. 18, American Mathematical Society, Providence, RI, 1996, ISBN 0-8218-1018-9. MR 98f:58070

[107] Jürgen Moser, *On the volume elements on a manifold*, Trans. Amer. Math. Soc. **120** (1965), 286{294. MR 32 #409

[108] James R. Munkres, *Topology: a first course*, Prentice-Hall Inc., Englewood Cliffs, N.J., 1975. MR 57 #4063

[109] ———, *Elements of algebraic topology*, Addison-Wesley Publishing Company, Menlo Park, CA, 1984, ISBN 0-201-04586-9. MR 85m:55001

[110] Charles Nash, *Differential topology and quantum field theory*, Academic Press Ltd., London, 1991, ISBN 0-12-514075-4. MR 93c:58002

[111] Richard S. Palais, *The Morse lemma for Banach spaces*, Bull. Amer. Math. Soc. **75** (1969), 968{971. MR 40 #6593

[112] J. Palis, *On Morse-Smale dynamical systems*, Topology **8** (1968), 385{404. MR 39 #7620

[113] J. Palis and S. Smale, *Structural stability theorems*, Gapspmvxbclobal nalysis (roc. ympos. ure ath., ol. iv, erkeley, alif., 1968), Amer. Math. Soc., Providence, R.I., 1970, pp. 223{231. MR 42 #2505

[114] Jacob Palis Jr. and Welington de Melo, *Geometric theory of dynamical systems*, Springer-Verlag, New York, 1982, ISBN 0-387-90668-1. MR 84a:58004

[115] M. M. Peixoto, *Structural stability on two-dimensional manifolds*, Topology **1** (1962), 101{120. MR 26 #426

[116] _____ , *Structural stability on two-dimensional manifolds. A further remark.* , Topology **2** (1963), 179{180. MR 26 #6533

[117] Lawrence Perko, *Differential equations and dynamical systems*, Texts in Applied Mathematics, vol. 7, Springer-Verlag, New York, 1991, ISBN 0-387-97443-1. MR 91m:34001

[118] Everett Pitcher, *Inequalities of critical point theory*, Bull. Amer. Math. Soc. **64** (1958), 1{30. MR 20 #2648

[119] S. Ramanujam, *Topology of classical groups*, Osaka J. Math. **6** (1969), 243{249. MR 41 #6246

[120] S. Ramanujan, *An application of Morse theory to certain symmetric spaces*, J. Indian Math. Soc. (N.S.) **32** (1968), 243{275 (1969). MR 41 #2708

[121] _____ , *Application of Morse theory to some homogeneous spaces*, Tôhoku Math. J. (2) **21** (1969), 343{353. MR 40 #3576

[122] Joel W. Robbin and Dietmar Salamon, *Dynamical systems, shape theory and the Conley index*, Ergodic Theory Dynam. Systems **8*** (1988), 375{393. MR 89h:58094

[123] H. L. Royden, *Real analysis*, Macmillan Publishing Company, New York, 1988, ISBN 0-02-404151-3. MR 90g:00004

[124] Walter Rudin, *Principles of mathematical analysis*, McGraw-Hill Book Co., New York, 1976. MR 52 #5893

[125] _____ , *Real and complex analysis*, McGraw-Hill Book Co., New York, 1987, ISBN 0-07-054234-1. MR 88k:00002

[126] David Ruelle, *Elements of differentiable dynamics and bifurcation theory*, Academic Press Inc., Boston, MA, 1989, ISBN 0-12-601710-7. MR 90f:58048

[127] Dietmar Salamon, *Connected simple systems and the Conley index of isolated invariant sets*, Trans. Amer. Math. Soc. **291** (1985), 1{41. MR 87e:58182

[128] _____ , *Morse theory, the Conley index and Floer homology*, Bull. London Math. Soc. **22** (1990), 113{140. MR 92a:58028

[129] _____ , *Lectures on Floer homology*, Sgtpcuymplectic eometry and opology (ark ity, t, 1997), IAS/Park City Math. Ser., vol. 7, Amer. Math. Soc., Providence, RI, 1999, pp. 143{229. MR 2000g:53100

[130] Dietmar Salamon and Eduard Zehnder, *Morse theory for periodic solutions of Hamiltonian systems and the Maslov index*, Comm. Pure Appl. Math. **45** (1992), 1303{1360. MR 93g:58028

[131] Matthias Schwarz, *Morse homology*, Birkhäuser Verlag, Basel, 1993, ISBN 3-7643-2904-1. MR 95a:58022

[132] P. Seidel, π_1 *of symplectic automorphism groups and invertibles in quantum homology rings* , Geom. Funct. Anal. **7** (1997), 1046{1095. MR 99b:57068

[133] Michael Shub, Albert Fathi, and Rémi Langevin, *Global stability of dynamical systems*, Springer-Verlag, New York, 1987, ISBN 0-387-96295-6. MR 87m:58086

[134] Ana Cannas da Silva, *Lectures on symplectic geometry*, Lecture Notes in Mathematics, vol. 1764, Springer-Verlag, Berlin, 2001, ISBN 3-540-42195-5. MR 2002i:53105

[135] S. Smale, *Stable manifolds for differential equations and diffeomorphisms* , Ann. Scuola Norm. Sup. Pisa (3) **17** (1963), 97{116. MR 29 #2818b

[136] Stephen Smale, *Morse inequalities for a dynamical system*, Bull. Amer. Math. Soc. **66** (1960), 43{49. MR 22 #8519

[137] _____, *On gradient dynamical systems*, Ann. of Math. (2) **74** (1961), 199{206. MR 24 #A2973

[138] Edwin H. Spanier, *Algebraic topology*, Springer-Verlag, New York, 1966, ISBN 0-387-94426-5. MR 96a:55001

[139] Michael Spivak, *A comprehensive introduction to differential geometry. Vol. I*, Publish or Perish Inc., Wilmington, Del., 1979, ISBN 0-914098-83-7. MR 82g:53003c

[140] Robert E. Stong, *Notes on cobordism theory*, Mathematical notes, Princeton University Press, Princeton, N.J., 1968. MR 40 #2108

[141] Arne Strøm, *Note on cofibrations*, Math. Scand. **19** (1966), 11{14. MR 35 #2284

[142] _____, *Note on cofibrations. II*, Math. Scand. **22** (1968), 130{142 (1969). MR 39 #4846

[143] W. Szlenk, *An introduction to the theory of smooth dynamical systems*, PWN—Polish Scientific Publishers, Warsaw, 1984, ISBN 83-01-03798-9. MR 86f:58042

[144] Clifford Henry Taubes, *Casson's invariant and gauge theory*, J. Differential Geom. **31** (1990), 547{599. MR 91m:57025

[145] René Thom, *Sur une partition en cellules associée à une fonction sur une variété*, C. R. Acad. Sci. Paris **228** (1949), 973{975. MR 10,558b

[146] Pierre Tremblay, *The Unstable Manifold Theorem: A proof for the common man*, Master's Paper in Mathematics, The Pensylvania State University (1988).

[147] Karen K. Uhlenbeck, *Connections with L^p bounds on curvature*, Comm. Math. Phys. **83** (1982), 31{42. MR 83e:53035

[148] Frank W. Warner, *Foundations of differentiable manifolds and Lie groups*, Graduate Texts in Mathematics, vol. 94, Springer-Verlag, New York, 1983, ISBN 0-387-90894-3. MR 84k:58001

[149] John C. Wells, *Invariant manifolds of non-linear operators*, Pacific J. Math. **62** (1976), 285{293. MR 54 #6206

[150] George W. Whitehead, *Elements of homotopy theory*, Graduate Texts in Mathematics, vol. 61, Springer-Verlag, New York, 1978, ISBN 0-387-90336-4. MR 80b:55001

[151] J. H. C. Whitehead, *Combinatorial homotopy. I*, Bull. Amer. Math. Soc. **55** (1949), 213{245. MR 11,48b

[152] _____, *On simply connected, 4-dimensional polyhedra*, Comment. Math. Helv. **22** (1949), 48{92. MR 10,559d

[153] Edward Witten, *Supersymmetry and Morse theory*, J. Differential Geom. **17** (1982), 661{692 (1983). MR 84b:58111

[154] Catalin Zara, *Personal communication*, Penn State Altoona (2004).

[155] Andreas Zastrow, *The second van-Kampen theorem for topological spaces*, Topology Appl. **59** (1994), 201{232. MR 95j:55028

Symbol Index

$(C_*(f), \partial_*)$	Morse-Smale-Witten chain complex, 198	
(M, g)	Riemannian manifold, 93	
(N, L)	index pair for an isolated compact invariant set, 208	
(N_q, L_q)	regular index pair for q, 209	
(X, A)	topological pair, 16	
$(\underline{C}_*(X, A; \Lambda), \underline{\partial}_*)$	CW-chain complex, 3, 27	
$(r_1, \ldots, r_n)$	Schubert symbol, 247	
B	Banach space of bounded sequences in E, 102	
$B^k(X; \Lambda)$	k coboundaries, 20	
$B_k(X; \Lambda)$	k boundaries, 16	
$C^k(X, A; \Lambda)$	k^{th} relative singular cochain group, 20	
$C^k(X; \Lambda)$	k^{th} singular cochain group, 20	
C_f	mapping cone, 36	
$C_k(X, A; \Lambda)$	k^{th} relative singular chain group, 16	
$C_k(X; \Lambda)$	k^{th} singular chain group, 15	
E	Banach space, 98	
$E(u)$	energy of u, 275	
E_s	stable subspace of the Banach space E, 98	
E_u	unstable subspace of the Banach space E, 98	
F	curvature, 282	
$FI_*(N)$	instanton Floer homology, 283	
F_i^k	i^{th} face map for Δ^k, 15	
$G \cdot x_0$	orbit, 229	
G_{x_0}	isotopy group, 229	
$HF_*(L, L')$	Lagrangian Floer homology, 281	
$HF_*(M, \omega, H, J; \mathbb{F})$	symplectic Floer homology, 279	
$H^k(X; \Lambda)$	k^{th} singular cohomology group, 20	
$H_*(X; \Lambda)$	graded singular homology, 16	
$H_k(D^n, \partial D^n)$	homology of a disk rel its boundary, 19	
$H_k(S^n)$	homology of a sphere, 19	
$H_k(X, A; \Lambda)$	k^{th} relative singular homology group, 16	
$H_k(X; \Lambda)$	k^{th} singular homology group, 16	

$H_p(f)$	Hessian, 46
$H_p^\nu(f)$	Hessian normal, 81
I	unit interval $[0,1]$, 31
$I(N)$	the invariant set of N, 207
$I(f, Z)$	intersection number, 146
J	almost complex structure, 273, 281
$L(f)$	Lefschetz number of f, 148
Lx_f	Lie derivative, 272
M^t	$f^{-1}((-\infty, t])$, 60, 63, 213
M_f	mapping cylinder, 36
$M_p(f)$	Hessian matrix, 46, 96, 137
M_t	$f^{-1}([t, \infty))$, 213
$N_x(G \cdot x_0)$	normal space to the orbit $G \cdot x_0$, 233
$O(n)$	orthogonal group, 304
$SO(n)$	special orthogonal group, 266
$SU(n)$	special unitary group, 266
$S^s(p)$	stable sphere, 181, 197
$S^u(p)$	unstable sphere, 181, 197
$Sp(n)$	symplectic group, 267
T	linear automorphism of $E_s \times E_u$, 98
T^*M	cotangent bundle, 58, 95
T_*M	tangent bundle, 45, 58
T_pM	tangent space at p, 45
T_s	T restricted to E_s, 98
T_u	T restricted to E_u, 98
$U(n)$	unitary group, 228, 266
$W(q, p)$	$W^u(q) \cap W^s(p)$, 158, 172, 175, 196, 271
$W^s(p)$	stable manifold of p, 94, 158
$W_r^s(\varphi)$	local stable set of φ, 100, 101
$W^u(p)$	unstable manifold of p, 94, 158
$W_p^s(\varphi)$	global stable manifold, 111
$W_p^u(\varphi)$	global unstable manifold, 113
$X^{(n)}$	n-skeleton of X union A, 21, 23
X_f	Hamiltonian vector field of f, 272
$Z^k(X; \Lambda)$	k cocycles, 20
$Z_k(X; \Lambda)$	k cycles, 16
$[S^n]$	fundamental class of $(S^n, *)$, 26
$[Y, x]$	vector tangent to the orbit $G \cdot x_0$, 232
Δ^k	k-standard simplex, 15
$\Gamma(g)$	graph of g, 148
Λ	commutative ring with unit, 15
$\mathcal{M}(q, p)$	moduli space of gradient lines, 197
Φ_n	inverse of Ψ_n, 199
Φ_n	inverse of isomorphism from Lemma 2.11, 26
Ψ_n	isomorphism from Lemma 2.11, 26, 199
Σ	Riemann surface, 284
δ_*	connecting homomorphism of a triple, 18, 211
δ_k	connecting homomorphism, 17

$\frac{\partial}{\partial \bar{x}} \nabla f$	matrix of $d\nabla f$, 96
$\frac{\partial}{\partial \bar{x}} \varphi_t$	matrix of $d\varphi_t$, 96
$\frac{\partial}{\partial x_1}, \ldots, \frac{\partial}{\partial x_m}$	local basis for $T_* M$, 95
γ_n	universal n-plane bundle, 268
λ_C	index of C, 84
λ_p	index of the critical point p, 46, 158, 196
$\mathbb{C}^*$	$\mathbb{C} - \{0\}$, 23
$\mathbb{R}^*$	$\mathbb{R} - \{0\}$, 22
$\mathbb{R}_+$	non-negative real numbers, 99
$\mathbb{Z}_+$	non-negative integers, 15
$\mathcal{A}$	space of connections, 281, 284
$\mathcal{G}$	gauge group, 281, 284
$\mathcal{J}(M, \omega)$	complex structures compatible with ω, 273
$\mathcal{L}M$	smooth contractible loops in M, 270, 274
$\mathcal{M}(L, L')$	smooth paths from L to L', 280
$\mathcal{M}(N)$	$\mathcal{A}/\mathcal{G}$, 281
$\mathcal{M}(x^-, x^+, H, J)$	finite energy gradient flows of a_H, 275
$\mathcal{P}(H)$	periodic solutions, 273, 274
$\mathcal{R}$	gauge equivalence classes of flat connections, 282, 284
$\mathcal{X}(M)$	Euler characteristic, 74, 76
$\mathcal{X}_F(N)$	instanton Euler characteristic, 283
$\mathfrak{g}$	Lie algebra, 229
$\mathfrak{su}(n)$	skew-Hermitian matrices with trace zero, 266
$\mathfrak{u}(n)$	skew-Hermitian matrices, 228, 266
a_H	action functional, 270, 274
$\mu_{CZ}(z)$	Conley-Zehnder index, 276, 277
∇f	gradient vector field, 58, 93, 95
νZ	normal bundle, 131
ν_k	number of critical points of index k, 73
$\bar{\partial}_k$	boundary operator on $C_k(X, A; \Lambda)$, 16
∂_k	boundary operator on $C_k(X; \Lambda)$, 15
ϕ	local coordinate chart, 46
ϕ_n	homomorphism such that $\phi_n[S^n] = 1$, 26
σ	singular simplex, 15
$\mathrm{Cr}(f)$	set of critical points of f, 116, 173, 196
$\mathrm{Diff}^r(M)$	diffeomorphisms of class C^r, 164
$\mathrm{Lip}(\varphi)$	Lipschitz constant, 99
$\tilde{V}, \tilde{W}$	vector fields, 46
$\underline{C}_n(X, A; \Lambda)$	free Λ-module generated by the n-cells not in A, 24, 27
$\underline{\partial}_n$	boundary operator on $\underline{C}_n(X, A; \Lambda)$, 27, 199
φ_s	$p_s \circ \varphi$, 98
φ_t	1-parameter group of diffeomorphisms, 94
φ_u	$p_u \circ \varphi$, 98
$b_k(F)$	k^{th} Betti number, 73
d_k	coboundary operator, 20
df	differential of f, 58, 95
$dx^1, \ldots, dx^m$	local basis for $T^* M$, 95
$e(\sigma)$	Schubert cell, 253

$f \pitchfork g$ f is transverse to g, 131

f_σ characteristic map, 21, 23

$f_{\partial\sigma}$ attaching map, 23

g Riemannian metric, 58, 93, 95, 273

g^{ij} entries of $(g_{ij})^{-1}$, 95

g_{ij} components of the metric g, 95

i_* inclusion homomorphism, 16

p critical point, 45

$p(j)$ number of partitions of j, 258

p_σ collapsing map, 26

p_s projection onto E_s, 98

p_u projection onto E_u, 98

Index

5-Lemma, 42
λ-Lemma, 8, 166
ω-limit sets, 223
$\varepsilon\, C^1$-close, 165
C^r topology, 140
J-holomorphic curve, 275

action functional, 270, 274
additivity property, 18
adjoint, 229
adjoint representation, 284
admissible, 277
algebraic automorphism, 120
almost complex structure, 227, 273, 281
anti-self dual connection, 282
Arnold Conjecture, 273
Arnold conjecture, 269
associative, 230, 267
Atiyah-Guillemin-Sternberg Convexity Theorem, 91
attaching map, 21
attractor, 223

Baire space, 159
Baire's Theorem, 160
Banach manifold, 2
base, 177
Betti number, 73
Betti numbers, 270
Bott, 80
Bott's perfect Morse function, 49, 75, 118, 227
boundary operator, 15
Brouwer degree, 154
Brouwer Fixed Point Theorem, 153
Bruhat cells, 264
bubbling, 278, 283

canonical immersion, 127
Casson invariant, 283, 284
Cauchy-Riemann equation, 275

cell, 165
cell equivalent, 177
Cellular Approximation Theorem, 30
Center Manifold Theorem, 292
chain complex, 41
chain map, 41
characteristic map, 21, 23
Chern class, 273, 274
Chern-Simons functional, 271, 282, 284
classic approach, 2
closed cell, 21
closure finite, 22
coboundary operator, 20
cofibered pair, 31
cofibration, 3, 15, 31, 185, 208
collar, 60, 145
compact open topology, 139
compactification, 272, 281, 283
compatible almost complex structure, 273, 281
compatible with the Morse charts, 176, 179, 185
complexification, 263
conjugate, 125
Conley index, 208
Conley-Zehnder index, 276, 277
connected simple system, 223
connecting homomorphism, 17, 18, 42, 199, 210, 211, 213, 216
connecting manifold, 181
connection, 271, 281
conservative, 119
Constant Rank Theorem, 152
contracting, 98, 111, 123
Contraction Mapping Theorem, 122
cotangent bundle, 58, 95
critical point, 4, 45, 130, 137, 270–272, 274, 276, 278, 280, 282
critical set, 80
critical value, 4, 72, 130, 178
C^r topology, 140
cup length, 91

curvature, 282
CW-chain complex, 3, 27
CW-complex, 2, 21
CW-Homology Theorem, 3, 15, 27
CW-pair, 23, 30
CW-structure, 21

degree, 154
dense, 139
derivative, 107
determinant function, 88
diagonal, 148
dimension property, 18
dominates, 73

embedding, 123, 134
energy, 275, 282
$\varepsilon \, C^1$-close, 165
Euclidean neighborhood retract, 149
Euler characteristic, 5, 74, 76, 151, 283
Euler-Poincaré Theorem, 78
exact sequence, 17, 27
exactness property, 17
excision property, 18
existence and uniqueness for O.D.E.s, 97, 114, 116
existence of Morse functions, 50
exit set, 208
expanding, 98, 111, 123

face maps, 15
fibered product, 152
fibration, 41
first fundamental form, 237
5-Lemma, 42
fixed point, 98, 100, 111, 113, 149, 269, 270, 272
flat connection, 282, 284, 285
Floer, 269, 270, 278, 279
Floer homology, 3, 269, 271
Floer homology for Lagrangian intersections, 270
flow, 294
flow line, 59
focal point, 236
framed cobordant, 185
framed submanifold, 180
Frankel's Theorem, 91, 261
Fredholm operator, 2
Frobenius inner product, 306
Fundamental Theorem of Algebra, 152

gauge equivalence, 271, 281
gauge group, 284
general position, 7, 134
generalized eigenvectors, 111
generic, 159
globally stable, 132, 138

gluing, 278
gradient, 4, 58, 93, 95, 274, 281, 282
gradient system, 125
gradient-like, 179, 183
graph, 121, 148
Grassmann manifold, 227

Hamiltonian diffeomorphism, 269, 270
Hamiltonian system of differential equations, 119
Hamiltonian vector field, 90, 264, 272
Handle Presentation Theorem, 73
Hartman-Grobman Theorem, 125, 184
Hessian, 4, 46, 81, 137
Hodge star operator, 282
holonomy, 284
homology, 41
homology 3-sphere, 283
homology sphere, 283
homology, fundamental properties, 3, 17
homotopy extension property, 31, 41
homotopy groups, 43
homotopy lifting property, 41
homotopy property, 17
homotopy type, 4, 64, 69
Hurewicz homomorphism, 273
Hurewicz Theorem, 43
hyperbolic, 5, 98, 111, 113, 166

immediate successor, 180
immersion, 6, 123, 127, 132
inclination, 167
index, 4, 46, 94, 155, 196, 271
index filtration, 184, 193
index pair, 207, 219
induced homomorphism, 30
inner product, 58
instanton, 282
instanton Floer homology, 271
intermediate critical point, 180
intersection number, 146, 197
invariant subset, 207
Inverse Function Theorem, 47, 108, 128–130, 271, 290
Inverse Image Theorem, 6, 131, 153
inward, 145
isolated compact invariant set, 207
isolating neighborhood, 207
isotopic, 134
isotopy, 189
isotropy group, 229

J-holomorphic curve, 275
Jordan form, 123

Kähler metric, 263
Killing form, 231

Klein bottle, 43
Kostant-Kirillov-Souriau structure, 262
Kupka, 159

Lagrangian Floer homology, 270
Lagrangian submanifold, 270, 280
λ-Lemma, 8, 166
Lefschetz Fixed Point Theorem, 269
Lefschetz map, 150, 155
Lefschetz number, 148
Lie algebra, 229, 281
Lie bracket, 227–229, 233
Lie derivative, 272
Lipschitz, 99
Lipschitz Inverse Function Theorem, 5, 101
Lipschitz Local Stable Manifold Theorem, 100
Ljusternik-Schnirelmann category, 91
local diffeomorphism, 133
local Lefschetz number, 155
locally Lipschitz, 108
locally stable, 132, 138
long exact sequence, 42

mapping cone, 36, 186
mapping cylinder, 36
mapping torus, 285
matrix group, 230
maximal invariant subset, 207
Mayer-Vietoris sequence, 42
Mean Value Theorem, 108, 112
metric space, 35
Milnor, 63
minimal Chern number, 274, 278
modern approach, 2
moduli space, 197
moment map, 265
monotone, 278
monotone symplectic manifold, 273
Morse chart, 89, 176
Morse decomposition, 224
Morse function, 4, 47
Morse Homology Theorem, 198, 270
Morse inequalities, 4, 73, 270
Morse Lemma, 4, 52, 88
Morse polynomial, 74
Morse sets, 224
Morse-Bott function, 80, 81, 164
Morse-Bott index, 84
Morse-Bott Inequalities, 85, 306
Morse-Bott Lemma, 83
Morse-Smale function, 158
Morse-Smale transversality, 7, 157
Morse-Smale-Witten chain complex, 9, 198, 270
multiplicity, 236

natural transformation, 17, 211, 216

NDR-pair, 34
neighborhood deformation retract, 34
non-degenerate, 4, 5, 46, 58, 94, 97, 137, 151, 269, 273, 280, 284
non-degenerate 2-form, 272
nullity, 237

ω-limit sets, 223
open cell, 21
orbit, 229
orbits, 190
ordinary point, 125
orientation, 143, 196, 201, 278, 283
orientation conventions, 27
orthogonal group, 304

Palis, 8, 166, 171
parabolic subgroup, 263
partial ordering, 8, 172
partition, 258
path method, 53, 54
Peixoto, 164, 184
perfect Morse function, 49, 76, 228
phase diagram, 173
Poincaré polynomial, 74, 268
Poincaré-Hopf Index Theorem, 155, 269, 284
Polar Decomposition Theorem, 276
positive definite, 58
positively invariant, 207
Preimage Theorem, 6, 130, 153
principal curvatures, 240
properties of CW-complexes, 22

quaternionic projective space, 43

Reeb, 61, 89
regular, 130
regular index pair, 208
regular pair, 279
regular point, 295
regular value, 130, 153
relative attaching map, 28, 181
relative homology, 16
relative index, 271
repeller, 223
residual, 159
retract, 32, 149
Riemann surface, 275, 284
Riemannian manifold, 2, 93
Riemannian metric, 4, 58, 95, 273, 274, 280, 282

Sard's Theorem, 3, 51, 136, 139
Schubert cell, 253
Schubert symbol, 247, 253
second fundamental form, 237
self-indexing, 89, 179
semisimple, 232

separable, 159, 160
simplex, 15
singular *k*-simplex, 15
singular cohomology, 3, 20
singular homology, 3, 15
singular homology group, 16
skeleton, 3, 21
skew-Hermitian matrices, 227, 228, 235, 266
Smale, 8, 159, 171
smoothly homotopic, 132
Sobolev space, 2
special orthogonal group, 266
special unitary group, 266
spectral flow, 284
spectral radius, 111, 123
sphere, CW-structure, 22
sphere, homology of, 19, 30
stable, 6, 132, 138, 172
stable manifold, 2, 5, 7, 94
Stable Manifold Theorem, 94
stable set, 100
stable sphere, 9, 181, 197, 215
stably trivial, 185
standard form, 176, 223
Strøm structure, 34
strong deformation retract, 34
strong Morse inequalities, 74
structural stability, 164
structurally stable, 164
subcomplex, 22
submersion, 6, 129, 133
succeed, 8, 173
symmetric, 58
symplectic diffeomorphism, 272
Symplectic Floer homology, 269

symplectic Floer homology, 270, 278
symplectic form, 272
symplectic group, 267
symplectic manifold, 262, 269, 272
symplectic vector field, 264

tangent bundle, 58
tangent space, 45
tangent vector, 45
Thom-Pontryagin construction, 180
topological pair, 16
topologically conjugate, 125, 164
topology of compact convergence, 139
torus, 48, 118, 158
trace form, 227, 230, 235
transverse, 2, 6, 131, 133, 145, 166, 280
triple, 17, 18, 27
tubular flow, 184
Tubular Flow Theorem, 125, 162, 163, 184
tubular neighborhood, 136, 180

uniform topology, 139
unitary group, 228, 235, 266
universal *n*-plane bundle, 260, 268
Universal Coefficient Theorem, 74
unstable manifold, 2, 5, 7, 89, 94
Unstable Manifold Theorem, 94
unstable sphere, 9, 181, 197, 215

weak Morse inequalities, 73
weak topology, 21
wedge, 40
Weyl chamber, 263
Whitehead Theorem, 44

Kluwer Texts in the Mathematical Sciences

1. A.A. Harms and D.R. Wyman: *Mathematics and Physics of Neutron Radiography.* 1986
    ISBN 90-277-2191-2
2. H.A. Mavromatis: *Exercises in Quantum Mechanics.* A Collection of Illustrative Problems and Their Solutions. 1987
    ISBN 90-277-2288-9
3. V.I. Kukulin, V.M. Krasnopol'sky and J. Horácek: *Theory of Resonances.* Principles and Applications. 1989
    ISBN 90-277-2364-8
4. M. Anderson and Todd Feil: *Lattice-Ordered Groups.* An Introduction. 1988
    ISBN 90-277-2643-4
5. J. Avery: *Hyperspherical Harmonics.* Applications in Quantum Theory. 1989
    ISBN 0-7923-0165-X
6. H.A. Mavromatis: *Exercises in Quantum Mechanics.* A Collection of Illustrative Problems and Their Solutions. Second Revised Edition. 1992   ISBN 0-7923-1557-X
7. G. Micula and P. Pavel: *Differential and Integral Equations through Practical Problems and Exercises.* 1992   ISBN 0-7923-1890-0
8. W.S. Anglin: *The Queen of Mathematics.* An Introduction to Number Theory. 1995
    ISBN 0-7923-3287-3
9. Y.G. Borisovich, N.M. Bliznyakov, T.N. Fomenko and Y.A. Izrailevich: *Introduction to Differential and Algebraic Topology.* 1995   ISBN 0-7923-3499-X
10. J. Schmeelk, D. Takacqi and A. Takacqi: *Elementary Analysis through Examples and Exercises.* 1995   ISBN 0-7923-3597-X
11. J.S. Golan: *Foundations of Linear Algebra.* 1995   ISBN 0-7923-3614-3
12. S.S. Kutateladze: *Fundamentals of Functional Analysis.* 1996   ISBN 0-7923-3898-7
13. R. Lavendhomme: *Basic Concepts of Synthetic Differential Geometry.* 1996
    ISBN 0-7923-3941-X
14. G.P. Gavrilov and A.A. Sapozhenko: *Problems and Exercises in Discrete Mathematics.* 1996
    ISBN 0-7923-4036-1
15. R. Singh and N. Singh Mangat: *Elements of Survey Sampling.* 1996   ISBN 0-7923-4045-0
16. C.D. Ahlbrandt and A.C. Peterson: *Discrete Hamiltonian Systems.* Difference Equations, Continued Fractions, and Riccati Equations. 1996   ISBN 0-7923-4277-1
17. J. Engelbrecht: *Nonlinear Wave Dynamics.* Complexity and Simplicity. 1997
    ISBN 0-7923-4508-8
18. E. Pap, A. Takaci and D. Takaci: *Partial Differential Equations through Examples and Exercises.* 1997   ISBN 0-7923-4724-2
19. O. Melnikov, V. Sarvanov, R. Tyshkevich, V. Yemelichev and I. Zverovich: *Exercises in Graph Theory.* 1998   ISBN 0-7923-4906-7
20. G. Călugăreanu and P. Hamburg: *Exercises in Basic Ring Theory.* 1998
    ISBN 0-7923-4918-0
21. E. Pap: *Complex Analysis through Examples and Exercises.* 1999   ISBN 0-7923-5787-6
22. G. Călugăreanu: *Lattice Concepts of Module Theory.* 2000   ISBN 0-7923-6488-0
23. P.M. Gadea and J. Muñoz Masqué: *Analysis and Algebra on Differentiable Manifolds: A Workbook for Students and Teachers.* 2002   ISBN 1-4020-0027-8; Pb 1-4020-0163-0
24. U. Faigle, W. Kern and G. Still: *Algorithmic Principles of Mathematical Programming.* 2002
    ISBN 1-4020-0852-X
25. G. Călugăreanu, S. Breaz, C. Modoi, C. Pelea and D. Vălcan: *Exercises in Abelian Group Theory.* 2003   ISBN 1-4020-1183-0

Kluwer Texts in the Mathematical Sciences

26.   C. Costara and D. Popa: *Exercises in Functional Analysis.* 2003   ISBN 1-4020-1560-7
27.   J.S. Golan: *The Linear Algebra a Beginning Graduate Student Ought to Know.* 2004
                                                                            ISBN 1-4020-1824-X
28.   J. Caldwell and D.K.S. Ng: *Mathematical Modelling.* Case Studies and Projects. 2004
                                                                            ISBN 1-4020-1991-2
29.   A. Banyaga and D. Hurtubise: *Lectures on Morse Homology.* 2004   ISBN 1-4020-2695-1

KLUWER ACADEMIC PUBLISHERS – DORDRECHT / BOSTON / LONDON